THE POLITICS OF
GENDER

THE POLITICS OF **GENDER**

A SURVEY

FIRST EDITION

Editor: Yoke-Lian Lee

LONDON AND NEW YORK

First Edition 2010
Routledge
2 Park Square, Milton Park, Abingdon, Oxfordshire OX14 4RN
711 Third Avenue, New York, NY 10017

Routledge is an imprint of the Taylor & Francis Group, an informa business

First issued in paperback 2014
© Routledge 2010

All rights reserved. No part of this publication may be photocopied, recorded, or otherwise reproduced, stored in a retrieval system or transmitted in any form or by any electronic or mechanical means without the prior permission of the copyright owner.

ISBN 978-1-85743-493-4 (hbk)
ISBN 978-1-85743-757-7 (pbk)
ISBN 978-0-203-85052-7 (ebk)

Development Editor: Cathy Hartley

Typeset in Times New Roman 10.5/12

The publishers make no representation, express or implied, with regard to the accuracy of the information contained in this book and cannot accept any legal responsibility for any errors or omissions that may take place.

Typeset by Taylor & Francis Books

For Jian Nin, Han Ben and Kora-Lee

Foreword

This book, *The Politics of Gender*, is intended to aid researchers, academics, students, the media, and all those interested in the relationship between gender and politics. It is informed by no single perspective and encompasses a broad range of topics in order to offer an extensive overview of the issues concerned.

The volume is divided into three sections. The first comprises original essays, each exploring some dimension of the politics of gender. There are many ways of looking at gender politics, and the authors do not share a common approach or framework for analysing the politics of gender. Hence, each essay reflects the contributor's specific research interests and theoretical commitments. Topics cover a broad range of issues within the realm of gender politics: the editor offers a general introduction on gender politics, while the following chapters offer overviews, analyses and case studies that include Romani women in Eastern and Central Europe, Hobbes, masculinity in Hollywood films, the gender politics of international relations, gender and climate change, female trafficked migrants, travel writing and journalism, gender and genocide, gender and the African Union, and sovereignty, subjectivity and human rights in terms of gender politics.

An extensive A–Z glossary of key concepts, issues, personalities and terms follows. Entries include affirmative action, Simone de Beauvoir, borders, Commission on the Status of Women, contraception, International Bill of Rights, intertextuality and women in development, offering a wide-ranging overview of gender politics world-wide.

The book concludes with an extensive bibliography for further research, partly drawn from the chapter contributors.

Yoke-Lian Lee
Stone
February 2010

Contents

List of Figures

Acknowledgements

In the course of producing this volume, I have become indebted to a number of people, to whom my expression of gratitude is owed. First and foremost, I would like to thank the Europa Development Editor, Cathy Hartley, for her enthusiastic engagement and patient guidance. Without her this book would not have been brought to completion.

Thanks are also due to the Production Editor, Paola Celli, who skillfully oversaw the entire production of this book. I am also grateful to Alison Neale, for her expert copy editing skills and help with tracking down incomplete references.

Thanks are especially due to the individual contributors, many of whom were already overwhelmed with their own tight schedules, for their work in advancing the study of gender and politics. I have learnt a lot from them. A considerable debt is also owed to several authors and editors for their permission to reproduce excerpts of some length from their works in the A–Z glossary section. Their names and full publication details are acknowledged with each entry.

Last, but not least, I would like also to thank John Horton, Jennifer Hulme and A.J.B. Holmes, who offered many helpful suggestions for improvements. Thank you also to Matthew Street for his IT support.

The Editor and Contributors

Dr Yoke-Lian Lee's research focus is on poststructuralist feminist reading of sovereignty and subjectivity, with specific interest in international and human rights law and the discursive practice of law in the construction of gendered subjectivities in world politics. She teaches at the University of Keele and is a member of the editorial board of the *Journal of Balkan and Near Eastern Studies.*

Tawia Ansah, LLB, PhD, is Professor of Law at New England Law, Boston, Massachusetts, USA. He writes at the intersection of international law, critical theory and theology. Currently he is working on a monograph on genocide discourse.

Sharron A. FitzGerald is Lecturer in Critical Feminist and Legal Geography at the Institute of Geography and Earth Sciences, Aberystwyth University, Wales. Her research focuses on the relationship between neoliberal governance and the state's attempts to regulate women's transborder mobility. She has written extensively on sex trafficking, state-formation and securitisation in the European Union. She is the editor of *Regulating the International Movement of Women: From Protection to Control* (2010).

Vassilis K. Fouskas is the Founding and Managing Editor of the *Journal of Balkan and Near Eastern Studies* (quarterly, Routledge). His publications include *Cyprus: the Post-imperial Constitution* (with Alex O. Tackie, 2009), *The New American Imperialism* (with Bülent Gökay, 2005), *Zones of Conflict* (2003) and *Italy, Europe and the Left* (1998). He is Professor of International Relations at the University of Piraeus, Greece.

Corinne Fowler is Lecturer in Twentieth-Century Postcolonial Literature at the School of English at Leicester University. Corinne is author of *Chasing Tales: journalism, travel writing and the history of British ideas about Afghanistan* (2007), co-editor of the forthcoming *Travel and Ethics. Theory and Practice* (2010) and editor of *Beatrice Grimshaw: From Fiji to the Cannibal Islands* (2008).

Lena J. Kruckenberg studied history and sociology at the University of Bielefeld, Germany. She holds an MA in International Relations from Keele University. Lena worked extensively on Roma human rights activism. She recently has completed an ethnographic case study of the UN

Committee on the Elimination of Racial Discrimination. Her general interests lie in the analysis of social microstructures as they constitute transnational and global 'macro' phenomena.

Dr Sherilyn MacGregor is a lecturer in environmental politics in the School of Politics, International Relations and Philosophy at Keele University. She is author of *Beyond Mothering Earth: Ecological Citizenship and the Politics of Care* (2006) and an editor of *Environmental Politics* journal.

Dr Tim Murithi is the Head of Programme with the Institute for Security Studies Office in Addis Ababa, Ethiopia. He has held posts at the Department of Peace Studies at the University of Bradford; the Centre for Conflict Resolution, University of Cape Town, South Africa; and the Programme in Peacemaking and Preventive Diplomacy at the United Nations Institute for Training and Research (UNITAR), in Geneva, Switzerland. He has worked as a Consultant with the African Union. He is the author of two books: *The Ethics of Peacebuilding* and *The African Union: Pan-Africanism, Peacebuilding and Development*.

Glen Newey is Professor of Politics and International Relations at Keele University. He is the author of *Virtue, Reason and Toleration* (1999), *After Politics* (2001) and *Hobbes and Leviathan* (2007), as well as numerous articles on toleration, political obligation, freedom of speech and political deception.

Jill Steans is Senior Lecturer in International Relations Theory at the University of Birmingham. She is the author of a number of books and articles on gender in international relations and international political economy, including *Gender in International Relations* (2006).

Abbreviations

AU	African Union
CAR	Central African Republic
CEDAW	Committee on the Elimination of Discrimination Against Women
CEE	Central and Eastern Europe
CWPE	Committee on Women, Population and the Environment
DRC	Democratic Republic of the Congo
ECCR	Economics Education and Research Consortium
ECOSOC	Economic and Social Council
EM	ecological modernization
ES	environmental security
EU	European Union
FAS	Femmes Africa Solidarité
FEMNET	African Women's Development and Communication Network
GAATW	Global Alliance Against Traffic in Women
GAD	gender and development
GATW	Coalition Against Trafficking in Women
GED	gender, environment and development
GMEF	Global Ministerial Environment Forum
ICC	International Criminal Court
ICCPR	International Covenant on Civil and Political Rights
ICESCR	International Covenant on Economic, Social and Cultural Rights
ICPD	International Conference on Population and Development
ICTR	International Criminal Tribunal for Rwanda
ICTY	International Criminal Tribunal for the former Yugoslavia
INSTRAW	United Nations International Research and Training Institute for the Advancement of Women
IPCC	Intergovernmental Panel on Climate Change
IPE	international political economy
HRC	Human Rights Committee
MDGs	Millennium Development Goals
NAFTA	North American Free Trade Agreement
NGO	non-governmental organization
NIDL	New International Division of Labour
OAU	Organization of African Unity

SD	sustainable development
UDHR	Universal Declaration on Human Rights
UN	United Nations
UNAIDS	Joint United Nations Programme on HIV/AIDS
UNCED	UN Conference on Environment and Development
UNEP	United Nations Environment Programme
UNESCO	United Nations Educational, Scientific and Cultural Organization
UNFPA	United Nations Population Fund
UNGA	United Nations General Assembly
UNICEF	United Nations Children's Fund
UNSAS	United Nations Stand-by Arrangements System
UNSC	United Nations Security Council
WEN	Women's Environmental Network
WHO	World Health Organization
WID	Women in Development
WSSD	World Summit on Sustainable Development

Essays

Introduction: the politics of gender

Yoke-Lian Lee

> 'It is important that we engage in dialogue with colleagues who work on gender issues from outside a feminist perspective … '
>
> (Baden and Goetz 1997)

The title of this volume, *The Politics of Gender: A Survey,* obviously suggests an analysis of politics from gender perspectives. The purpose of this collection of essays is to engage with the question of to what extent gender, as a key principle, can help us to understand the politics of everyday life. In addition, it is about whether a gendered and, more specifically, feminist epistemology is capable of transforming the dominant concept of knowledge-making that has traditionally excluded women as subject of knowledge. These questions raise two immediate sorts of concerns. First, conceptual difficulties: 'gender' as an analytic concept is neither straightforward nor easy to define. It is located within a wide diversity of theoretical and philosophical traditions and approaches and, hence, can be read in multiple ways. Moreover, feminists also come from diverse social groups, each with their own social and political experiences and, thus, their conceptual frameworks are also distinctively different. Gender politics is typically rooted in masculine and feminine identities. However, some recent critical feminist theorists have suggested that not only are there no fixed gender identities—a point that most gender theorists would now accept—but that the very idea of a coherent identity is, after all, a cultural construction. This raises problems for the very idea of gender as a key principle of analysis.

Second, gender invisibility: there is also the larger issue of long-standing gender blindness and male monopoly of knowledge production, where feminist scholarship continues to be marginalized. Typically, feminist scholarship employs a 'gender lens' to 'see' different social and political realities by drawing on different epistemologies that seek to escape from the gender bias of conventional masculine theoretical assumptions. Although much mainstream academic scholarship continues to exclude gender from its conceptualizations, in the last two decades or so there has been a surge of feminist scholarship casting doubt on the role of traditional methodologies in the production of objective knowledge. Thus, feminists ask whether it is possible to produce, in Ann Tickner's words, 'a universal and objective foundation of knowledge'. Rather than viewing the traditional systems of knowledge production as an objective scientific inquiry, feminists urge us to think of knowledge production as a social construction, formed mostly by men, that is 'variable across time, place, and culture' (Tickner 1992, 30; Tickner 2001, 4). So, then, to what extent can the increasing visibility of women scholars and the development of feminist epistemologies and methodologies influence our conceptions of knowledge? These are some of the questions, among others, that this volume seeks to address.

The essays collected here emerge from a rather straightforward formulation: the authors were simply asked to address, with a degree of explicitness, the gender aspects of their own specialized fields of study or disciplines, and to interrogate and challenge dominant analytical categories and conventional methodologies from this point of view. No specific themes or topics were imposed on the contributors. This collection thus takes on a multidisciplinary approach that transgresses disciplinary boundaries. One of the key objectives of the volume has been to expand the boundary of gender analysis; not necessarily to exclude men and masculinity ideas, for the politics of gender, as Steans writes, 'need not exclusively be about women, nor need it only be by women' (see Steans's chapter in this volume). Critical theory, for instance, warns against the knowledge claim made in the name of 'women'. Inevitably, we have to confront the question of how we categorize women and whether there is an inscribed gendered identity. Who is included and excluded from such characterizations? One interesting point about grounding claims about women is that, as soon as the category of women is invoked, positing an 'essence' of women, arguments soon arise among feminists questioning that supposed essence. For example, some feminists claim that there is an ontological specificity: women as mothers. This categorization must form the foundation for a specific common legal and political representation. However, not all women are or want to be mothers, and not all mothers see that as the proper foundation for their legal and political identity. Moreover, women are always divided along a variety of class, ethnic, religious, national, racial and sexual lines. Gender politics, therefore, is heterogeneous and 'should involve the analysis of power relations between men and women and the discursive and cultural construction of hegemonic masculinities and femininities' (see MacGregor's chapter in this volume). The inclusion of four male-authored chapters in this edition affirms the commitment to a more inclusive politics.

The analysis of gender politics can expose the structural instability of sexual identity, making possible a multiplication of possibilities in social and political life. Gender analysis enables us to understand how 'man' and 'woman' are not categorically separated, independent entities, but rather are mutually constituted and interdependent. This argument is crucial in challenging essentialism, reflecting instead the complexity, contingency and often contradictory process of identity formulation. The key to a gender analysis of politics, on this view, is to disrupt the dominant conception of the feminine within the gender dichotomy as a crucial step towards the displacement of gender hierarchy and undermining the role of gender distinctions in structuring the socio-political order. A common theme that emerges from this volume is that of subjectivity;

to a considerable extent the various chapters seek to explore the many social and legal practices through which the subject is constructed.

There are, of course, differences among the contributors. Most noticeable, perhaps, is that between Jill Steans and Glen Newey; while they both discuss what they take to be 'orthodoxy', they have entirely different points of view. Steans's critique of 'malestream' orthodoxy in international relations (IR), defines orthodoxy as a particular (male) mode of knowledge production; more precisely, as a disciplinary tool to tame feminist scholarship. Newey, on the other hand, makes an attempt to right orthodox feminists' countless wrongs in misreading Hobbes's political theory. In this sense, he turns feminist critique of male orthodoxy on its head. In defence of Hobbes, especially Hobbes's view on women, family, sexual politics and rape, Newey argues that Hobbes's political theory actually consistently displays 'a basic commitment to original gender equality', which is far more egalitarian than that which those dominant modern liberal political theorists, such as John Rawls, have managed to achieve. Newey charges that orthodox feminism, notably Carole Pateman, has misread the Hobbesian conceptions of the liberal sovereign individual and patriarchy. In defence of Hobbes, Newey asserts that Hobbes is 'not a woman hater' and he is wholly misunderstood with orthodox feminist reading of his work.

GENDER AND SUBJECTIVITY

One of the main objectives of the interrogation of the gendered subject has been to challenge some theorists' underlying assumptions that there is a pre-social, pre-cultural and pre-existing *natural* person waiting for law and politics to take cognizance of it, while masking the constitutive function of the social order in the formation of its subject. The process of production or construction of the gendered subject, as some essays in this volume suggest, can be viewed as a regulatory practice—a kind of cultural and political disciplinary practice/mechanism that helps to classify the human subjects in their 'place' within the social order.

The claim to a gendered identity, as Butler explains, is an exclusionary act that 'is invoked with the unmarked dimensions of class and racial privilege intact in practice'. Further, 'the insistence upon the coherence and unity of the category of women has effectively refused the multiplicity of cultural, social, and political *intersections* in which the concrete array of "women" are constructed' (Butler 1990, 14, my emphasis). Several contributors in this book take up the theoretical analytical framework of intersectionality to address the structural tension and conflict between race, class and gender. Kruckenberg argues that intersectionality is a useful concept for examining gender politics in the complexity of Romani women's diverse and marginalized position. Her chapter challenges the politics of identity by exploring the politics of difference through Romani women's activism. She argues persuasively for an inclusionary gender politics 'to find innovative ways to transcend simple identity politics by recognizing multiplicity'. Thus, Kruckenberg asserts that the use of intersectionality as a theoretical framework can help us to understand diverse and yet intertwined accounts of human subjectivity, 'such as race, ethnicity, gender, class or sexual orientation and other categorizations'.

The intersection of subjectivity and body informs Fouskas's chapter, highlighting in particular the construction of masculinity. He offers a neo-Marxian meta-narrative on the tragedy and vulnerability of the masculine body in and through three contemporary Hollywood action movies, namely *Raging Bull*, *Heat* and *Miami Vice*. For him, masculinity is embodied in the body that is not wholly secured: it is capable of becoming 'fat and fit' and, thus, 'lives in a state of permanent insecurity inside and outside'. He shows how the *intersection* of violence, sexism, class and masculinity *works* in a dramatic male-dominant capitalist world. However, as Fouskas observes, male-dominant movies often provoke a paradoxical 'hegemonic presence' of multiple and fragmented female figures. Scenes and lines of action are often punctuated and complicated by the entry or exit of these women. Domestic disputes and gender–relationship feuds—the different forms of conflict and battle between men and women—are a common theme. However, who gets to win the fight? In *Raging Bull*, Fouskas sees the complex inter-relationship of power and control between man and woman in the relationship between La Motta and Vickie: 'Vickie was getting a beating, yes, but La Motta could not win this fight'. Women have figured largely in Hollywood action films alongside violence and aggressive masculinity: this illustrates how the gendered subjects are not categorically separated, but are mutually constituted and codependent.

GENDER AND KNOWLEDGE

There is considerable consensus among Steans, MacGregor, FitzGerald and Fowler that the construction of normative knowledge is gendered. Steans makes an important feminist intervention in the study of IR in noting a repression of feminist scholarship within the discipline. Her chapter offers a dialogical discourse analysis between feminists and mainstream (positivist) scholars through an assessment of the impact and status of feminist scholarship. Steans's chapter shows us how the politics of gender is at work in the construction of IR knowledge; more precisely, the ways in which the 'discipline' has marginalized and continues to marginalize feminist work. Steans argues that the core assumptions of IR, noticeably realism and neo-realism, are deeply gendered in ways that serve to reify social structures, and which mask the 'historical character of social and political relations'. She contends that IR as a discipline needs to take a more inclusive approach to broaden its scope of studies to accommodate feminist concerns and feminist modes of inquiry.

What, it might be asked, has gender got to do with climate change? MacGregor looks through the lens of gender politics and finds that 'the concept of gender is almost completely absent in policy documents and research reports on climate change at all levels'. The exclusion of gender issues thus creates gender-blind analyses and responses in the ways that we 'see' environmental issues. Masculine environmental discourses play a decisive role in framing and shaping normative understanding of climate change. However, MacGregor argues that climate change is not simply a gender neutral issue, but it is a discursive cultural construct that is based on deeply-entrenched dominant gender discourses. The hegemonic construction of masculinity and femininity effectively legitimizes the masculine subject as subject of knowledge. Under the 'current climate of gender politics', MacGregor argues, 'climate change has brought about a masculinization of environmentalism', which in turn helps to shape the ecological modernization and environmental security agendas. So, what can the discourses of climate change tell us about contemporary gender relations? Fundamentally, MacGregor claims that '[t]he lens of gender politics brings into acute focus the process, norms and power relations through which we can recognize the working of hegemonic masculinities and hegemonic femininities in all social phenomena'.

Sharron FitzGerald questions the liberal feminist production of knowledge about trafficked female sex workers as the 'victim subject'. She offers a critical analysis of the intersection of racialized class and gender in the construction of the non-Western female migrants as subjects of legal regulation. The politics of gender and feminist identity politics as mechanisms of knowledge and power help to sustain power relations between liberal (dominant) feminism and the female trafficked migrant. This, in turn, as FitzGerald argues, 'reinforces a hierarchy of womanhood buttressed through colonial stereotypes of non-Western women. This neo-colonialism places Western feminists at the centre of the spaces of authority to advocate for all "Others"'. Consequently, feminist anti-trafficking initiatives and measures designed to protect and advance the human rights of non-Western female trafficked migrants become repressive law enforcement mechanisms—seeing all consensual women migrants as trafficked and thus rendering their border crossing as illegal. In conclusion, FitzGerald argues that unreflective neoliberal feminists' 'contemporary anti-trafficking measures are another cog in the transnational structures of governance and regulation'.

Fowler asks: what is the similarity between a 19th century woman travel writer, such as Beatrice Grimshaw, who writes in racial/colonial discourse, and John Miller, a Channel 4 liberal feminist reporter who writes about women's rights? The similarity between them is their tendency to construct their narrative structure in simple, stark imperialistic and racial terms. Fowler is especially critical of the role of gender politics in the production and dissemination of 'knowledge' about the postcolonial female subject as part of an industry of 'knowledge production' in the publishing and media industries. Gender assumptions embedded in the 'lucrative industry of Iranian and Muslim women's memoirs', often cast non-Western women as victims of cultural and religious oppression. Fowler shows how advocating equality and freedom as remedies for postcolonial women has unintended political consequences: progressive gender-specific legal measures that have been expressly designed to provide objective remedies are often, in practice, not responsive to women's needs and experiences. Thus, Fowler shows how the emancipatory promise of women's human rights can often have the unintended consequence of framing the legal status of women in narrow and unhelpful terms. Fowler's analysis suggests that colonial women's cultural imperialism has had the consequence of removing postcolonial women's complex subjectivities.

GENDER AND POLITICS

What gets to count as politics? Many would argue that politics is not simply a given; it is a social process of construction, often the result of contestation and ideological struggle: an act that is historically specific and contingent in accordance to social order (Donald and Hall 1986). Social order is established by *power* because 'the existence of power capable of obtaining generalised obedience and allegiance implies a certain type of representation [...] concerning the legitimacy of the social order' (Lefort 1986, 282). Legitimate sovereign authority, at least in Western democratic states, it can be argued, resides in bureaucratic techniques of governance (power) expressed through organizational institutional structure of state processes. Once the bureaucratic structure of politics has been established, it is extremely difficult to remove. In terms of gender politics, it could be argued that masculinity and femininity are created out of societal and legal structures through techniques of governance.

Tawia Ansah takes up the theme of 'governance' to examine the politics of gender in relation to the crime of genocide. He argues that the deployment of a gendered analysis as a way of 'seeing' constitutes a form of 'governance'. The reason for this is that gender analysis informs and shapes the meaning of genocide and, thus, influences how we come to view both gender and genocide. To demonstrate this, Ansah offers a comparative analysis of three gender modalities in relation to genocide. First, the 'gendered lens' approach conflates gender violence as genocide; second, the intersectional approach sees each form of violence as separate, but compounded in relation to the other; and, finally, the critical approach that takes a non-gendered view of genocide. Ansah advocates a non-gendered view of genocide in line with Janet Halley's call in favour of 'taking a break' from feminist legal discourses. That is to say 'taking a break' from the operations of the governmentality of law which favours military intervention to punish the perpetrators of violence as its operational strategy. Ansah warns of the inherent violence of humanitarian intervention and the violence of the law's institutional structure.

Tim Murithi begins where Ansah leaves off. He uses the idea of gender mainstreaming to examine the African Union Peace and Security Agenda with a special focus on gender-based violence, in particular rape as an instrument of genocidal civil war. He questions the effectiveness of gender mainstreaming in the formulation and implementation of peace and security programmes and projects. Murithi observes that, although the African Union has adopted a gender equality approach, there is an acute lack of commitment to implement practical initiatives to address gender concerns. As such, gender mainstreaming agendas have only very limited ways of changing for the better the lives of ordinary African women in conflict situations. He identifies two key problems with current approaches to gender mainstreaming: the lack of political will and commitment to implement gender-sensitive projects, and the cultural and conceptual difficulties in using the term 'gender' as a tool for programmatic reforms because it is extremely difficult to transmit the notion of gender into a workable technical model. Murithi calls for a normative shift to ensure that women play a key role in socio-political affairs.

Yoke-Lian Lee is interested in the 'structural' subjugation of 'women' in international law and, in particular, in international gender-specific human rights law. Her chapter discusses the contradiction between claims about state sovereignty and claims about human rights and the question of the relationship between gender and subjectivity. The principal objective of her analysis is to examine the impact of various discourses and practices of sovereignty and their political and legal effects on international law, particularly the international human rights of 'women'. Lee seeks to extend the problematization of the discourses of sovereignty that has become familiar over the past decade in order to politicize women's human rights law. The analysis thereby also develops a reading of struggles over the gendered identities expressed in human rights law. It analyses how international women's human rights tend to decontextualize, ahistoricize and, thus, depoliticize the multiple subjectivity of women. It further explores the normative consequences of treating the state as sovereign in relation to international law; how sovereignty is reinscribed within international law; and how the individual inscribed as 'woman' is constituted within the discourse of international human rights law.

CONCLUSION

As can be seen, much of the material in this volume seeks to question the idea of a pregiven unitary sovereign individual as the subject of politics and knowledge. The unsettling of the subject that takes place across different disciplines in these chapters and, consequently, the rethinking of the subject leads us to understand that the process of producing the subject of politics is itself political. The political, therefore, can be seen as the constitution of the social order where gendered structural social processes intersect with race, class, ethnicity, sexuality and violence to produce political hierarchies and structural subordination—to keep gendered subjects in their place. To a certain extent, this book contributes to the theorization of subjectivities through research that engages with a broad range of topics. In addition, it displays the continuing force with which the various contributors continue to fuel debate about the construction of gendered knowledge through contesting the dominant mode of knowledge. We hope this project is a modest contribution to the politics of gender.

Not a woman hater: Hobbes's critique of patriarchy

GLEN NEWEY

In recent years, Hobbes's views about women have been the subject of increasing discussion and dispute. In his own life, Hobbes moved in an exclusively masculine intellectual milieu, and the same goes, more or less, for what is known of his social contacts. Hobbes's friend John Aubrey famously remarked in his *Brief Lives* that Hobbes was 'not a woman hater' (Aubrey 1898, I, 390), an observation which might be taken to repudiate any imputation of misogyny, or again of homosexuality. Hobbes of course never married, and rumours that he may have fathered an illegitimate daughter have never been substantiated.[1] At any rate, nothing in Hobbes's biography gives colour to the notion that his views on women would have differed in any significant respect from the patriarchal opinions held by his contemporaries. Very few of the parties to his surviving *Briefwechsel* were female, an exception being Margaret Cavendish, Marchioness of Newcastle.[2] However, even here the contact between Hobbes and the Marchioness was limited: according to Margaret herself, she and Hobbes never exchanged more than 20 words (Martinich 1999, 317).

HOBBES, THE FAMILY AND IDEOLOGY

In some respects, Hobbes's theory of the family offers a more thoroughgoing egalitarianism than that achieved by modern liberalism, at least in its dominant Rawlsian version. As feminist critics of Rawls have long noted, a striking feature of *Theory of Justice* is its exclusion of the family from the purview of distributive justice.[3] Rawls devises general principles of justice, assumed to operate at the level of the sovereign state, but the benefits and burdens of domestic labour are ignored when it comes to devising both the entitlements due to contractors in the Original Position and the legitimate claims they can make against society's stock of resources. The effect, accordingly, is to eliminate domestic labour and more generally the terms of the marriage contract from the remit of distributive justice, though Rawls made some effort to address these issues in his later work *Political Liberalism*.[4]

For feminist critics of Rawls, the major problem with the theory was that it at least did not, and on some views could not, accommodate gender within the terms of distributive justice.[5] However, when we turn to Hobbes's works on political theory we find a remarkably consistent advocacy of a basic commitment to original gender equality, against the overwhelmingly patriarchal attitudes espoused by his contemporaries. It can be traced directly through all of Hobbes's major works in English on political theory, from *The Elements of Law* (1640) to *De cive* (1642; English translation 1647) to *Leviathan* (1651).[6] By stark contrast with the Rawls of *A Theory of Justice*, Hobbes does not exclude the family from the terms of the social contract—or Hobbes's version of it, namely the contract of civil association.

For some writers, Hobbes remains a mouthpiece for patriarchal orthodoxy. For example, R.W.K. Hinton argues (Hinton 1967; Hinton 1968) that Hobbes's theory offered the 'ablest exposition' of patriarchalism (Hinton 1967, 300). Leo Strauss contended that far from attacking patriarchy, as some of Hobbes's contemporary critics alleged, patriarchy 'is itself defended by Hobbes' (Strauss 1952, 103). More recently, from a feminist perspective, a qualified version of the same reading has been defended by Carole Pateman (Pateman 1988; see also Pateman 1991). For Pateman, Hobbes defends a form of patriarchy—not, like many of his contemporaries, as a feature of the state of nature, but as an irreducibly 'political' state (Pateman 1988, 50).

However, some writers have cast doubt on the picture of Hobbes as a defender, if in certain respects an unorthodox one, of patriarchy. For example, some time ago R.A. Chapman pointed out the resources for radical critique that Hobbes's account of the family contains, although Chapman regards the evidence for Hobbes's anti-patriarchalism as ambiguous (Chapman 1975). Similarly, Joanne H. Wright argues that Hobbes laid 'the foundation for a thorough critique of the notion of natural inequality between the sexes' (Wright 2002). However, it remains the case for these writers that the basis for such a critique confines itself to the state of nature, to be superseded once the commonwealth arrives.

This raises the question of how *ideological* we should regard Hobbes's account of the family as being. As Raymond Geuss points out (Geuss 2008, 52–53), ideology is a set of beliefs (and associated practices), whereby the existing configurations of power make certain contingent features of human existence appear to be universal or 'natural', as a way of expressing the interests of those holding the beliefs and, in the case of those who enjoy dominion, of entrenching those interests. It is clear from this definition how 'patriarchy', as an attempt to naturalize men's domination over women, is ideological. Of course, Hobbes does not understand his own enterprise as that of *Ideologie-kritik* in the Critical Theory sense. But if the position defended by Pateman and others is correct, Hobbes was not merely—in fact not even—providing a critique of patriarchy as ideology, but rather producing a further example of it. In this way, Hobbes could be seen not just as deceived, but delusional.

None the less, I shall try to show how far Hobbes goes in removing the shackles of patriarchal ideology in his own theory. For the family in Hobbes is not just under the power of the sovereign, but a microcosm of it.[7] At both the micro and the macro levels, Hobbes makes express allowance for female domination over males. That extends not merely to men-children, but male adults. Because of the homology between the family and commonwealth, the possibility of matriarchy in the one institution implies its possibility in the other.[8]

The main lineaments of the political theory that Hobbes puts forward in *Leviathan* are sufficiently familiar to require no recapitulation here. However, in understanding his *theoretical* appraisal of gender relations and their linkage to political power, we can notice two salient features of the theory. First, the theory of 'absolute' sovereignty that Hobbes advanced in *Leviathan* is gender-neutral, for more than one reason. It is gender-neutral because the entity that wields sovereignty in the theory need not be a natural person at all: it may be a committee or other corporate body, and it is in the nature of such bodies to be genderless. In addition, even given that in *Leviathan* chapter XIX Hobbes makes explicit his preference for a polity in which the sovereign *is* a natural person rather than a corporate body (Hobbes 1996, 131–33), at no point does Hobbes make any suggestion that this person must or should be male. This issue is simply not addressed. The conclusion follows, that power for Hobbes is not gendered. It is not simply that Hobbes fails to identify political or domestic power as essentially masculine: he goes to some lengths in all his three major political works to deny this.

RAPE IN THE STATE OF NATURE

Before we address the specific theory that Hobbes offers in support of this view, we need to examine one issue, which immediately arises from his account of the state of nature, though surprisingly commentators on Hobbes's theory of sexual politics have sometimes ignored it.[9] A fundamental issue in investigating Hobbes's view of gender relations is his implicit analysis of *rape* in the state of nature—implicit, because Hobbes does not address this issue directly. Hobbes does not, for instance, use the term 'rape', though this was current at the time he was writing.[10] Indeed, in comparison with the other familiar features of the state of nature, Hobbes's treatment of rape might be thought to be sanitized, even quaint. For although the institution of matrimony clearly cannot exist in the state of nature, Hobbes is happy to countenance the idea that families do, with the conditions necessary not only to raise children but to dispose of the question of guardianship and rights to obedience by contract (Hobbes 1996, 140).[11]

As far as this goes, then, it looks as though Hobbes's view of the matter will involve his usual application of the principle *volenti non fit injuria*.[12] A woman faced with a man who is intent on having sex with her and who fails to resist only through fear of worse consequences if she resists, is deemed to consent—just as, in Hobbes's famous example in his discussion of freedom and necessity, a passenger who throws her belongings overboard to keep the ship afloat 'does it nonetheless very willingly' (Hobbes 1996, 146). Since for Hobbes 'liberty and necessity are consistent', he does not have to deny that the woman submits to sex out of necessity; he denies only that it follows from this claim that the woman does not also consent to it. It then follows, defining 'rape' as being subjected to sexual intercourse without one's consent, that rape does not occur in such a situation. However, even allowing for this, rape remains a possibility in the state of nature wherever the woman has no opportunity to resist, regardless of whether she would avail herself of it if the opportunity presented itself.

However, it certainly does not follow that men have a natural right to commit rape in the state of nature. This is obscured by the frequently-made but false claim that Hobbes says, simply, that every person has a right to everything in the state of nature.[13]

> And because the condition of man, (as has been declared in the precedent chapter) is a condition of war of every one against every one; in which case every one is governed by his own reason; and there is nothing he can make use of, that may not be a help unto him, in preserving his life against his enemies; It followeth, that in such a condition, every man has a right to every thing; *even to one another's body.*
>
> (Hobbes 1996, 91; emphasis added)

Similar remarks occur in *Elements of Law* XIV §7 (Hobbes 1994, 79) and in *De cive* I §§7–8 (Hobbes 1998, 27). The point is that one's having a right to something depends on whether that thing could be of use in preserving one's life, to which each person also has a natural right. For whoever has the right to an end, has *ipso facto* the right to the necessary means to that end:

> Because where a man has right to the end, and the end cannot be attained without the means, that is, without such things as are necessary to the end, it is consequent that it is not against reason, and therefore right for a man, to use all means and do whatsoever action is necessary for the preservation of his body.
>
> (Hobbes 1994, 79)

The right of everyone to 'all things' then follows from the additional stipulation that there is nothing which may not be of use in securing the preservation of one's life.[14]

Hobbes is quite explicit, of course, in the *Leviathan* passage quoted above, that one of the 'things' that may be of use to one in self-preservation, is another person's body. However, it is not clear that it follows from Hobbes's account that just any *action-description* φ will be such that everybody has a right to φ, in virtue of the fact that φ-ing may in certain circumstances conduce to self-preservation. In some cases, there are no conceivable circumstances in which φ-ing will further the project of self-preservation. An obvious example is committing suicide: there are no circumstances in which killing oneself can be conducive to one's *preservation* (though of course it may promote other aims one has, such as escaping from unbearable pain). If so, the right to 'all things' already needs qualification.[15]

It might be thought that suicide constituted a special case, since it is *definitionally* incompatible with self-preservation. However, this is not the only case in which a limitation of right by the presumption end of self-preservation enters the picture in qualifying the headline right to all things. In the famous discussion of the 'Fool' in *Leviathan*, Hobbes does not assert that the Fool acts as of right in failing to perform what he has agreed when the other party has performed first. As Hobbes elaborates, the 'keeping of covenant is a rule of reason, whereby we are forbidden to do anything destructive of our life; and consequently a law of nature' (Hobbes 1996, 103). As Hobbes had already argued in the previous chapter, 'right consists in liberty to do or to forbear, whereas law determines, and binds to one of them; so that law and right differ as much as obligation and liberty, which in one and the same matter are inconsistent' (Hobbes 1996, 91).

Hobbes has a supplementary line of argument here. First, the basis of the right of nature is reason, since 'that which is not against reason, men call "right"' (Hobbes 1994, 79). It is not against reason for human beings to do everything within their power to preserve their own lives. Hence, as a particular human being, doing everything within one's power to preserve one's life is called 'right'. That is, one has a 'right' to do whatever is within one's power to preserve one's life. However, this says nothing about what happens when an action *is* against reason, since it cannot figure in any foreseeable rational project to preserve one's own life.

In view of this, Hobbes's headline position that everybody in the state of nature has a right to all things has to be qualified. There is an *ex ante* right to all things which may conceivably be of use in preserving one's life, though this must mean that there is no even *ex ante* right to commit suicide. Hobbes also reserves the discretion to each individual to decide whether or not, in given circumstances, a given action will conduce to his own preservation. However, even with this qualification in place, it still does not follow that anyone has a right to φ when he does *not* believe that φ-ing will help to preserve his life. In the case of rape, though, there is in general no reason for men to think that this will help preserve their lives.

In view of this, there is no reason to think that the state of nature includes a natural right to rape. It is hard to imagine circumstances in which a man's exercise of his right to self-preservation will require him to rape a woman. There is the marginal situation where he is ordered to do this on pain of death by a third party who has dominion over him, and such situations do occur. However, a right to second-party rape is hard to derive from the general basis for assigning rights that Hobbes sets out not only in *Leviathan* but also in his other works of political theory.

Hobbes has a further argument, which bears on the permissibility of rape within the commonwealth. This is that the sovereign is presumed to have a duty to 'increase the people', i.e. promote population growth (Hobbes 1994, 173). Accordingly, the sovereign is bound, in Hobbes's view, to issue 'ordinances concerning copulation', including a ban on 'such copulations are against the use of nature' (presumably homosexuality, bestiality and so on), on polyandry, on consanguineous unions, and 'forbid[ding] the promiscuous use of women'. Such acts are deemed to be 'so prejudicial' to 'the improvement of mankind' that the sovereign's failure to prohibit them is itself 'against the law of natural reason' (Hobbes 1994, 173). The claim is puzzling, in that, if the aim is demographic growth, it is not clear why rape would hinder it. As with some of the other examples of deviance from heterosexual monogamy, one could take it that Hobbes is offering a rule consequentialist style of justification: the end of population increase will be achieved most effectively by the general observance of monogamy, even if in specific cases the end will be breaching it. One might equally take it that when a thinker's conclusions are markedly under-supported by the presented justification of them, his adherence to those conclusions must be strong.

Two further qualifications are in order. First, Hobbes does not allow for the possibility of rape in marriage, and his contractual apparatus militates against making such an allowance. Indeed, Hobbes makes the same stipulation with respect to sub-matrimonial unions such as concubinage (Hobbes 1994, 131–32).[16] The fact of having entered into a marriage or concubinage is assumed to render consent to sexual intercourse. However, even here there is nothing in the nature of the matrimonial contract that prevents making sex contingent on subsequent consent. Second, Hobbes apparently assumes a uniform marital contract. Where patriarchy obtains, marriage, as the institutionalized dominion of men over women, will reinforce it. This will be true, for example, if relations within marriage, between either both spouses or one spouse and a third party, are not subject to the regulation of the criminal law. However, the scope exists for the sovereign to regulate these matters to the degree deemed appropriate for civic peace.

HOBBES AND MATRIARCHY

Naturally, Hobbes had to hand historical illustrations of how supreme political power could successfully be held by a woman. Until Hobbes was in his 15th year he was the subject of a highly successful female ruler, Elizabeth I; among the successes that Elizabeth could claim was to have seen off the dynastic and military threat posed by another woman monarch, her cousin, Mary Queen of Scots, who was executed in the year before Hobbes was born. Most of Hobbes's adult life was spent in the service of the Cavendish family, where Hobbes occupied a position subordinate to figures such as Margaret Cavendish, and to the wives of his successive pupils in the Cavendish household (Martinich 1999, e.g. 316–17). Other prominent female potentates at Hobbes's time or immediately before it included Catherine de' Medici, Anne of Denmark, the consort of James I, and, later, the dowager Queen Henrietta Maria, with whose circle Hobbes was linked in the years of exile in France following the execution of her husband, Charles I. As Hobbes noted in *De cive*, 'there are several places today where women have sovereign power' (Hobbes 1998, 108). The presence of females in positions of dominance was, then, for Hobbes, already a fact of political life.[17]

It is now a well-established feature of Hobbesian commentary that the extended account of the state of nature that Hobbes gives in *Leviathan* allows explicitly for matriarchal dominion. His principal statement on the subject comes in chapter XX, 'Of Dominion Paternal, and Despotical'. The discussion occurs as part of Hobbes's distinction between commonwealths by 'institution', or agreement, and 'acquisition', i.e. conquest. That distinction applies first and foremost to the conditions in which political power or 'dominion' comes into existence. Hobbes makes it clear—in this, if not in much else, following an Aristotelian template—that the household is synecdochic of the political community as a whole. Clearly one might regard the household as being historically or ontologically prior to the political community, so that it sets the terms for the legitimate existence of the latter. Hobbes does not take this course. His view is that it is not the mere fact of procreation that creates the right of dominion which parents enjoy over their offspring, but 'from the child's consent, either express, or by other sufficient arguments declared' (Hobbes 1996, 139).

In fact, Hobbes goes further in this passage, and provides a *reductio ad absurdum* argument from the indivisibility of sovereignty or 'dominion', to show that procreation *could* not form the basis for the parent's dominion over his or her child. For both the father and mother are equally the parent of the child, since 'there be always two that are equally parents'; but, since 'no man can obey two masters', it would follow, if parenthood conferred the right of dominion, that no such right could exist. Hence we find Hobbes attempting to refute a traditional argument for patriarchy, which had of course already been drawn upon by Sir Robert Filmer in *Patriarcha*. Filmer argues, for instance, that not only 'Adam, but [also] the succeeding patriarchs had, by right of fatherhood, royal authority over their children' (Filmer 1991, §3); he goes on to infer from this that the basis for political power lies in the Adamic right conferred by fatherhood (rather than parenthood) over one's offspring. '[M]any a child, by succeeding a king, hath the right of a father over many a greyheaded multitude, and hath the title of *Pater Patriae*' (Filmer 1991, §8). For if no right of dominion follows from the fact of parenthood then, in particular, no such right follows from paternity, and not only any attempt to ground domestic authority in this fact, but *a fortiori*, any attempt to ground political authority in it, must fail.[18]

Hobbes had already made this argument in *De cive* IX, where he argues that 'since dominion, i.e. sovereign power, is indivisible, so that no one can serve two masters, but generation requires the cooperation of two persons, a male and a female, it is impossible for dominion to be wholly acquired by generation alone' (Hobbes 1998, 108). In fact, Hobbes had already staked out this position in *The Elements of Law*, chapter four of which, 'Of the Power of Fathers, and of Patrimonial Kingdom', closely foreshadows the arguments of his later works of political theory on the subject of patriarchy:

> Because generation gives title to two, namely, father and mother, whereas dominion is indivisible, they therefore ascribe dominion over the child to

> the father only, *ob praestantium sexus* [i.e. on the grounds of men's being the more excellent sex] but they show not, neither can I find out by what coherence, either generation infers dominion, or advantage of so much strength, which, for the most part, a man has more than a woman, should generally and universally entitle the father to a property in the child, and take it away from the mother.

In the *De cive* treatment of this subject, Hobbes argues also that it relies on a similar appeal to self-evidence as does arguing that it is self-evident that a triangle equals the sum of two right angles (Hobbes 1998, 107–8). However, the polemical force of this second argument, which is not repeated in *Leviathan*, is blunted by the fact that it is after all true, if not self-evident, that a triangle equals two right angles. Hobbes needs to show that it is not even true that paternal dominion results from generation, not simply that it is not self-evident.

That is, he needs an argument to show that the right of paternal dominion is in no sense one simply conferred by nature, as orthodox patriarchal writers like Filmer supposed. Hobbes makes this explicit in the passage immediately following that just quoted. Hobbes gives short shrift to the idea that because men are 'the more excellent sex' they therefore enjoy a natural title to dominion over others, specifically women and their offspring: 'And whereas some have attributed the dominion to the man only, as being of the more excellent sex; they misreckon on it' (Hobbes 1996, 139). Here again Hobbes follows the position of *De cive* IX, §3 (Hobbes 1998, 108–9), as well as *The Elements of Law* XXIII, ii (Hobbes 1994, 130). The effect of Hobbes's account of the sexes in the state of nature is systematically to undermine arguments from nature to specific configurations of gendered power.

In the *De cive* passage and elsewhere Hobbes elaborates on these claims. In *De cive*, Hobbes says that in the state of nature:

> By right of nature the victor is master of the conquered; therefore by right of nature dominion over an infant belongs first to the one who first has him in their power. But it is obvious that a new-born child is in the power of his mother before anyone else, so that she can raise him or expose him at her own discretion and by her own right.
>
> (Hobbes 1998, 108)

Thus the *Leviathan* treatment of maternal dominion can in no sense be regarded as an aberration or new departure in Hobbes's thought, unlike, for instance, the novel theory of sovereignty developed in *Leviathan* chapters XVI–XVIII, which has no counterpart in his earlier political writings, including *De cive*. In that work, Hobbes had already argued that 'master is not in the definition of father' (Hobbes 1998, 107). It is clear that Hobbes here has the state of nature in mind: 'in the state of nature every woman who gives birth becomes both a mother and a mistress (*domina*)' (Hobbes 1998, 108).

As it happened, the question of sexual dominance, both in the mutual relations of adult progenitors, and as regards the engendering of offspring, had enjoyed a lengthy history before Hobbes. This view, and views like it, attributing generational potency to the male who stamped his sexual identity on the product of his sexual activity, remained current until well into the 18th century.[19] By contrast, Hobbes takes it as read that the only means by which dominion can be conferred on a parent, or somebody who finds herself *in loco parentis*, over a neonate is by the fact and recognition of overwhelming power—the selfsame means, of course, by which the sovereign dominion over political subjects is conferred.

In the *De cive* discussion, as well as in *Leviathan* and the *Elements*, Hobbes also knocks down other pretended justifications for paternal dominion over children. The most important is Hobbes's argument, central to his characterization of the state of nature, that human beings in the natural state are equal, or as near equal as makes no difference: 'the inequality of natural strength is too small to enable the male to acquire dominion over the female without war' (Hobbes 1998, 108).

Hobbes famously makes the same argument in *Leviathan* chapter XIII, where the natural equality between human beings is central to his explanation of why the state of nature must be a state of war. He is, in fact, a thoroughgoing egalitarian. He not only contends that humans are equal in respect of pre-emptive killing power, that each person in the state of nature possesses the same bundle of natural rights, and is equally beholden to the laws of nature; human beings are, according to Hobbes, also more or less equal in intelligence, since as he says in chapter XIII of *Leviathan*, 'there is not ordinarily a greater sign of the equal distribution of anything, than that every man is contented with his share' (Hobbes 1996, 87). All this Hobbes applies, without qualification, to the comparative abilities, duties and entitlements of men and women.[20]

Hobbes also furnishes a more specific argument in *De cive* to show why natural dominion must lie in the mother. That is, in essence, that (as the old saying goes) it is a wise man that knows his own father. As Hobbes remarks, 'in the state of nature it cannot be known who is a child's father, except by the mother's pointing him out; hence he [i.e. the child] belongs to whomever the mother wishes, and therefore belongs to the mother' (Hobbes 1998, 108). He is also clear that maternal dominion can persist from the state of nature into the commonwealth, citing in *De cive* as he does elsewhere the precedent for matriarchy in the society of the Amazons (Hobbes 1998, 108; 109–10).[21] In the *De cive* treatment, Hobbes clinches the point by noting that 'among men no less than other animals, the offspring goes with the womb' (Hobbes 1998, 108).

More generally, Hobbes eschews one of the obvious resources available to orthodox theorists in their efforts to legitimate patriarchy. This is to naturalize the family as a patriarchal unit, on which political and legal effect is then conferred by the foundation of the commonwealth. One version of this strategy was to take a primordial family as *being* the commonwealth, at least *in nuce*. As Filmer had argued in his *Patriarcha*:

> [...] not only Adam, but the succeeding patriarchs had, by right of fatherhood, royal authority over their children [...] That the patriarchs [...] were endowed with kingly power, their deeds do testify; for as Adam was lord of his children, so his children under him had a command and power over their own children, but still with subordination to the first parent, who is lord-paramount over his children's children to all generations, as being the grandfather of his people.
>
> (Filmer 1991, §4).

Filmer further contended that 'God at the creation gave sovereignty to the man over the woman, as being the nobler and principal agent in generation' (cf. Locke 1988, §55). In Filmer's view, then, the patriarchal authority of Adam, as the head of the first human family, simply was coterminous with that which he enjoyed as first earthly ruler. Even this structure was open, in principle, to subversion, since nothing in it required that the head of family in any given generation had to be male. Of course, Filmer assumed that a female head of family would not be the case.

MATRIARCHY AND THE COMMONWEALTH

So much, then, for Hobbes's arguments to show that the default condition in the state of nature is for maternal dominion over offspring. The question now is how he describes the transformations of natural right that mark the civil condition, as compared with the state of nature. In the orthodox feminist view of Hobbes defended for example by Pateman (1988), he at this point abandons the primitive egalitarianism by which, as we have seen, he characterizes the state of nature. 'Dominion' in Hobbes's understanding involves the right to demand obedience. Hence any shift in dominion must involve a transfer of right or title.[22] In fact, this can already take place in the state of nature, as he points out, since the mother can relinquish her dominion over a child by abandoning it, the vacant title then being transferred to whomever may subsequently decide to foster it. Hence the title conferred by dominion is not inalienable, as some writers who had claimed that the fact of procreation confers title had asserted.

In the state of nature, if a man and woman enter a partnership (*societas*) in which neither is subject to the other's power, dominion over any resultant offspring remains in the hands of the mother (Hobbes 1998, 109–10). However, as Hobbes notes, the terms of any agreement between them may allow for the man or some third party to take over dominion, as he tells us the Amazons did in respect of their male offspring. On the face of it, this marks a difference with life under a commonwealth, where dominion belongs to the man (Hobbes 1998, 109–10). This provides the strongest evidence for Pateman's reading of Hobbes, whereby the primitive sexual equality of the state of nature is firmly superseded once the commonwealth is in place. She notes, for instance, that 'in *Leviathan* he states that a family "consists of a man and his children, and servants together; wherein the father or master is the sovereign"'

(Pateman 1988, 47), and quotes a similar passage from *De cive*. On this view, Hobbes dallies with sexual equality in his description of the state of nature, only to jettison it once society is a going concern: 'Hobbes assumes that, in civil society, the subjection of women to men is secured through contract; not an enforced "contract" this time, but a marriage contract. Men have no need forcibly to overpower women when the civil law upholds their patriarchal political right through the marriage contract' (Pateman 1988, 48).

There is no doubt that Hobbes assumes that, in the specific form of social organization sustained by the institution of marriage, men will be the heads of families. To this extent they will stand in a relation to their spouses which is analogous to that between the sovereign and his or her subjects: that is, they—the sovereign or, again, the paterfamilias—will 'bear the person' of the relevant community, be it a family or a commonwealth. This relation does indeed involve a form of subsumption by the author or represented person in that of his or her representative. However, Hobbes makes it clear that such arrangements are *specific* to societies that incorporate the form of social arrangement known as 'marriage'. That is, it is open to those who contract to form a society to waive this arrangement in favour of others if they so choose. Indeed, this remains a possibility *within* societies that maintain the institution of marriage, as Hobbes states in *De cive*:

> In a commonwealth, if a man and a woman make a contract to live together, any children born belong to the father [...] and such a contract, if made in accordance with the civil laws, is called a 'marriage'. But if they only agree to concubinage, the children belong to either mother or father according to the different civil laws of different commonwealths.
> (Hobbes 1998, 109–10)

Thus the relation of 'concubinage',[23] namely cohabitation (*con* 'with', and *cubare* 'to lie') without the explicit sanction of the civil law, remains a possibility for Hobbes, as is fully consistent with his wider account of the grounds on which obligations arise. Again, in *Leviathan* Hobbes argues that the question of which sex is to be dominant in commonwealths 'is decided by the civil law: and for the most part, *but not always*, the sentence is in favour of the father, because for the most part commonwealths have been erected by the fathers, not by the mothers of families' (Hobbes 1996, 139–40; emphasis added).

Pateman's quotation of this passage omits this qualification, making it look as though in Hobbes's view a patriarchal institution of marriage was an invariable feature of life in civil society. That this is not so is made clear in *Leviathan* chapter XX, where Hobbes notes that '[i]f the mother be the father's subject, the child is in the father's power'. But he immediately adds that 'if the father be the mother's subject (as when a sovereign queen marries one of her subjects), the child is subject to the mother, because the father also is her subject' (Hobbes 1996, 140). Again, when Hobbes considers the possibility that a child is conceived by parents each of whom is the monarchical sovereign in his or her own land, dominion 'passes by contract'. In the absence of any such contract, 'the dominion follows the dominion of the place of his [i.e. the child's] residence', since the child will necessarily be subject of whomever is there sovereign.

These remarks testify to the very considerable coherence and consistency of Hobbes's thinking about the relations between men and women, and the abrogation by him of anything that might be construed as a principled defence of patriarchy. In this reading of Hobbes, I depart from Pateman's influential interpretation of his views on the relation between the sexes.[24] As Pateman notes, Hobbes departed from all previous contractarian theories in regarding all relations, *including* relations between the sexes, as political (Pateman 1988, e.g. 44). Pateman's point is that in naturalizing as a 'fact' the dominion of males over females, pre-Hobbesian writers removed this dominion from the political agenda by essentializing as given what in fact was merely the result of social convention.

HOBBES, PATRIARCHY AND THE CRITIQUE OF IDEOLOGY

Of course, the social constructivist analysis, both of the reification of 'gender' as 'natural', and of the concomitant gendered power relationship that marks patriarchy, is itself open to a form of immanent critique. After one has denatured supposed biological (or otherwise 'natural') facts of human behaviour as gendered constructions, the question immediately arises: from where do these constructions come? If the answer appeals to the claim that the constructions are impositions by a naturally dominant group, namely men, to provide an ideological rationalization of the brute fact of their dominion over another group, namely women, the argument will have gone round in a circle. The apparent alternative, to regard patriarchy as simply an accident, is not obviously satisfactory. Pateman seeks to corral Hobbes—his clear and radical differences from conventional patriarchal writers notwithstanding—into a defender of the established order of gender relations.

However, Hobbes's analysis can by-pass the explanatory difficulties facing social constructivist critiques of these relations. His theory has the resources needed to explain why supposedly 'natural' gender relations are, in fact, wholly conventional in their origins. Because it does not have to rely on—indeed it is subject to a radical critique of—the idea that gender relations are given by nature, Hobbes's theory can cope easily with the possibility that patriarchy arises only contingently. His own historical counter-examples to patriarchy make this contingency plain. He can locate, in the spurious universalism that sees a Western patriarchal model as dispositive for human societies everywhere and always, evidence of the same naturalization of contingency that marks the reification of gendered power relations themselves as 'natural' and hence 'universal'.

In other words, Hobbes furnishes the materials for what Critical Theory would call an *Ideologiekritik*: he can explain not only why the naturalization move is erroneous, but also why the very discourse of naturalism arises conventionally in such a way as to promote the illusion that the basis for extant power relations eludes critique, because it is 'natural' and therefore given prior to politics. This is not because Hobbes rejects naturalism across the board. Rather, the radical egalitarian naturalism that he espouses undercuts the pseudo-naturalistic basis for conventional gendered power relations. In its place, Hobbes puts agreement which, though structured by power, is explicitly *not* structured by gendered power. Of course, the critique of naturalism is implicit in the philosophical programme that Hobbes sets out in the opening chapters of *Leviathan*.

At this point, a Critical Theorist may smell a rat. If Hobbes is genuinely practising a form of *Ideologiekritik*, how come the theory that emerges is one of agreement structured by *power*? After all, the very concept of ideology can hardly be freed from the notion that the belief-structures in which people find themselves express or reflect attitudes that are functional for extant power relations, whether or not the holders of the beliefs benefit from those relations. So how can Hobbes tell a story in which the powerful achieve dominion that is not, in the pejorative sense, ideological? The answer is that the theory operates at a higher level of generality than that of specific ideological discourses. It provides the materials that can explain how particular value-laden constructions might come to be dominant in certain societies, without itself endorsing the *content* of these constructions—of prime importance among which has been, historically, the notion of the natural as putting extant power relations beyond question.

That of course does not preclude an ideological reading of Hobbes's theory itself, of the kind that C.B. Macpherson provided in *The Political Theory of Possessive Individualism* (Macpherson 1962). Macpherson reads *Leviathan* as seeking to legitimate the commercial relationships characteristic of early capitalism, centrally including those of contractual freedom on which Hobbes's theory of political legitimacy rests. Whatever may be said for this reading—and it has more going for it than some recent commentary has been willing to acknowledge—its truth would not in itself negate the claim that the theory provides the wherewithal to deconstruct patriarchy. The fact, if it is one, that in so doing the theory itself proves to instantiate ideology as well as to undermine some versions of it may show merely that no Archimedean point exists wholly beyond the reach of ideology—which does not prevent a given theory from unmasking the ideological content of other discursive practices.

CONCLUSION

I have argued that Hobbes's account of gender relations and the family derives entirely from his general account of contracting and submission as supervenient on natural equality. As such, Hobbes is largely free of the patriarchal assumptions dominant in his time, and in some respects provides the raw materials for a more thoroughgoing liberal critique of patriarchy than those to be found in works of some prominent modern liberals.

NOTES

1 This allegation is recorded in Bishop White Kennett's memoir of Hobbes. See Newey 2008, 18.
2 See for example Hobbes's letter to Margaret Cavendish, Marchioness of Newcastle, of 9 February 1662, in *Thomas Hobbes: the Correspondence*, N. Malcolm (ed.) 1997, Vol. II, 524–25. The Marchioness aspired to be a thinker and *littérateuse* in her own right. She wrote plays, as well as a work of philosophy, *The Description of a New World, Called the Blazing World.* Hobbes contributed a short piece to a collection of poems and letters in her honour in 1662.
3 See Rawls 1972; for the most influential statement of the feminist critique of Rawls, see Okin 1987, and Okin 1989. Also Martha Nussbaum 2003, 'Rawls and Feminism', in Freeman (ed.).
4 Some feminist respondents to this statement of Rawls's theory have doubted whether his efforts to accommodate the family as a site of distributive justice are compatible with his championing of 'political' liberalism, or whether it relapses into a form of 'comprehensive' liberalism that Rawls is avowedly seeking to avoid.
5 On a more radical feminist critique of Rawls, the very project of conceptualizing 'justice' in terms of distributive principles devised in the abstract by the thought experiment of the Original Position, betrays gendered attitudes. See Gilligan 1993.
6 In subsequent citations from these works I modernize spelling.
7 Cf. *Leviathan*, ch. XX: 'a great Family if it be not part of some commonwealth, is of itself, as to the rights of sovereignty, a little monarchy' (Hobbes 1996, 142).
8 See also the important works on this subject in Schochet 1967; also Schochet 1975, and Schochet 1990.
9 The issue is, for instance, not addressed in Schochet's full-length work on patriarchalism (Schochet 1975)
10 The *Oxford English Dictionary* entry for 'rape' cites the earliest occurrences of the word from the 15th century. It was in current usage in 17th-century England.
11 Cf. Hobbes 1994, 131–32, where Hobbes also seems to have in mind covenanting outside the framework of civil law.
12 Hobbes cites this principle at various points. For example, Hobbes 1994, 120: 'it is against reason for the same man, both to do and complain; implying this contradiction, that whereas he first ratified the people's acts in general he now disallows some of them in particular. It is therefore said truly, *volenti non fit injuria*'.
13 See, for example, Martinich 1999, 157–58.
14 Cf. Hobbes 1998, I §8: 'But because it is in vain for a man to have a right to the end, if the right to the necessary means be denied him; it follows, that since every man hath a right to preserve himself, he must also be allowed a right to use all the means, and do all the actions, without which he cannot preserve himself.'
15 For further discussion see Stoffell 1991.
16 Even here, again, Hobbes explicitly allows that the children of a concubine may belong to her.
17 For these reasons I demur from Stanlick 2001 and Hirschmann 2007, both of whom suggest that for Hobbes, men's dominion over the family is the paradigm for their dominion over the polity.
18 Schochet (1975, 241) rightly notes that Hobbes was innovative in giving familial dominion a basis in consent rather than in nature, but assumes that the consent must be directed towards the father, Hobbes's explicit denials of this notwithstanding.

19 It informs, for instance, Aristotle's hylomorphist account of sexual reproduction in his *De Generatione et Corruptione.*

20 A failure to recognize the force that this consideration had for Hobbes underlies, in my view, the mistaken interpretations offered in Schochet 1975, 235 and Hirschmann 2007, 44, that because the basis for both political power and domestic power lies for Hobbes in acquisition (rather than institution), both the polity and the family will be dominated by men.

21 Cf. Hobbes 1996, 140; Hobbes 1994, 131.

22 A.P. Martinich (1999, 157–58) argues that Hobbes is confused in granting to the mother a right over her offspring, since that right follows trivially from the fact that, in the state of nature, everybody has a right to all things. However, in his discussion of maternal rights, Hobbes is not talking of the pure liberty-rights of the state of nature, but the claim-right which is thought to arise from the infant's *acceptance* of the mother's dominion over it.

23 Historically, female concubines were often of inferior social status to their male partners and to that extent comprised relations of unequals. However, Hobbes says nothing that indicates that he regards this connotation as intrinsic to 'concubinage', which simply signifies a long-term cohabitative sexual relationship that is not given formal legal effect.

24 For further criticism of Pateman, see Wright 2002.

Politics of difference and activism at intersections: Romani women in Eastern and Central Europe[1]

LENA J. KRUCKENBERG

'There is not one single definition of a "true" Roma woman.'

(Joint Statement of European Roma Women Activists, in ODIHR et al. 2006, 13)

'Romani women in Romania, as elsewhere in Europe, are stereotyped as illiterate, loud, lazy, and irresponsibly burdening the state by bearing too many children, too early, and too often.'

(Surdu and Surdu 2006, 5)

Romani women[2] in Central and Eastern Europe (CEE) face enormous challenges. Their experiences of intertwined forms of racial and gender discrimination demonstrate in which ways structures of social domination and subordination *intersect* in everyday lives. Such structural patterns do not have an impact independently of each other: they cross social boundaries and reach from the public well into the private sphere. It is the complexity of the experiences of Romani women that is the subject of this chapter. These experiences are also the source of Romani women's activism, which addresses persistent inequalities and patterns of subordination within their own communities as well as within the societies of CEE where the majority of them live.[3] This chapter contributes to discussions on politics of difference, intersectionality and *activism at intersections* as they arise from the experiences of Romani women.

As the Joint Statement of European Roma Women Activists points out, not one single definition of a 'true' Romani woman exists. In fact, there does not exist a single definition of a 'true' or 'authentic' woman: this term fails to be exclusive as, 'if one "is" a woman, that is surely not all one is' (Butler 1990, 3). However, women are as a matter of course and subconsciously identified and addressed as being 'women'—by their neighbours, siblings, colleagues or the police. Women who are recognized as Romani are confronted with the repercussions of stereotypes such as those in the quotations above. Even if there is no single definition of a Romani woman as such, there are, none the less, their distinct experiences of what it means to be perceived as being a Romni by non-Roma as well as by their fellow Roma. Every Romani woman may define her identity in a different way by emphasizing different aspects of culture and heritage, social status or activism. However, in doing so she simultaneously positions herself within collective understandings of what it means to be a Romni.[4]

In the introduction to her seminal study on gender divisions and gender politics in international relations, Cynthia Enloe writes that she had started with Pocahontas, the Native American princess, and had ended up 'mulling' over Carmen Miranda (Enloe 1989, xi), the Brazilian singer and Hollywood star of the 1940s and 1950s. This chapter shares her perspective, as it reflects on a broad range of women's biographies and the conflicting forces that shaped them. Some of these forces originate in families, others in particular communities; some relate to more encompassing social structures, some arise from the international sphere. Each single biography rests on different social identities that are situated within different contexts.[5] Stuart Hall argues that individuals speak 'from a particular place, out of a particular history, out of a particular experience, a particular culture' (1988, 258). However, it is important to note that some of these places, or social positions, seem to be more coherent, more obviously distinct, and closer to prototypes as they are represented in narratives passed down by generations. 'White man', for example, can be understood as a reference point for many typical categories of social relationships. However, it goes without saying that choosing a particular 'prototype' or starting point is rather a question of conventions and their dominant narratives than of pertinence or even necessity. Why not start with the Romani woman?

Discrimination is considered to arise from segregation and stigmatization. Analysing these implicitly suggests as starting point the dominant, 'normalized' position of e.g. a 'white man' and the relative difference to the subordinate, 'exceptionalized' one of e.g. a Romani woman. Instead of shedding light on how those in the superordinate position achieved their advantages, most narratives of inequality and discrimination focus on the accrual of disadvantages in the subordinate position, leaving dominant groups often unacknowledged (Warner 2008, 456; Valentine 2007, 14). None the less, it seems justified that much of the literature on intersectionality starts from the abhorrent abuse that women of colour still suffer all over the world, and focuses on the deep-rooted patterns of discrimination that strap them to the bottom of their societies. In the European context, the history and current situation of Romani women can be regarded as exemplary of the devastating consequences of long-lasting and intersecting patterns of discrimination related to race and gender. The exploration of underlying mechanisms and resulting experiences of discrimination leads on to more general questions about ways in which patterns of inequality and discrimination emerge and are stabilized through politics of difference. From this perspective, reflecting on intersectionality implies to identify and examine patterns of socially constructed distinctions such as between men/women and majority/minority and the processes through which these distinctions *interlace.* Within the analytical framework of 'intersectionality' (Crenshaw 1991), the complex and 'mutually constitutive relationships' between two or more devalued facets of a multiple identity are articulated (Shields 2008, 301; Krane et al. 2000, 3). Intersectionality is also constitutive of the *lived experiences* of Romani women. As their lives emerge from ethnographies, blogs, articles, interviews and reports, Romani women themselves reveal a clear understanding of intersectionality.

Notwithstanding the difficult situation in which Romani women find themselves, it is important to transcend mere description and examination of the causes and effects of their subordinate position. Romani women's activism has its roots in the tensions between intersecting identities of race and gender; it seeks to transcend exclusion to open up possibilities of inclusion. An exclusive focus on patterns of victimization and experiences of oppression loses sight of these possibilities. This perspective tends to ignore the struggle for rights and recognition by many women in their multiple capacities (Kapur 2002, 16). In this contribution I want to highlight ways in which intersectionality is addressed—and not merely represented—by Romani women activists and advocates of their cause. Rather than seeing gender and ethnic movements as mutually exclusive, Romani women's activism stands a chance to overcome these divisions.

This chapter starts with an exploration into the current situation of Romani women in CEE. It is decisive to understand their situation, before turning to the next section in which a theoretical framework of politics of difference, and of persistent inequalities is developed based on the work of Charles Tilly (1999) on 'durable inequality'. Within this framework, the characteristics and effects of 'structural intersectionality' (Crenshaw 1991, 1,245) are analysed. The following section explores more complex forms of intersectionality. I argue that perceiving intersecting categories as fundamentally different and separate not only constitutes a central problem of intersectionality as lived experience, but also reflects on the ways intersectionality is addressed and responded to. Many Romani women find themselves in the 'empty space' between Romani and feminist discourses. However, more and more Romani women in CEE make their voice heard *and* understood. Their activism demonstrates that their intersecting identities function as a seedbed for outreach and enables practices of inclusion. Finally, the role of Romani women's experiences and activism in changing our understanding of intersectionality and politics of difference will be explored.

ROMANI WOMEN IN CEE: PATTERNS OF DISCRIMINATION AND STRUCTURAL INEQUALITIES

Romani women are in a particularly disadvantaged situation as they suffer from exclusionary practices both as women and as members of a discriminated minority. Belonging to Europe's largest and also most vulnerable (Gil-Robles 2006, 5) 'archipelago' of minorities (Marushiakova and Popov 2001, 33), many find themselves at the bottom of Europe's societies, in particular in CEE where the majority of the approximately 10 million European

Roma live (Bancroft 2001, 146; European Commission 2004, 9). Only a few of them are still nomadic (UNDP 2002, 13).[6] Describing the Roma as a single, unified ethnic group rather reflects the perceptions of the majority populations than those of many Roma themselves. 'This imagined community shares no common language [...], culture, religion, identity, history or even ethnicity' (Kovats 2003, 5). Accordingly, most Roma derive their individual sense of identity from subgroups and extended families (see Liégeois 2007; Marushiakova and Popov 2001), which does not imply that there is no sense of an *overarching* community. Since their arrival in the Balkans nearly 1,000 years ago (Marushiakova and Popov 2001, 33), Roma have been subjected to slavery, prosecution and several attempts of genocide (Crowe 1991, 155). Being perceived as 'strangers' who are simultaneously 'close' and 'remote' (Simmel 1971, 143), they experienced exclusion and oppression throughout their history (Guy 1975, 203), and they continue to do so. After the fall of communism much of the safety net of social provisions was abolished, including health care, and, in many regions, housing (Pogány 2004, 58). Roma, in particular Romani women, often were the first who were laid off, and they stood little chance in post-communist labour markets (ERRC 2007a). Large proportions of the region's Romani minorities suffer from severe poverty; in some regions, poverty rates are more than 10 times higher than those of non-Roma (Ringold et al. 2005, xiv). This 'racialization of poverty' (Kligman 2001, 64) coincides with a disproportionately high number of women in the ranks of the poor.[7] Here, intersecting patterns of racist and gender-based discrimination create a 'glass-box' (ERRC 2007a) from which Romani women can hardly escape.

Discrimination in the labour market corresponds to widespread discrimination in education. High proportions of Romani children in the region have been systematically assigned to segregated, inferior schools or 'special schools' for children with learning difficulties (ERRC 1999). In some regions of Slovakia more than 80% of Roma children were placed in such 'specialized institutions', while only 3% reached secondary education (Gil-Robles 2006, 21). Many Roma live in segregated settlements with limited access to basic medical care, clean water and sanitation (Pogány 2004, 1), and their life expectancy is estimated at 10 years below that of the majority populations in the region (Ringold et al. 2005, 48).They are over-represented in all groups in need of special protection, namely 'the very poor, the long-term unemployed, the unskilled, the uneducated, members of large families, individuals without resident permits, identity documents or citizenship papers' (Asylum Aid 2002, 13). Until recently, neither the governments of the region, nor (pro-)Romani non-governmental organizations (NGOs) paid much attention to the particularly difficult situations many Romani women face. This neglect is even more glaring, as 'it is usually the Roma/Gypsy women who keep the contact with the majority communities' (Bitu 1999).

Economic exclusion coincides with social marginalization in an environment where nationalism is on the rise. Prejudice in the general public and institutional racism are widespread, and both create a vicious cycle of constant repetition and mutual reinforcement.[8] Roma suffer from racist attacks in the streets, pogroms, evictions from their communities, and abuses by the police. Roma are denied equal access to public services (Pogány 2004, 5) and they are treated less favourably by state authorities in general (Pogány 2004, 13). Romani women are confronted with specific forms of discrimination in public institutions, such as segregated maternity wards, inferior or degrading treatment in the health care system, and violent abuse such as coerced sterilization (Council of Europe and EUMC 2003).[9] Poverty and marginalization make Romani women highly vulnerable to trafficking and other types of exploitation such as forced begging (Bitu 1999).

Romani women also find themselves in a subordinate position in their communities and families, which limits their choices for their lives. As Pogány points out: 'It is almost as if the rigid social hierarchies that have confined most Roma to a lower social status than almost all non-Roma in Central and Eastern Europe, have been mirrored by the gender-based stratification characteristic of Romani culture(s)' (2004, 116). In many communities the purity and chastity of women are understood as defining the honour of the family and are thus of distinct importance (Bitu 1999). These beliefs provide the background for practices such as arranged marriages of children or juveniles and virginity testing.[10] Romani women find themselves not only at the intersection between racial and gender discrimination, but also between the traditional beliefs of their community and those proclaimed by the majority population surrounding them.

These conditions within their own communities as well as the majority society all contribute to the vulnerability of Romani women to domestic violence. Incidents of domestic violence occur generally more often in communities that experience serious economic and social problems, and the humiliation and discrimination that many Romani men experience outside their community certainly is a serious factor. Patriarchal patterns prevent Romani women from reporting their victimization at the hands of their partners, in-laws or other male family members (Cahn 2007). Confronted with neglect and ignorance by public institutions—the police, the judicial system and even women's shelters—Romani women who try to fight against domestic violence encounter mounting obstacles. Still, Romani women live in societies where domestic violence is widespread and acceptable to an extent that a 2000 Romanian edition of *Playboy* magazine included an article explaining 'how to beat one's wife without leaving marks' (Asylum Aid 2002, 165).

It is important to bear in mind that these problems arise from multiple contexts, and depend, amongst other factors, on a Romni's marital status, the particular group to which she belongs, whether she has children, or lives in an urban or rural area, her social class and citizenship status, her religion, and the conditions in the majority society (Bitu 1999). There is the mother of eight living in a remote Romani village in Romania, a single local government administrator in Sofia, an unemployed community nurse and grandmother in rural Slovakia, or an NGO activist studying in Budapest. Considering the diversity of these lives it is all the more remarkable that it is, none the less, possible clearly to identify patterns of disadvantage, exploitation and oppression that apply to the majority of Romani women, as they are perceived to be 'different', both as women and Roma.

POLITICS OF DIFFERENCE OR THE RELATIVE ORDER OF INTERSECTING CATEGORIZATIONS

How can we make sense of the situation of Romani women? What politics of difference are at the roots of a social order in which being a Romni significantly affects all minutiae of life? Current debates on multiculturalism and cultural difference often address these questions within a framework that focuses on state governments as they interact with (their) ethnic minorities. Young (2007, 74) describes this approach as '*politics of cultural difference*' insofar as it establishes a perspective that critically addresses the 'absolutist impulses' of the majority populations in CEE as these seek to segregate and subordinate this culturally distinct group. Politics of cultural difference emerge around issues of *recognition* and *accommodation*. This approach frames the discrimination of Romani *women* in terms of a 'minority-within-a-minority' problem, nested in the particular traditions of Romani minorities as well as the overall 'societal culture' (Kymlicka 1995, 18) of CEE states. However, is a severe deficiency in terms of recognition and accommodation the main cause of their problems—though without doubt both account for many of them? Indeed, in many cases these problems seem not to stem from the failure to recognize and appreciate Romani ethnicity and culture, but from a failure to recognize Roma's *humanity*. If cultural differences actually were the core problem, culturally assimilated Roma would not encounter racism—which apparently is not true. Rather, deeply engrained norms and recurrent practices of stereotype and segregation that exclude Roma from high status positions and assign them to unemployment and poverty, all point to a *racial* hierarchy (Mills 2007, 94).[11] In addition, processes of social discrimination pervade all kind of social institutions, the state as well as non-governmental actors outside Romani communities, their own minority institutions as well as private networks and families. Consequently, only fragments of the experience of a Romani woman can be understood in terms of her membership in a minority (women) *within* a minority (Roma). The framework of the politics of cultural difference might be appropriate for issues such as representation, political accommodation or minority rights, but it diverts attention from structural manifestations of racism and sexism (Young 2007, 81).

In order to address such multi-faceted subordination, '*politics of positional difference*', as Young (2007, 64) termed this approach, seems more suitable. Here, experiences of discrimination and inequality are understood as structural social processes: 'Some institutional rules and practices, the operation of hegemonic norms, the shape of economic and political incentives, the physical effects of past actions and policies, and people acting on stereotypical assumptions, all conspire to produce systematic and reinforcing inequalities between groups' (Ibid.). Patterns of social difference manifest themselves in unequal access to, and distribution of, resources and opportunities, and thus delineate social boundaries (Lamont and Molnár 2002, 168). These boundaries confine Romani women to circumscribed spaces in economic, social and political hierarchies and prevent them from developing and exercising their capabilities. The politics of positional difference approach theorizes gender as well as race as a 'set of structural social positions' and equally addresses both racial and gender boundaries (Young 2007, 87). Further, *structural inequality* can account for the *persistence* of the subordinated positions in which most Romani women find themselves, in as much as these are sediments of the social history of the region, its politics of difference and racialization.

A decade ago Charles Tilly (1999, 6) asked 'How, why, and with what consequences do long-lasting, systematic inequalities in life chances distinguish members of different socially defined categories of persons?' He developed the concept of 'durable inequalities', i.e. inequalities that are persisting through social interactions, the life course, or even generations—exactly like those that can be observed for Romani women in CEE. Tilly argues that significant inequalities of chances and advantages mainly correspond to 'bounded categories' such as black/white or man/woman, and thus depend on the institutionalization of such categorical juxtapositions and the social boundaries that they create (Tilly 1999, 8). The mechanisms that make inequalities 'durable' operate through *social interactions* and the *collective experiences* of these (Tilly 1999, 100). Transactions and communication across boundaries generate advantages for one group and disadvantages for the other (Tilly 2000, 782), encourage 'opportunity hoarding' as well as 'exploitation', and hence perpetuate 'durable inequality' (Tilly 2005, 73).

The asymmetrical nature of the Roma/non-Roma categorization is obvious in the many inferiorizing descriptions of their culture and appearance, foremost their skin colour. Recent surveys show that for official purposes such as registration many Roma in CEE seek to 'pass' as non-Roma (Cahn 2007) in order to avoid further stigmatization. However, non-Romani Europeans try to expose and reject these efforts: 'The effort to "pass" is exposed where attempted, if the person concerned is not "white" enough to pass. Or, more likely, it is not openly exposed, but the person is quietly treated differently: denied work, rejected school placement, excluded from various benefits' (Ibid.). The boundary between Roma and non-Roma is never to be violated, and thus becomes *relevant* in all interactions 'across'—such as seeking employment, school registration or claiming benefits. This similarly applies to gender relations. Being perceived as a Romni does make a difference when communicating with fathers and husbands, as well as with police officers, teachers or welfare administrators. 'Doing difference' (West and Fenstermaker 1995) and scripts of communicating or transacting across boundaries differ between social situations, but persist across wider contexts. Distinctions of race and gender instruct interactions in all realms of life, in work and welfare, as well as in restaurants or even clubs. Signs indicating that Roma are not admitted to a club, or discriminating attitudes by doormen, demonstrate the relevance of the Roma/non-Roma boundary. Patriarchal traditions prohibiting young women from going out and dancing (ERRC 2000) add a gender-based barrier, as possibly does the behaviour of men who party in these places.

The scripts of encounters across boundaries are related to each other and coalesce into coherent stereotypes and patterns. The transfer of boundaries from one setting, such as a school, to another, such as a job centre, generalizes the influence of categorical distinctions, a process that Tilly terms 'emulation' (Tilly 1999, 11). When applied in various settings, the distinctions between Roma/non-Roma and man/woman are 'naturalized' and more readily available. They lend themselves to be used as organizing principles across contexts, become habitual, and hence less likely to be questioned. For Tilly, the type of the distinction, whether in terms of gender or race, does not make a difference with respect to the mechanisms that drive the formation of 'durable inequalities' (Tilly 1999, 64; see also West and Fenstermaker 1995, 22). However, their predominance varies with the specific social setting. For Romani women racial discrimination mainly operates within the public realm, and at the boundary between majority and minority population, while gender discrimination cuts through households, communities, class, etc., and is thus to be found in public as well as private spaces.

Within his framework of bounded categories, Tilly has little to offer for our understanding of *intersecting* identities and discrimination. Is the experience of being a Romani woman just the sum of two distinctive forms of discrimination associated with both her identities? Tilly observes that boundaries have to remain partial as not every distinction is operative in each situation (Tilly 2005, 73). Crenshaw uses here a concept of 'overlapping systems of subordination' (Crenshaw 1991, 1,265). Her conceptualization of 'structural intersectionality' (1,245) implies that subordinate identities are related in a synergistic way when intersecting at the crossroads of race, gender, class and other categorizations. However, the image of a road junction where race and gender cross each other rests on the tacit assumption that these categories are in principle independent outside of their junction (Walgenbach 2007, 49). It is impossible to experience gender without simultaneously experiencing race (West and Fenstermaker 1995, 9, 13), as everyone has a multiple identity emerging from various intersections. Race and gender do not have interdependent relations, but are interdependent in themselves (Walgenbach 2007, 61). Such a conceptualization of 'structural intersectionality' implies that the experience of intersecting *devalued* identities (e.g. 'Romani' and 'women') is fundamentally different from the experience of intersecting *valued* identities (e.g. 'white' and 'man').

Patterns of discrimination such as domestic violence, early marriage and deprivation of education are closely linked and reinforce each other, thus creating a dynamic process of devaluation. In order to achieve a deeper understanding of the politics of difference involved here, it is important not to start from the dominating position and the practices of domination and active discrimination that define it. As proposed in the introduction, the focal point of the analysis is the subordinate position and the experience of devaluation which is inevitably linked to it. Figure 2.1 represents this approach, with the Romani woman in the crosshairs of two axes. The axes represent *two analytical dimensions* along which distinctions are being made with respect to gender (man/woman) and race (Roma/non-Roma). These dimensions span a social space which is divided into four fields each defined by a race/gender intersectional group. This creates four exemplary *situational boundaries* for the Romani woman, juxtapositioning her to the four gender/race groups. The shading indicates social distance between her and those groups. This figure shows how each of the relationships between a Romani woman and other Romani women, Romani men, non-Romani men and non-Romani women entail particular, immediate boundaries that are attached to, but do not exclusively consist of gender or racial distinctions.

Let us now explore this representation of structural intersectionality for a problem that is seminal in the production of 'durable inequality': education. Along the race axis particular patterns are easily identified. A disproportionate number of Romani girls are sent to schools for children with learning difficulties, where they experience racial discrimination by non-Romani students (Surdu and Surdu 2006, 52). They are also likely to receive less attention and support from teachers, not the least because these expect them to

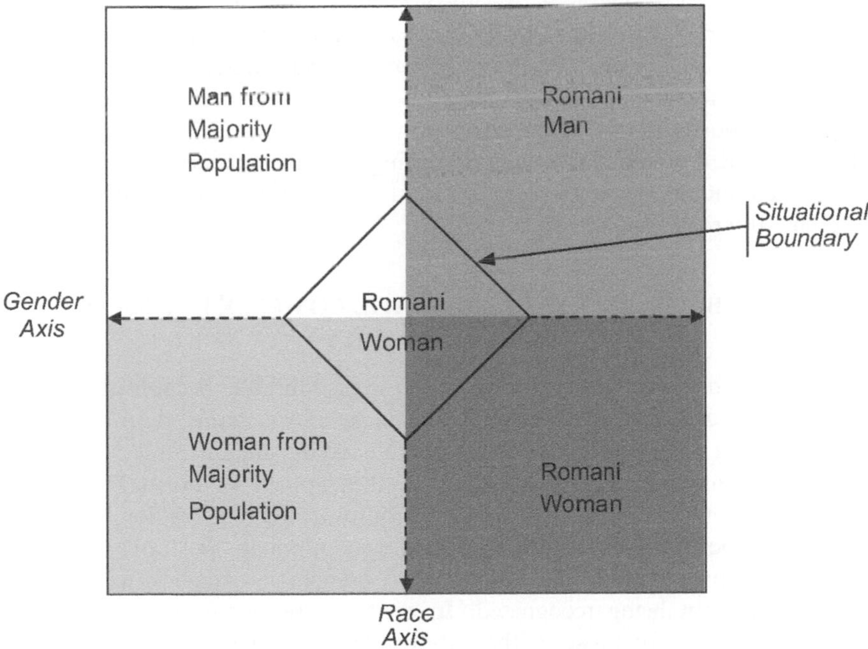

Figure 2.1 Structural intersectionality and situational boundaries.

drop out at an early age. A high number of early and child marriages as well as teenage pregnancies amongst Roma account for this attitude. Authorities in CEE states often turn a blind eye on child marriages even if they are legally prohibited. This situation is exacerbated by the lack of provisions facilitating re-entry of young mothers into education. Taken together, these conditions severely diminish the chances of Romani girls achieving even basic levels of education. Furthermore, many young girls are burdened with domestic chores and have to take care of their siblings from an early age (Surdu and Surdu 2006, 38). Further, Romani girls suffer from a lack of support not only in the educational system but also at home due to patriarchal attitudes within Romani communities that deem education as predominantly important for the male breadwinner. This position is reflected in a statement of a Romani woman from Romania: 'We don't have any expectations when it comes to girls […] It is likely that tomorrow she'll get married and then follow her man, and she doesn't need to study anymore. Plus, her husband probably won't let her get a job, anyway.' (Surdu and Surdu 2006, 46). In some traditional communities Romani women who achieve higher education are considered as non-Romani and abandoning their own traditions. Combined with discrimination in the labour market, these attitudes negatively affect the motivation of girls (Kligman 2001, 69) who have to rely on their often even less educated mothers for help with homework and relating to teachers (Surdu and Surdu 2006, 48). In summary, the less than sufficient education of Romani girls and women is the result of complex interactions between situational boundaries within the community on the one hand, and the majority society on the other hand. 'Positions, identities, and differences are made and unmade, claimed and rejected' (Valentine 2007, 14) each time a Romani girl encounters mostly non-Romani teachers, fellow students, her Romani father, later husband, her mother or others. Living at these intersections confronts her with conflicting demands from different social orders, and positions her at different *situational boundaries.*

FALLING BETWEEN THE CRACKS—THE COMPLEX LAYERS OF INTERSECTIONALITY

Tilly argued that inequalities are so surprisingly 'durable' because they often come disguised as lack of individual talents or efforts rather than structural phenomena (Tilly 1999, 6). As they operate across innumerable situational boundaries embedded in all realms of life, they appear as 'normal' and consequently become *invisible.* However, invisibility is different for dominant and subordinate positions. Intersecting identities in dominant positions are often perceived as prototypes and collapsed into *one* single category, its intersectional nature not being recognized. In contrast, the artificial *separation* of multiple identities for those in the subordinate positions implies a focus on one *or* the other dimension. Experiences of intersectionality in these positions are often ignored when academics, politicians or activists engage with intersectionality; they separate 'women's experiences' from 'Romani experiences', as if these were mutually exclusive. However, not all Roma are men, not all women in CEE belong to the majority population. Separating race from gender misses the point, as it unduly reduces the complexity of experiences of intersectionality. This phenomenon is quite common in what can be termed the 'intersectional fallacy' of official statistics. As Oprea (2003) demonstrates, statistical data are often collected either according to gender *or* racial 'differences' from the dominant 'norm'. This renders Romani women invisible in many statistical reports. Most information on the situation of Romani women, therefore, is contained in very few reports that are written on their particular experience, whereas reports on women in the CEE region or on Roma in general rarely provide disaggregated data or more specific information on their lives (Ibid.).

A similar fallacy can be observed for social and political movements that seek to address different forms of structural inequality and discrimination. Notwithstanding that Romani women are *formally* represented by both the Romani movement and the women's rights movement, until rather recently they were left without any representation proper. The patriarchal organization of the Romani movement marginalizes them as women, as does their Romani identity within the women's rights movement. Crenshaw identifies this problem as 'political intersectionality' (1991, 1,252) and argues that 'the failure of feminism to interrogate race means that the resistance structures of feminism will often replicate and reinforce the subordination of people of color' (Ibid.)—and vice versa for the Romani movement. Enisa Eminova, a Romani woman activist had the following experience:

> An American woman was doing an e-mail introduction between me and another Canadian woman to talk about gender studies in Canada. She introduced me as: 'Enisa is a young Roma quasi feminist'. I opened the dictionary to check definitions for quasi and it says: appears to be something but not really so; partly, almost. I did not ask her what she meant by that because I did not want to listen to any possible explanations coming from a Western, white, privileged perspective telling me if I was a real feminist or a quasi feminist in my context [....] If I was to define myself and my work [....] I would probably say I am [a] feminist activist challenging embedded cultural prejudice in my community.
>
> (Eminova 2006, 35)

Romani women activists encounter analogous problems in the Romani movement. When a Romni raised women's issues at a roundtable discussion among Romani activists, a male Romani activist advised her that the problems of all Roma had to come first, before women's issues could be tackled in the community: 'To be clear and short, for now, in the eyes of the gadje, you are not a girl or woman first, you are a GYPSY' (Martin Demirovski in ERRC 2000).[12] His position is shared by many of the predominantly male leaders in the Romani movement. For Romani women activists the problem of political intersectionality arises from the fact that though gender and race intersect in their own identities, the major movements that address gender and racial discrimination do not. In this situation the impact of asymmetrical relations becomes manifest: 'Structural intersectionality' remains suppressed, invisible and left unaddressed by the gap that is created by the two movements. Romani women are left in limbo as Crenshaw notes: 'Existing within the overlapping margins of race and gender discourse and the empty space between, it is a location whose very nature is telling' (Crenshaw 1992, 403).

The resulting 'acute social invisibility' (Purdie-Vaughns and Eibach 2008, 381) at the intersections of race and gender is reflected in a range of social and political institutions. Most democratic governments have established departments for minority issues and separate ones for gender issues; most judicial systems treat claims of alleged racial and gender discrimination as independent and separate claims. The international human rights system separately established women's rights and minority rights, a convention on the elimination of gender discrimination and a different one on the elimination of racial discrimination; each of these is monitored by a different body of the UN. Notwithstanding current efforts to increase communication between these agencies and to unify anti-discrimination legislation, institutionally separated sets of norms are the rule. This reinforces perceptions that they are independent of each other.[13]

Figure 2.2 represents this situation of Romani women activists who again take centre stage. They are situated in the empty space of 'intersectional invisibility' (Purdie-Vaughns and Eibach 2008) created by the division between two distinct movements and human rights regimes. They operate on at least three levels of intersectionality: *structural intersectionality* designates their identities as women and Roma; *political intersectionality* defines the

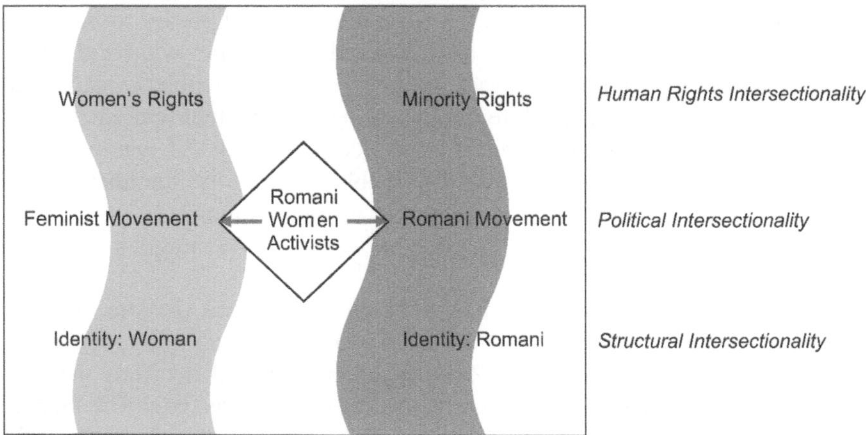

Figure 2.2 Intersectional invisibility: Romani women activists caught 'in between'.

space of their activism; the *intersectionality of human rights* delineates the space of normative orientations and claim-making.

ACTIVISM AT INTERSECTIONS

None the less, Romani women activists increasingly raise their voice in order to be heard and seen from both sides of the divide. Their activism has its roots in and takes up their *experiences of intersectionality at all levels.* These experiences are embedded in their own biographies and the environment of their families and communities (Schultz 2005, 265). They emerge from working in and with Romani and human rights NGOs, from contacts with women's organizations and government authorities, and rarely from lobbying international and regional organizations. Romani women's activism is a recent and complex development. It is, therefore, impossible to give an encompassing account of all its different initiatives (see e.g. Schultz 2005; Mihalache 2003; Izsak 2008; Bitu 1999, 2003, 2005). Instead, I focus on its particular characteristics, its self-definition, its outreach and, importantly, its own understanding of intersectionality.

Individual activist voices can be traced back well into the early 1990s (Bitu 1999). Only at the turn of the century did they become stronger and triggered an ongoing debate within the CEE Romani movement. Romani women activism finds its most visible expression in the form of networks of activists, which create flexible, horizontal channels for communication and debate, rather than existing as representative umbrella organizations. Members of these networks have highly diverse visions and approaches, and their experiences in the Romani movement differ widely. They organize meetings and conferences and seek support for networks that are based on internet and e-mail communication. They hope that the new media will help them to break up the isolation in which many Romani women at the grassroots level find themselves (see Metodieva 2003). Numerous Romnia participate in these networks in their capacity as primarily *individual* activists (Bitu in ERRC 2000), while others speak for small groups within Romani organizations. Finally, there is a small but rising number of Romani women's organizations and initiatives.

In contrast to other, highly organized women's movements, Romani women's activism does not aim at a homogeneous movement and a safe 'women's corner' (Bitu 2003), that protects female activists from challenging debates with male colleagues or family members. Instead, its objective is to achieve a broader understanding of *intersectionality*, namely *as a linked fate embedded in social relationships*: '[W]hen we speak about gender, we don't speak about women but about the relations between men and women in all aspects of life' (Memedova 2003). Romani women activists insist on their social, economical and emotional interdependence with their husbands, fathers, sons or male neighbours. Many of the initiatives and objectives of Romani women's activism target the situation of the *Roma in general* (Izsak 2008, 3). Participation of men (Romani and non-Romani) in meetings and activities is mostly welcome. Romani women activists explicitly want to *include* the experiences of male members of their communities. Their struggle focuses on the integration of a *Romani women's perspective* into the Romani movement and into all programmes, policies and initiatives targeting CEE Roma at large. In their endeavours, Romani women activists share the vision of 'bring[ing] the margins to the centre', as Azbija Memedova stated at the Roma Women's Forum, referring to bell hooks (NWP/OSI 2003, 35).

Even if their numbers are estimated at a few hundred at best (Ovalle 2006, 1) Romani women activists seem to be everywhere. They approach the Organization for Security Co-operation in Europe (OSCE) and the Council of Europe (see e.g. ODIHR et al. 2006; Mihalache 2003), which are both particularly active in dealing with Roma issues. A few Romani women activists also engage with national governments and the European Union. Lívia Járóka, a Romani activist and Member of the European Parliament for Hungary, presented a report on the situation of Romani women that was adopted in 2006, including several policy recommendations.[14] Other Romani women activists make their way on committees that design Roma policies throughout the region (Bitu 2005). Some lobby international human rights bodies and seek support from transnational human rights and development NGOs. Activists also speak to regional women's organizations and participate in major conferences of the global feminist movement.[15] Romnia increasingly are present in many organizations of the Romani movement and challenge its predominantly male leaders. They set up action research programmes within their own communities like surveys on sensitive gender issues, such as virginity testing, and thus start debates throughout the region (see e.g. Eminova 2005; Centre for Roma Initiatives 2006). Activists conduct their own research on the situation of women in their communities and provide urgently needed information (see e.g. Surdu and Surdu 2006; Centre for Roma Initiatives 2005). They deliberately utilize their unique position to get access to this knowledge.

In these ways Romani women's activism consciously uses and fills up the empty space created through political intersectionality. Figure 2.3 captures the very nature of this *activism at intersections*. It shows the web of relationships in which Romani women's activism emerged (dashed grey lines). Romani women's activism develops within and becomes part of existing networks, but also establishes new links of its own (black lines). Through Romani women's activism pro-Romani NGOs and initiatives are linked to women's rights activists. Romani women activists provide international women's organizations with information on Romani women at the grassroots level and, in turn, Romani women are informed about women's rights and the global feminist movement. In addition, Romani women's activism reinforces established relationships. In contrast to male Romani leaders, most Romani women entered the Romani movement from the local level (NWP/OSI 2002, 42) and, while moving into the regional or even global field of activism, 'constantly

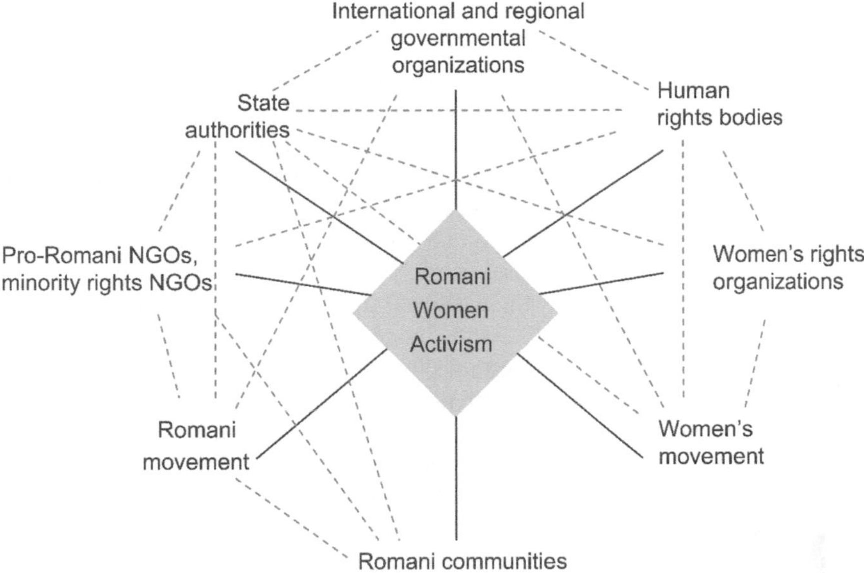

Figure 2.3 Activism at intersections.

interrogate their own privilege and challenge themselves to be accountable to grassroots communities' (Schultz 2005, 249). In this way they seek to bridge the hierarchical divide that had limited the male-dominated, 'top-down' Romani movement, as it had created a legitimacy deficit between 'official' Romani leaders and their own constituencies (Kovats 2003, 4).

However, an exclusive focus on the relatively few 'formal', English-speaking, and often university-educated, full-time activists omits a decisive section of Romani women activism. As Ovalle (2006, 2) rightly notes, many Romani women at the grassroots level launch their own local initiatives or become active in their individual capacities as mothers, home workers, carers or mediators. They express the feeling that 'something is wrong with the present state of inequality between Romani women and Romani men and the majority society' (Ibid.). In Pecs (Hungary), 16 young Romani women from a traditional community founded their own women's rights organization; some of them had been married as 14 year-olds, they were busy young mothers and none of them had enjoyed much education (Szakács 2003). Their activism had its seedbed in their experience that, 'obligated to serve their husbands and children, Roma[ni] women are not permitted to tend to their own dreams, desires, or even health' (Pálmai in NWP/OSI 2003, 29). Romani women are at the centre of their families and simultaneously function as contact points for the majority population and state authorities (Schultz 2005, 259). Their activism opens their communities to others and starts debates about women's roles in their communities. In this way, Romani women's activism is 'as much about personal change and self-empowerment as it is about collective and social change' (Ovalle 2006, 2).

'Romani women activists are a very diverse lot' (Schultz 2005, 246), as they come from various backgrounds, with different capacities, and act in a multiplicity of contexts. In addition, calls for women's rights challenge the perceptions of many Romani women. The older generation of Romnia are reluctant to question the patriarchal order of their communities, which they perceive as an integral part of Romani culture and tradition (NWP/OSI 2002, 45). A female Romani activist stated at a roundtable discussion: 'I do not like to talk about the issue of women's rights [....] Possibly these activists do not consider this damage to the community' (Sztojka in ERRC 2000). None the less, she supported women's activism, for example in the establishment of community centres where women could meet and discuss their problems (Ibid.). These traditional positions are criticized by a generation of young, university-trained women who do not hesitate to challenge Romani traditions from a feminist perspective, or the global women's movement from a Romani perspective. They are versed in academic discourses on feminism, multiculturalism and intersectionality.

Notwithstanding the considerable tensions between different generations of activists, they continue to work together. In 2006 the main Romani women networks, the International Roma Women Network and the Joint Romani Women Initiative, joined forces in order to integrate Romani women's perspectives into the agenda of a European conference on the implementation and harmonization of national policies for Roma. Romani women activists from 10 European countries discussed employment, housing, relationships with the police, arranged marriages and virginity testing. Even though 'a clear gap between the thinking of the younger and the older women' (Izsak 2008, 7) became obvious, in the end they adopted a 'Joint Statement of European Roma Women Activists', which acknowledges the differences amongst them in its opening statement:

> There is not one single definition of a 'true' Roma woman. Roma women across Europe are diverse [....] We are aware of our differences, we accept them and we fully appreciate them [....] We want to preserve our Romani culture but also acknowledge that there are harmful practices which violate the human rights of Roma women. Roma women recognize the existence of double standards in the movement for human rights of Roma. Double standards should not be present [....] One cannot fight racism in a society while discriminating others on the basis of gender in their community.
>
> (ODIHR et al. 2006, 13)

The statement furthermore includes a declaration on early and arranged/forced marriages as well as virginity testing as 'harmful to young women and men', and demands that they should 'be eliminated' as they 'are not Roma practices but [...] rather exist in every patriarchal society/community'.

This statement represents a milestone in the brief history of Romani women's activism. Its importance lies in their understanding of intersectionality. It disapproves of existing gender differences and all *asymmetrical relationships* in which Romani women find themselves, but neither rejects nor denies *differences* among them; instead these are 'appreciated' as expressions of diversity. The statement demonstrates an understanding of mechanisms of gender and racial discrimination as not essentially different. It transcends the notion that patriarchal inequality is a particular feature of Romani culture, and defines it as a gendered hierarchy irrespective of its cultural framing. In acknowledging 'men' as victims of practices such as forced marriages, the authors understand intersectionality as a linked fate between those whose identities are devalued in similar as well as different ways.

The 'Joint Statement of European Roma Women Activists' thus forwards an *inclusive* understanding of intersectionality. 'Bringing the margins to the centre' does not imply establishing a new centre (and thus new margins), but rather interrogating the conditions through which marginality is created. This understanding implicitly addresses the mechanisms of durable inequality as identified by Tilly (1999). Romani women activists do not predominantly seek essential representation of their particular intersecting identities, but rather an encompassing understanding of their perspectives. They frame multiple social identities as what they are: interdependent. They transform intersectionality from an unequal, seemingly static position into a series of social relationships emerging from interaction and communication across boundaries. Romani women activists utilize their intersecting identities and their experience of multiple boundaries as assets for inclusionary activism instead of being trapped in exclusion. A prominent Romani, Nicoleta Bitu, summarizes: 'some of us appear to be taking advantage of our multiple roles of women and our opportunity to have access to more roles than men do' (ERRC 2000).

The best way to characterize Romani women's activism is as *representation through mediation.* Instead of engaging in narrowly framed politics of identity and difference, Romani women activists *join forces with others across differences*, and thus Romani women's activism slowly moves from the margins to the centre. Figure 2.4 shows how Romani women's activism assumes a central role in mediating across boundaries between men and women in Romani communities, between traditional culture and a culture of change, between their communities and the majority population. Mediation allows for the creation of new links between the Romani movement and the women's movements, and recalibrates links like those between Romani organizations and state authorities. Finally, they mediate between abstract notions of intersectionality in academic discourses and a concept of intersectionality that is brimming with lived experience.

MOVING ON?

Walker (2004, 991) notes that 'the goal of understanding intersectionality is difficult to the extreme'. This chapter has demonstrated time and again how

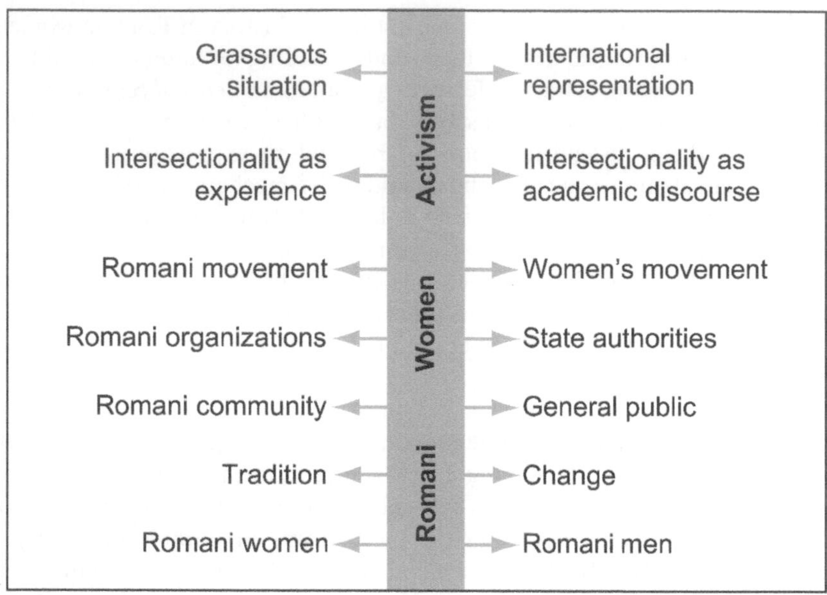

Figure 2.4 Representation as mediation: 'Bringing the margins to the centre.'

prevalent understandings of social categorizations obscure the very nature of intersecting identities. The ways in which these conceptualizations ensure order at the 'centre' entail rather diffuse and biased understandings of the experiences accumulated at the 'margins'. To a certain extent such framing reflects the subterranean adherence to seemingly independent boundaries of race and gender as they are enacted in everyday life. Even though race and gender are simultaneously experienced by everyone, current theorizing often artificially separates them, and thus contributes to the problem that intersectionality poses for us. Studies have as their subjects either gender or racial issues and identities. Many reports that implicitly address Romani women's issues fail even to recognize them in their intersecting identities as they approach their 'subject' from the perspective of the dominant position.

However, any attempt at understanding intersectionality requires us to move on and 'see things from the worldview of others and not [merely] from our own unique standpoints' (Walker 2004, 991). I followed this approach and let Romani women take centre stage. This was decisive for unearthing the interdependent character of race and gender. I showed in which ways domination and subordination emerge from interactions across the manifold situational boundaries of everyday life, and how these define asymmetrical social relationships that produce social exclusion. However, the numerous boundaries that Romani women activists encounter also increase chances for outreach, mediation as well as inclusion through recognition and redistribution. Romani women's activism lends itself as a rich source for theory-building on intersectionality and the politics of difference. Their activism challenges the dominance of prevalent narratives of race and gender within their own communities, the majority societies of CEE, the feminist movement and in academic discourses on intersectionality. A prominent activist of the younger generation asks: 'So, if we are told [...] that Western culture is more progressive, more democratic, more inclusive, then we also have to ask: where are we [...] represented, seen, heard, in the so-called West? Why is it that there is an invisibility that shrouds us [...]?' (Eminova 2006, 36). Her question implicitly reminds us that the recognition of dominance and privilege is as essential to any understanding of intersectionality as it is for the 'construction of inclusive politics' (Oprea 2004, 38).

Notwithstanding its vigorous and hopeful beginnings, it is more than likely that Romani women's activism will be a long, uphill struggle. A severe lack of resources inhibits their activism. Romani women activists know that they will not be able to change their situation significantly only by virtue of their own efforts. Azbija Memedova summarizes this feeling: 'Building a better society and integration of Roma women cannot happen by itself; it is an interactive process with at least two players—the one who has power and the other who does not. The basic precondition for a better society is the willingness of both to listen and understand' (NWP/OSI 2003, 34). *Representation through mediation* offers itself as a forward strategy for Romani women to be heard, but it does not guarantee that those who are addressed do listen and understand. This would require 'those with relatively more power to critically rethink the ways they had worked in the past' (Cole 2008, 448). Tilly (1999) reiterates this point in his analysis of durable inequalities. As asymmetrical relationships foster 'exploitation' and 'opportunity hoarding', it is clear that those in dominant positions have vested interests in these relationships. However, durable inequalities become invisible as they are emulated in and adapted to multiple categorical boundaries (Tilly 1999, 10). This process is uncovered and challenged by Romani women activists. They show that 'politics of difference' include more than mere *interest* politics—and that they also do not simply translate into *identity* politics.

Romani women activists understand intersectionality as a linked fate at the individual and collective level. In seeking representation through mediation they challenge both interest and identity politics. Like other women activists they 'find innovative ways to transcend simple identity politics by recognizing multiplicity within these categories' (Cole 2008, 444) and promote new notions of difference—and similarity.[16] In this way they move gender politics further into the realm of the politics of difference. Their activism demonstrates that 'gender politics begin to look very different if there is no essential woman at the core' (Zinn and Dill 1996, 322). One should add that politics of difference start to look different, too, once it is recognized that categorical differences do not exist independently of each other. In paraphrasing Zinn and Dill, intersectionality can only provide a framework for understanding the experiences of Romani women, if the experiences 'of all women, and men, as well' are included (Zinn and Dill 1996, 331).

NOTES

1 I owe a great debt of gratitude to Susanne Karstedt who commented on earlier drafts of this chapter. I am grateful to the editor Yoke-Lian Lee for her trust and support. Special thanks go to all the Romani women's organizations whose impressive and inspiring activism taught me a lot about intersectionality—and life in general.

2 'Roma' is an umbrella term for an archipelago of different groups all having their own names. As a term, 'Roma' is predominantly used in Central and Eastern Europe for groups who speak (or spoke) one of the languages of *Romanes*. As the term 'Gypsy' is often considered to be pejorative, it seems appropriate to use 'Roma' even if some Romani groups reject this name (Klímová-Alexander 2005, 31). I use 'Romani' as adjective, 'Roma' as the plural noun, 'Rom' as the singular noun for men, 'Romni' for a woman. 'Romnia' is the term for Romani women. For background information on the Roma see A–Z.

3 Being neither a Romni myself, nor of Eastern and Central European origin, this might look like a presumptuous undertaking. However, I do believe it is worthwhile to cross boundaries in an analysis of this kind. The argument that only Romani women are best placed to understand Romani women's issues is problematic in itself. This concerns the 'essentialist trap' that there is a core Romani experience to which only members of this minority are able to give voice. Further, rejecting cross-boundary analyses of this kind would imply that it is only appropriate for researchers to research their own individual experiences. This would severely limit social research and would restrict human experience in general.

4 It is of utmost importance to avoid essentialist assumptions by framing Romani identity and culture as closed, clearly bounded, internally uniform and having a true and authentic core. However, rejecting essentialism does not rule out the assumption that particular cultures and collective identities do exist (Mason 2007).

5 For the concept of identity used here see A–Z.

6 For overviews of the history of Roma in Europe see Hancock (1987), Marushiakova and Popov (2001), Liégeois (2007), Fraser (1992) and Crowe (2007).

7 Many stereotypes link being Roma with being poor, and make the one characteristic an essential part of the other. Kligman (2001) quotes from her interviews: 'How could I be a Gypsy when we have such a big house?' (70), or 'We are so poor that we barely survive […] if you are as poor as we are, in the eyes of others you are a Gypsy' (74).

8 Representatives of public and political life are no exception. Degrading speech can address Romani origin as well as gender: on 19 May 2007, the (male) President of Romania, Traian Basescu, addressing Andreea Pana, a female journalist, stated, 'You pussy, don't you have anything to do today?' and then said privately while being recorded, 'How aggressive that stinky gypsy was' (Scicluna 2007). These statements are indicative of public discourse across Europe.

9 Coerced sterilizations are documented for a number of Eastern European countries such as the Czech Republic and Slovakia (ERRC 2008) and are currently debated and addressed on the national and international level. See Centre for Reproductive Rights and Poradna (2003); Commission on Security and Co-operation in Europe (2006); Bond (2004). Detailed information can also be found on the websites of the European Roma Rights Centre (www.errc.org) and the Centre for Reproductive Rights (www.reproductiverights.org).

10 It is important to note that practices vary between different groups and even families, and statements like 'the Roma community itself openly embraces juvenile arranged marriage' (Timmerman 2004, 479) are misleading at best. However, the probability of being exposed to practices of child marriage and virginity testing is higher for Romani girls than for other girls in CEE. See Oprea (2005) on juvenile arranged marriages in the context of prevailing anti-Romani racism.

11 I use the term 'race', instead of 'ethnicity'. Ethnicity is predominantly perceived in cultural terms, while race refers to an assumption of 'ineluctable biological hierarchy' (Mills 2007, 101). The term ethnicity is often used 'as a way of circumventing the racist history of "race" and [is] associated with apparent cultural choices' (Gunew 2004, 21). Most Roma do not have a choice given the racist projections that majority populations link to their visible differences. The resulting process of racialization, albeit resting on perceived differences in culture, centres on a racist determination of what is *normal*, and what is *not*.

12 As Oprea notes: 'When put to the test of logic, this assertion proves to be false [....] The rhetoric of a "global struggle" against a uni-dimensional form of racism would only make sense if one were treating the Romani community as consisting only of heterosexual Romani men' (2004, 34).

13 Bond analyses violations of the reproductive rights of Romani women and the ways in which the international human rights regime reacted. She concludes that Romani women 'who experience multiple forms of human rights abuses operating simultaneously cannot fully benefit from a remedial system that artificially fragments human rights violations' (2004, 916).

14 Romani women are still dramatically under-represented in both state and local administrations (Mihalache 2003). Only a few have ever made it into party politics and in most CEE countries they are not represented in parliament at all. Lívia Járóka's position as MEP is exceptional.

15 As Nicoleta Bitu, an experienced activist, says: 'For me it's very rich when I learn the experience of non-Roma women, like minority women, like Muslim women, like Dalit women from India. It was very rich for me to learn from them and to realize that we're not unique in this sense. I think that's why [....] I am sensitive to opening the Roma issue to others [....] I'm not afraid of making comparisons' (Schultz 2005, 272).

16 Cole's concept of coalition through which she theorizes 'an alternative to categorical approaches to intersectionality' (2008, 443) partially coincides with what I term 'representation though mediation'.

Class, violence and masculinity in Hollywood movies: a neo-Marxian meta-narrative[1]

VASSILIS K. FOUSKAS

INTRODUCTION

The aim of this chapter is to re-create and interpret the ways in which a particular genre of Hollywood movies brings history and social reality into arts. We will focus on action movies in which acting and directing are influenced, directly or indirectly, by Lee Strasberg's 'method' school. To be more specific, we are proposing an interpretative and sociological contextualization of three movies. The first, released in 1980 by Martin Scorsese, *Raging Bull*, captures the social milieu of the 1940s through to the 1960s. The second, *Heat*, and the third, *Miami Vice*, were released in 1995 and 2006, respectively, and were directed by Michael Mann. In these three motion pictures, we plan to identify and interpret, by way of sociological meta-narrative, issues of class, violence and masculinity. How do class and masculinity, as reified social entities in capitalism, assume such artistic forms that are potent enough to transform social malaise, exploitation and violence into artistic pleasure? We do not want violence in our lives, but we do not mind seeing it in a movie.[2] Exploring this question constitutes the guiding thread of our efforts here. This, it should be said, is an 'anti-Adorno and anti-Horkheimer' question, whose 1944 influential essay presented mass culture as 'stylized barbarity', produced by what they termed as 'bourgeois culture industry' (Adorno and Horkheimer 2005).

Choosing these three films was a difficult, but not entirely arbitrary, undertaking. One could argue that *Scarface*, for example, could be the substitute for *Raging Bull*, and *Goodfellas* or *Scent of a Woman* a substitute for *Heat*. However, a closer look at *Raging Bull, Heat* and *Miami Vice* offers something more convincing: it offers a *structuralist* interpretation of the themes under examination here, while at the same time *historicizing* them. We have detected that these three films move from the classic sexism of *Raging Bull*, to the globalist and libertarian perspective of *Miami Vice.* However, at the same time there are meanings that do not change in any of the three movies. *Heat* is in a position—as we shall see below—to present all the explosive elements of an *avant-garde masculinity* in the operatic art of the 1990s (the contemporary-historical dimension), while incorporating some age-old themes about class and masculinity that can be found in every artistic commodity, and not only movies, across the centuries (the structuralist dimension).[3] Similarly, we saw *Raging Bull* as a repository for a number of cultural and sub-cultural events of post-war American society. Overall, the choice of movies was influenced by this dialogue between *structuralism* and *historicism*: we dropped movies in which the dialogue between these two philosophical currents was unbalanced.

How do we use the terms 'meta-narrative', 'violence', 'masculinity' and 'class' here? Postmodern objections could be raised in that 'meta-narratives' constitute an all-encompassing, in fact, unrealistic, attempt to reconfigure holistic interpretations of the social reality, which is itself fragmented, compartmentalized and localized. 'I define post-modern', said Jean-Francois Lyotard in his classic, *The Post-modern Condition; A Report on Knowledge*, 'as incredulity toward meta-narratives' (Lyotard 1979, xxiv–xxv). We adhere to some penetrating criticisms of postmodernism (Callinicos 1991; Habermas 1981). We argue that only a post-Hegelian, neo-Marxian perspective can capture the meanings of social action in history. By 'meta-narrative' we mean the holistic structuring of theoretical and cognitive orders along the themes of 'violence', 'masculinity' and 'class', themes that are themselves structured in capitalism in reified forms. The de-reification of these themes from the researcher dictates exactly forms of meta-narrative, which is the only way to capture the fragmented, segregated and indeed reified social order caused by the advanced capitalistic forms of social/technical division of labour.

If 'meta-narrative' is the overall analytical framework, then 'class', 'violence' and 'masculinity' are ordered within it as interpenetrating, de-reified social categories subject to the defining dualism of structuralism/historicism. What is 'masculine' in *The Code of Hammurabi* (circa 1750 BC) as opposed to 'feminine', is not identical with the masculinity of paganism, or the one that is structured alongside the Fordist factory of the most part of the 20th century. History and technological advances are matters with which the intellectual has to reckon with. At the same time, there are certain constant features. In *Raging Bull* we will see that Jake La Motta is in a constant battle with himself, wanting to prove his masculinity in the ring, *in a male-male relationship of equality*, something which he could not do through his absurd relationship with Vickie (Cathy Moriarty). Vickie was getting a beating, yes, but La Motta could not win this fight. The ring was the solution.

Lastly, we define 'class' broadly as a condition of relative poverty for individuals and families caused by the social/technical division of labour under capitalism—a mode of production that separates producers from the objects and means of their labour. This system of unequal production and distribution of wealth, whether legal or illegal, cannot alleviate relative poverty in history; it can only re-produce it in an extensive form as capitalist history goes by. The roots of reification, in which social relations of exploitation appear to the individual as relations between objects, lay bare in the capitalist relations of productions.[4] Hence our duty: to de-reify the social meanings of the movies under examination here, unravelling the mystery of why violence and exploitation, when artistically presented, give us pleasure.

As we shall see, these themes cut across all three films chosen here, but it is also one which can be found in more pronounced forms in films such as *Mean Streets* (1973) or *The Deer Hunter* (1978). It should not go unnoticed that the directors themselves, Scorsese and Mann, were the sons of émigrés. Scorsese was the son of Sicilian immigrants (born in New York City in 1942). His father and mother worked in New York's garment district, his father as a clothes presser and his mother as a seamstress. Mann grew up in a tough neighbourhood in Chicago (born in 1943). His father, a Ukrainian immigrant of Jewish descent, had fled the Russian revolution, and later was wounded in the Second World War. In Chicago, he started an unsuccessful grocery store. To a great degree, these biographies determine the genre they produced decades later.

First, we shall focus on *Raging Bull*. We will then move on to *Heat* and, lastly, *Miami Vice*. These are three movies that encapsulate a lot more than three decades. In fact, the entire post-war American society is present here, in all its imperial/hegemonic greatness as a producer of rare and aggressive cultural commodities. However, as we shall see, these films remain three *different* and, at the same time, *equivalent*, meta-narratives about violence, class and masculinity.

GREY BLOOD ON THE ROPES

Jake La Motta, a white middleweight boxer, had made a good career in the 1940s, managing even to become a champion, not so much because of his boxing qualities but because he could take a punch. 'La Motta's great gift', Paul Shrader said, 'was his ability to take a beating; he took a beating almost better than anybody else. It's really all he could do' (Kelly 2004, 124). Although La Motta's chin has been celebrated as the toughest in boxing, the story has it that he had let unnecessary punishment be inflicted on him by his opponent. In 1970, the year his autobiography was published, and on which Scorsese (director), Robert De Niro (actor) and screenplay writers Paul Shrader and Mardik Martin based their film,[5] La Motta, in an interview with Peter Heller, admitted the following (La Motta 1970, xi):

> I wanted to get punished and I took unnecessary punishment when I was fighting. I didn't realize it but subconsciously I was trying to punish myself. Subconsciously—I didn't know it then, I realize it now when I know a little bit more about the mind and the brain—I fought like I didn't deserve to live.

Knowledgeable film critics and reviewers saw La Motta and his admirable personification by De Niro as someone who 'passes through successive stages of punishment, compromise and self-disintegration, due to numerous inner demons' (Dirks 2009). In many respects, whether a Cuban criminal deportee—the case of Tony Montana (Al Pacino) in *Scarface*—or a boxer, this theme of 'rise and fall' is bound up with the so-called 'American dream', where streams of refugees and immigrants 'invade' America from all over the world, the land of liberal opportunity and freedom, in order to work their way to the top.[6] No matter which means they use to reach the top (drugs-trafficking, the construction business, mafia networks, boxing or whatever), the fact remains that ambition is a matter of psyche, not a social condition encompassing both the ambitious individuals and the conditions of their existence. Thus, film reviewers are inclined to present social dynamics as psychological events, internalized by the individual in question, and that is how they interpret acting. However, a careful reading of La Motta's biography shows colourfully that it was the social condition in his immediate Bronx neighbourhood that pushed him out to fight. 'Fight', that is not boxing as such, but how to survive in general, 'how to make the dough'. La Motta, in his autobiography, puts all this in many passages discursively, unable to articulate a sociological proposition, a thesis, a meta-narrative even in 1970, 'when he knew a little bit more about the mind and the brain' (La Motta was born in 1922). Yet Scorsese has a way to show this in his first scenes when La Motta is still living in his poor neighbourhood in Bronx. He beats his wife because she overcooked his steak: 'Don't overcook it, you overcook it, it's no good, it defeats its own purpose'. The off-screen, urban neighbour, Larry, complains about the noise caused by the domestic violence ('What's the matter with you up there, you animals?'), but Larry himself is not immune to criticism. He has a dog that disturbs La Motta's sleep: 'He calls me an animal', La Motta says to his brother Joey (Joe Pesci) who just stepped in, and leaning over the open window, throws a carnivorous threat: 'I'm gonna eat your dog for lunch, you hear me Larry?', and then he goes on to complain to his brother that he will never be able to fight Joe Lewis, the world heavyweight champion, because he has 'girl's hands'. 'Masculinity', in that respect, is decisively the form of sexism created out of a class condition, a condition of relative societal poverty. Scorsese and De Niro were in a position to present this as artists. The reviewers of the film, which number in thousands all over the world, have dismally failed to do so.[7]

Let us reinforce this thesis further with a discursive passage from the recollection of La Motta himself (La Motta 1997, 5–6):

> The first thing you got to do if you want to be a fighter is to fight. I figured out once that by the time I got into amateurs [...] Some of the kids I grew up with, they were good kids, their fathers were something respectable, these kids went to school and wore clean clothes. With them it was a big deal if they got into a fight. That's no way to learn how to fight. That's the way to make sure you get your block knocked off [...] Along with learning how to fight, I also learnt how to steal. In fact, about the only things I didn't learn was what they teach you in school [...] Everybody has a temper, but mine was set on a hair trigger [...] I don't know

> where it came from. And what made it worse was that I had mastoiditis when I was eight, from the cold in the tenements, and one of my ears was bad.

This is the writing of someone who wants to say something simple and meaningful, namely that those who fight are, first and foremost, those who have the social need to do so ('in order to get out of the poor and cold tenements'). Temper and physique matter, but societal/survival needs come first. Violence is the means to survive.

In *Scarface*, Tony Montana, equally sexist and poor at the start as a Cuban 'political refugee who demands his human rights, like President Carter says so', supersedes poverty by making his way through Miami's drugs industry, but he never relinquishes sexism. Sexism dies with him, straight after his sister, Gina (Mary Elizabeth Mastrantonio), dies from the bullets of the Colombian drugs-dealer, while herself aiming at her brother, vengeance for all the suffering he inflicted on her, including the killing of her fiancé, Manolo (Steven Bauer). *Scarface* is thematized around the *problématique* of 'rise and fall'—an American dreamer who has reached the top, but lacks the ability to hold onto this achievement ('you reach the top but still you gotta learn how to keep it' are the lyrics of a 1970s disco song in one *Scarface* scene). *Raging Bull* runs on a similar track, but it is much more complicated.

La Motta chooses the ring to fight. This is a macho, violent act of two equal male fighters in which violence reigns supreme, institutionalized by way of assigning 'rules of the game', conveyed by the referee and, to some extent, respected by the fighters. In order to get a title shot, he throws a fight to Billy Fox at Madison Square Garden in 1947. La Motta beats even his nemesis, Sugar Ray Robinson, in 1943 and finally becomes a middleweight champion of the world in 1949, beating French champion, Marcel Cerdan (the Mafia's reward for throwing the fight to Fox). The moments in which the champion's belt tightens around La Motta's waist are presented in a 'state of the art' choreography: carefully crafted by Scorsese and his team, these moments make the point with sad music playing on the soundtrack as if this signifies already the beginning of the hero's downfall, which is true. However, La Motta's fall, as opposed to that of Montana, is discursive, multiple, multi-faceted and fragmented. In fact, we are aware that it begins well before 1949, the zenith of his career. The very opening scene of the movie presents La Motta with his face hidden in the leopard-skin hood and robe, shadow-boxing into the smoky air. There is a melancholy all over the place, transmitted by the soundtrack of the 'Intermezzo', from *Cavalleria Rusticana* (an opera by Pietro Mascagni). As the film is a flashback, it closes with the same music. La Motta's downfall, unlike Montana's, is present from the very beginning. More than a drama, *Raging Bull* could be visualized as a tragedy.

So, in the first place, *Raging Bull* is a flashback. The very first scene that follows the shadow-boxing shows a fat La Motta—to play La Motta, De Niro ballooned from 145 to 215 pounds—in New York City in 1964, entertaining people in the Barbizon Plaza Theatre. He recites bits of Shakespearean tragedy, blended with bits of his life, a hotchpotch that he calls 'entertainment':

> I remember those cheers
> they still ring in my ears
> and for years, they'll remain in my thoughts
> Cos one night
> I took off my robe, and what did I do?
> I forgot to wear shorts.
> I recall every fall, every hook, every jab
> the worst way a guy can get rid of his flab
> I know my life was a jab
> I'd rather hear you cheer when I delve into Shakespeare
> 'A horse, a horse, my Kingdom for a horse'
> I haven't had a winner in six months [he lights his cigar]
> And though I'm not Olivier
> if he fought Sugar Ray he would say that the thing ain't the ring
> but the play
> so gimme a stage where this bull here can rage
> and though I can fight, I'd much rather recite.
> That's entertainment!
> That's entertainment …

This is all very complicated. La Motta gets fat and then goes back on a diet in order to fight; he lives in a state of permanent insecurity inside and outside his family home, suspecting his wife of cheating on him; he even suspects his own brother, Joey, of having sex with Vickie, his second wife; unlike Montana, he has children; and he has to fight constantly to get punished as this is cathartic for him; but he also needs to fight to hope for a title shot. La Motta's is an agonizing and zigzag rise to the top, whereas Montana's own is more straightforward. When he loses the championship to his nemesis, Robinson, on Saint Valentine's day in 1951, La Motta believes he is still the winner. A mangled, beaten-to-a-pulp La Motta, staggers about to Robinson's corner to say: 'Hey, Ray, you never got me down, Ray, you see, I can walk […]', and as the ring announcer proclaims the winner by technical knock-out, a massive Scorsese close-up shows La Motta's grey blood dripping off the rope: Scorsese's film is black and white, denoting La Motta's own flashback in his autobiography: 'Now, sometimes, at night, when I think back, I feel like I'm looking at an old black-and-white movie of myself. Why it should be black and white I don't know, but it is' (La Motta 1997, 2).

Before Scorsese brought La Motta to his initial setting in the backstage dressing room, he took care to detail the hero's complicated downfall even further. He showed La Motta wrestled into a prison for soliciting clients with prostitutes in his nightclub; he showed a dull scene in which Vickie announces that she is divorcing him assuming custody of the kids. Now, in the closing scene, Scorsese films La Motta's swollen body facing a mirror, reflecting his own image: his body is his life. In *Raging Bull,* masculinity is embodied in the body of La Motta, whether fat or fit.

The fat La Motta speaks now about the famous 'I coulda been a contender scene' in *On the Waterfront* (Marlon Brando/Terry Malloy 1954). He, as Terry Malloy, was once an 'upper-and-comer', now turned into a 'down-and-outer'. There is no emotion or passion here. La Motta is cool, and just ready before he goes out to perform in a crowded hall, he rehearses:

> It wasn't him Charley. It was you. You remember that night at the Garden you came down in my dressing room and you said, 'Kid, this ain't your night; we're going for the price on Wilson?' Remember that? 'This ain't your night'. My night. I could've taken Wilson apart that night. So what happens? He gets a title shot outdoors in the ballpark, and what do I get? A one way ticket to Palookaville. I was never no good after that night Charley. It's like a peak you reach, and then it's downhill. It was you Charley. You was my brother. You should've looked out for me a little a bit. You should've looked out for me just a little bit. You should've taken care of me just a little bit instead of making me take dives for the short-end money. You don't understand. I coulda had class. I coulda been a contender. I coulda been somebody, instead of a bum, which is what I am. Let's face it. It was you, Charley. It was you, Charley.

This is not simply the lament of a broken man, of a man who becomes a puppet to his inner-psychological pathology with no escape on the horizon. Quite the opposite: this is a (fat!) man who keeps fighting, who sees the defeat as a lesson from which he learns that he should never allow himself to go back to those social conditions of deprivation and misery that marked his early life. La Motta's own autobiographical testimony is again very revealing. Reflecting on how he felt when in prison, he states the following: 'So here I was at thirty-five, right back where I was at fifteen—in the can. Except this was worse. Back when I was fifteen, who the hell ever heard of Jake La Motta? Now everybody in the world knew that Jake La Motta, once the world's middle-weight champion, was a real bum. A first-class bum' (La Motta 1997, 210).

Boxing taught him, though, certain reified meanings too. It is no accident that, in the final scene of the film, as he grunts off-screen while shadow-boxing with his suit on this time around, he mumbles the following lines: 'I'm the boss, I'm the boss, I'm the boss, I'm the boss, the boss, the boss [....]'

La Motta believes that only by becoming 'a boss again' a return to poverty could be avoided, which is a misleading, reified belief. The 'rise and fall' idiom is the 'American dream' in its extreme and is a class meta-narrative that embodies masculinity, sexism, violence, that is to say, the reified meanings of a capitalist society in its conceivable forms. Scorsese, unwittingly, portrayed all this in a master-film.[8]

HEAT AND THE CITY

Clearly, we are not exploring here the actors' performances, or the production careers of the directors of the movies in question. This is the job of film reviewers. Quite unashamedly, we are trying to offer a sociological interpretation—in the form of meta-narrative—based on the themes of 'class' and 'masculinity' as they appear in the films we have chosen. The crucial research question that we are called to answer is this: how do violence, social malaise and sexism, from reified objects get transformed into pleasurable signs artistically woven in the film? Having interpreted *Raging Bull*, we think that the answer to this comes from an artistic and aesthetically-informed combination of the grey picture used, of the soundtrack, of the fusion of sexism, class, violence and masculinity. This is, obviously, an explosive mix, which Scorsese took care to surrender to the extraordinary acting of De Niro and the other cast members in order to make sense as a sequential film with technical instances, emotional layers and necessary choreographic hiatuses.

With *Heat* we move on to a different type of artistic and aesthetic plateau, not because class and masculinity or, for that matter, great soundtrack, are absent. Quite the contrary: in *Heat*, class and masculinity are embedded in the *time* and the *space* of the city, radicalizing them to extremes by way of acting and mobilizing the highest forms of technology available at the time. Class and masculinity appear transformed within the *spatial* and *temporal* matrixes of the city, absorbing the rate of movement into the muffled voices of Elliot Goldenthal's music. Our task here, therefore, will be to decipher the way in which the major themes of this chapter, class, violence and masculinity, present themselves in the time/space artistic framework of Mann's understanding of the city.

The city is Los Angeles (LA). Mann loves LA, the entire movie is about LA, but it is Chicago that follows him in a very peculiar way. Neil McCauley, incarnated by De Niro in the film, was a real-time Chicago criminal in the 1960s, taken down by Chicago police officer Chuck Adamson (played by Al Pacino with the screen name Vincent Hanna). Mann conducted more than 10 years of research and experimented with *LA Takedown* before the idea reached a mature stage. The reviewers of *Heat* have so far concentrated on the following:

- Two great actors who, for the first time in the history of cinema, share a majestic scene together—*The Godfather Part II* does not count as they did not appear together—and realize that, although on the opposite sides of the law, they are virtually identical;
- The film is essentially a sophisticated remake of *LA Takedown,* which Mann initially produced for TV, but which ended up becoming a low-budget film in 1989;
- The film is essentially a game of survival, as we know almost from the beginning that either Neil or Hanna will fall, and we also know that the survivor will be damaged by the loss, as he will eliminate a piece of himself.[9]

Malicious reviews, such as that by Steve Rhodes (1995), saw in *Heat* nothing but 'four sets of characters: good guys (cops), bad guys (robbers), all their families, and the people doing the stunts'. According to the same reviewer, 'the music is bland, made up mostly of pseudo-haunting violin melodies'. This is nonsense.

Heat reconstructs fundamental structural features in America's psyche and social condition: it brings into a motion picture how poverty, race, violence and masculinity play out in the city's urban, high-tech landscape. The age factor should be noted here too: the complexity of masculinity in *Heat* is chiefly presented through two middle-aged men. These themes are not delivered by Mann in a formalistic way, as many postmodern productions have the tendency to do. For instance, the camera details the city not only in its external, urban settings, but penetrates deeply into the very constitutive materialistic parts of it. Mann shoots Hanna and Neil's flats from inside, as well as 'a refinery and a scrap yard', or when Neil looks over midnight LA from Eady's balcony (Amy Brenneman) to say: 'In Fiji they have these iridescent algae that come out once a year in the water. That's what it looks like out there'. The balcony becomes an organic extension of midnight LA, as it works out the sexual chemistry of the characters. Hanna, for his part, coming back from the massive shoot-out in the wake of the bank's semi-successful robbery, explains to the playmate of his wife, Ralph, that he is somewhat welcome to 'come and have dinner in her ex-husband's dead-tech post-modernistic house'.

How do we interpret *Heat*? We mainly see it as anything *but* a report on violence. We see it as how hallucinations of personal drive, male heroism, guns, gangs, wealth and ambition are the mystifying products of tensions generated by the geography of the great American city, global in nature and scope, then and now in permanent ascendance *and* decline. The two key characters, obviously, are superb: sharp personalities, high drive, leaders and loners interacting in a dramatic *space* in order to dramatize the city's *time.* Yet it is the city that commands and consumes time at their expense. They are great performers, thus a pair of complex transformers. They can transform everything, including the spectator and a building's geography, except the city's divine time.[10] Hanna and Neil have been picked up to be enemies by default: forced to kill each other not because of what separates them—the Law—but because of what unites them, that is, the city's space and architecture within which they move fast, relentlessly, professionally and vigorously, because this is what 'keeps them sharp, on the edge', because 'the action *is* the juice'. They promise each other that this lifestyle damns any tangling commitment that disables them from 'escaping in under thirty seconds flat if the heat sparks around the corner'. Masculinity seems to be portrayed by Mann as the loner's, middle-aged man's affair, unashamedly requiring you 'to do what you do best' before you die. However, the winner must not delude himself that he will be safer from the 'enemy' he beat, because no angel will be staring at him, but dead people, as Hanna's recurring dreams reveal in his conversation with Neil. In essence, both have to place their hopes on the Messiah. 'The Messiah', says Walter Benjamin, 'comes not only as a redeemer, he comes as the subduer of Antichrist […] *Even the dead* will not be safe from the enemy if he wins' (Benjamin 1992, 245, emphasis in original).

Hanna and Neil meet and confess to each other that they need more time 'to do what they wanna do', but they are puppets, mere pawns in the merciless hands and unpredictable dictates of the city, that is, the all-encompassing entity that produced for them the hallucinating signs of ambition, male heroism, and wealth—which they do not possess. Being workaholic is a passport to societal alienation, more than a pecuniary addiction. The following dialogue between Justine and Hanna is very revealing:

Justine: I guess the earth shattered.
Hanna: So why didn't you let Bosko take you home?
Justine: I didn't want to ruin his night too. What was it?
Hanna: You don't want to know.
Justine: I'd like to know what's behind that grim look on your face.
Hanna: I don't do that, you know it. Come on, let's go.
Justine: You never told me I'd be excluded.
Hanna: I told you when we hooked up baby that you were going to have to share me with all the bad people and all the ugly events on this planet.
Justine: And I bought into that sharing, because I love you. I love you fat, bald, young, old, money, no money, driving a bus, I don't care. But you've got to be present like a normal guy, some of the time. That's sharing. This is not sharing, this is leftovers.
Hanna: I see. So what I should do is come home and say, 'hi honey, guess what. I walked into this house today where this junkie asshole just fried his baby in a microwave because it was crying too loud', so let me share that with you, come on, let's share that and in sharing it we'll somehow cathartically dispel all that heinous shit, right? Wrong. You know why?
Justine: Because you prefer the normal routine: we fuck and you lose the power of speech.
Hanna: Because I've got to hold onto my angst. I preserve it, because I need it. It keeps me sharp, on the edge, where I've got to be.
Justine: You don't live with me. You live with the remains of dead people. You shift through the detritus, you read the terrain, you search for signs of passing, for the scent of your pray, and then you hunt them down. That's the only thing you're committed to. The rest is the mess you leave as you pass through. What I don't understand is why I can't cut loose of you.

It could have been a dialogue between Neil and Eady. The two men share a parroted quality. *Heat* echoes the lines from the original *Scarface* book, written in 1930 by Armitage Trail (the pseudonym for Maurice Coons) and upon which the black and white film of the 1932 Howard Hawks and Paul Muni screenplay was based: 'To Tony, the only difference between a policeman and a gangster was the badge' (Trail 1997, 7–8). What, though, is the case with their women? We need to refresh the film here. Justine and Eady appear to be better educated, but so does Hanna, the cop. Hanna has graduated and undergone all sorts of training, and Eady studied graphic design in New York City. Justine is highly sophisticated and nerdy, although we do not know much about her background, other than that she has a daughter and there is a father somewhere. However, everything is crystal clear when it comes to Neil: 'I'm a needle, going back to zero, a double blank'. Time and again, this takes us back to *Scarface*, when Montana (Pacino) confesses to Elvira (Michelle Pfeiffer) that 'he has no education, but that's OK'. Montana, as does Neil, thinks that it is enough both for him and Elvira that he 'knows the street' and he can 'make all the right moves' to bring in enough dough. This has proved to be a delusion in both films.

It is the imposing supremacy of the imperial city that makes Neil and Hanna totally drawn into the hopeless game of life and death. The city, full of cases that need to be sorted out, is a huge depository of beauty, ugliness, crime, poverty, money, sex, art and, above all, action, indulging their heroes to concede that 'all they are is what they are going after'. Time is of the essence. Time-wasters do not count, appear nowhere in the equation, as if Mann suggests that they should stay at home, or retire to the countryside.

In *Heat*, although violence has the typical features blended with speed and high-tech means, both machismo and sexism take on very peculiar forms. The only sexist scene in the movie—in the classic sense—is when Hanna, terrorizing Charlene's playmate, Alan Marciano (Hank Azaria), says that Charlene's 'got a big ass, and you [Alan] have your head all the way up it'. For Hanna, this is something 'ferocious', yet these words constitute plain sexism. Nevertheless, in the overall motion picture the theme presents itself in a totally different manner.

Gender meanings in *Heat* are transformed and border on a kind of 'unequal equality' paradigm, in which men are in a duel, in a male-male equal relationship, the classic theme we saw in *Raging Bull,* but at the same time women are elevated into a different, yet undefined, level of self-consciousness and sophistication. Mann does not tell us whether the woman is part of the male equation, that is to say, whether *she* is a constituting part, a kind of 'cause', of the man's high drive, speed and paroxysm in the city's unearthly landscape. We can assume, though, that it is so, particularly if we read carefully the context in which Chris Shiherlis (Val Kilmer) pronounces these words about Charlene (Ashley Judd): 'For me the sun rises and sets with her, man'. What is the context in which Chris pronounces these words?

Neil wants to lay down the work schedule for Chris ('take the cash delivery from Roger Van Zant [William Fichtner], give Kelso [Tom Nooman] the deposit, take the bank and then get out'), but Chris would consider going nowhere without Charlene. He insists on that, despite the fact that Neil reminds him of the Folsom dictum: 'If you wanna make moves on the street', Neil tells him bluntly, 'have no attachments, allow nobody to be in your life that you can't walk out on in thirty seconds flat when the heat sparks around the corner'. Yet Chris is the only one from Neil's crew who escapes death. To do that, he had to leave Charlene behind, but one should not fail to notice the soundtrack 'Armenia' in the background, with all its bearing sadness and grief of the lives lost in something that usually is described as 'genocide'. Because, from then on, Chris is virtually destined to live a lifeless life, that is a 'normal' life, without eroticism and action, but full of 'barbecues and ball-games'. In fact, he is worse off than his dead comrades: Neil, Trejo (Danny Trejo) and Michael Cheritto (Tom Sizemore).

Mann describes the city's social reality as it is, insinuating that mothers and fathers know that their sons and daughters will eventually glorify its heat, by not becoming time-wasters. They will do so even better than their dead: watch how Hanna hugs the mourning mother of a 16-year old 'prostitute', just killed by Waingro (Kevin Gage). Here, Italian-American Camille Paglia, socialist feminists and many others triumph over certain, rather revengeful, branches of radical and liberal feminism, that is 'a WASP, middle-to-upper-middle class discourse' (Paglia 1992, 73–82, 121, passim). Likewise, with class and race: 'take out the garbage, empty the bin, take a break later', orders the white boss to the African-American Don Breedan (Dennis Haysbert) in the diner. However, Neil saves the black man's pride. He walks to the counter and Don remembers him from Folsom D-wing. He offers him the job of driver in the bank robbery and he claims a straight answer, yes or no. It took Don a few seconds before he answered positively, taking the opportunity to shove his boss to the floor on his way out. Together with his ex-Folsom mate, Don preferred to do something brave and heroic in the city, than to stay on a miserable salary, slavishly obeying the racist attitude of his boss. For Don, going beyond his class and race immediacy was of paramount importance in order to rebuild pride and reclaim his woman as 'equal'.

By all standards, Mann's *Heat* is a very complicated motion picture. Among other things Mann teaches us, is that he tells us that masculinity is about knowing *how* to die. Remember *Zorba the Greek*? He died standing up, with his arms wide open, inhaling deeply in front of the open window, looking out to the sea, far off, gently moving. Neil and Hanna share a few quiet moments of reflection and affection as Neil dies, and it is a 'tribute to the dramatic force of the movie', as Nick James put it, 'that we do not find this image ridiculous' (James 2002, 83).

'I'M A BUSINESSWOMAN'

Heat's music sounds like an industrial, albeit discursive, electronic complex. Composer Elliot Goldenthal weaves masterfully the high-tech gloss of the film with the drones of the Kronos quartet, the synthscapes of Moby, the guitar of Terje Rypdal and the authority of Brian Eno/U2. When there is energy in the scene, there is Moby's 'New Dawn Fades'. When Neil enters the bank, his steps are captured by Goldenthal's 'Entrada & Shootout': t-t-toop-topa-topa-toop. In moments of sadness, Einstürzende Neubauten's 'Armenia' reigns supreme.

With *Miami Vice* we enter an even more elevated terrain, both musically and thematically. The sound is more pluralistic, flamboyant, global and energizing. Masculinity, violence, sexism and gender are presented in a far more sophisticated and complex way, as they are infused into a global underground network of drugs business. In many respects, this film, as a whole, moves beyond race perimeters too, and not only because one of the two protagonists is an African-American (Jamie Foxx).[11] *Miami Vice* shifts the boundaries of anti-racism in the film by diluting the identity of the female protagonist, Isabella (Gong Li), into a cosmopolitan perspective: she was born in Angola, where her mother died when she was 16; she is based in Latin America doing business in Miami, Malaysia, Hong Kong, Ukraine, Russia and Switzerland; in the movie she is Cuban-Chinese. This is the majesty of *Miami Vice.* Race, in a classical sense, appears in the explicit form of the Aryan Brotherhood gang, which kidnaps detective Trudy Joplin (Naomie Harris), Rico's girlfriend, in order to force the FBI to deliver the load to them. The theme also appears in the discourse of the actors, who talk of 'white supremacist gangs' and so on, but nothing more than that.

Likewise, class does not feature explicitly as a social condition of deprivation, poverty and discrimination: in fact, it is Mann's camera through its silence that delivers the power of this theme to us. When the protagonists, FBI agents Sonny Crockett (Colin Farrell) and Rico Tubbs (Jamie Foxx), walk through the dirty slums of Latin America to meet kingpins of drugs-trafficking, we are in a position to discern why the poor there can only be linked to the drugs business, or to a Guevarian, or to a Bolivarian, albeit populist, political perspective. The city, this time around, is Miami, but we only get snapshots of it, such as that in the beginning of the movie, when Sonny and Rico are on a roof and the nocturnal jewel box of Miami glitters behind them. More high-tech than *Heat*—we are in the mid-2010s—Mann couples Miami with its uneven counterpart: the crowded, polluted slums of Cali, Barranquilla and Ciudad del Este. However, *Miami Vice* deals with class also on another, more intensive and analytical level.

Mann moves the issue of class from the *problématique* of *agent* to that of *structure.* Class is part and parcel of two parallel economic and power/security structures: the FBI's state bureaucracy, which is the legal one, and that run by drug lord Arcángel de Jesús Montoya (Luis Tosar), which is the illegal, underground network based in Latin America but ramified globally. We need this methodological separation, as well as a class, holistic analysis to understand what is going on in the movie. We also need a global, deeply political, perspective to explain lines such as 'that's the type of stuff CIA does, in Baghdad', when Rico and Sonny's cell phones are jammed by Montoya and

Jose Yero's (John Ortiz) security apparatuses. How could a democrat intellectual in America, who opposed the US-led war in Iraq, come to terms with lines like this, and even come to like them *and* the movie? We return here to our original research question asked in the beginning, which attempts to formulate an answer, as to how art, by imitating reified realities, comes to transform them into visual pleasures.

Miami Vice is a new version of a 1980s TV series with the same name, also directed by Mann and written by Anthony Yerkovich. The film is dark and blue, the dialogue at times very technical, but few critics hailed that as 'a plus' for the movie, because they failed to understand how the technical FBI jargon links with the action, high-technology, the blue and dark film and the soundtrack as a whole. Even perceptive critics, such as *Sight and Sound*'s Graham Fuller (2006), isolate the parts of the film in order to describe their meanings. Instead, we argue that a holistic, 'Hegelian' interpretation of *Miami Vice*, that is to say, an anti-postmodern understanding of it, is the prerequisite to recognize the enduring strength of its composite parts, and that it is this that we come to like. As Mann transcends the postmodern formalism, it is our duty to reverse the 'Hegelian' presentation of the film in order to demystify its reified, unconscious message, which is our neo-Marxian perspective.

Violence of all sorts reigns supreme in the film. It is hemmed in with eroticism and an unprecedented, even hegemonic, presence of female characters, such as that of the financial adviser to Montoya, Isabella, who falls in love with Sonny, and of detective Trudy (Naomie Harris), who is captured by a neo-Nazi gang as a ransom for a drug deal. However, one should not fail to notice Elizabeth Rodriguez's contribution to the film. As Gina, she perhaps pronounces the most authoritative lines of the movie, aiming at the white supremacist neo-Nazi gorilla, who was threatening to press the button on the detonator, thus killing everybody, including the captured Trudy:

Gina: Drop the detonator!

[A small interval during which Rico gets rid of two other neo-Nazis.]

Neo-Nazi: Shoot me. She dies. Shoot me, go ahead. Fuck it. We can all go. That's cool.

Gina: That's not what happens. What will happen is I will put a round at 2,700 per second into the medulla at the base of your brain and you will be dead from the neck down before your body knows it. Your finger won't even twitch. Only you get dead. So tell me, sport, do you believe that?

Neo-Nazi: Hey … !

[Gina, cold-blooded, who leads the operation to free Trudy, shoots the neo-Nazi instantly; what happens is what she described.]

There are plenty of scenes in *Miami Vice* where female discourse dominates the artistic milieu. This, it should be noted, takes place not only in the form of isolated incidents, where male and female protagonists confront each other (the case of *Heat*). In fact, a woman's personality, at least as an equal, cuts across the whole film and appears as being an organic, leading member of a team. Isabella, after all, tells Sonny that she is 'a businesswoman, who does not need a husband to have a house to live in'. Later in the movie, she will soberly attack Sonny for being 'too protective, typical Cuban male thinking', she tells him. It is Trudy's presence, though, that one should discern in this respect. In front of a team of detectives dominated by males, she will have the following dialogue with Nicholas (Eddie Marsan), the FBI's contact to mediate to facilitate an initial meeting with the drugs lords:

Rico:	Call Colombia.
Nicholas:	Man, that's Jose Yero.
Rico:	Really?
Nicholas:	He is AUC, you know, Colombian right-wing paramilitaries. You know who they are? They are vertically integrated, they are …
Rico:	You mean they walk around with constant erections?
Nicholas:	No. They farm, process, produce, export …
Rico:	I know what it means.
Nicholas:	No, see, it gives them attitude. A player negotiates too hard and you never hear from him again, cos these guys kill everything! And [pointing at Trudy] I gotta know what's the skinny.
Sonny:	It's none of your fucking business.
Nicholas:	It can come back on me baby.
Rico:	It can't come back on you baby.
Trudy:	Hey, sunshine, when has Rico or Sonny ever lied to you, huh? I mean when has anything Rico told you not happened exactly like he said … ?
Nicholas:	It's just these variables, you know, randomness, see, that's why …
Sonny:	That's why what?
Nicholas:	This group goes from zero to high level violence like this!
Trudy:	You made 15% commission off three money laundering operations I put you into, you know, which is why you live in your $4m. condo. And you question Rico and Sonny? Fuck that. I will cap your skanky ass and throw it off that goddamn balcony …
Nicholas:	Why is this happening to me?
Trudy:	Because you lead a life of crime.
Gina:	Can't do time, don't mess with crime.
Sonny:	He's cool, he'll make the call.

This dialogue is so interesting. Mann is not portraying women as a stay-at-home sophisticated subject waiting for her partner—whether a cop or a criminal—

to turn up, which is the case with *Heat.* In *Miami Vice* he elevates women's presence at every conceivable level: the film shows women as being *both cops and criminals in real action*, while assigning to them leading roles to perform and dominant discourses to deliver.[12] This, as the dialogue above confers, turns classic sexism on its head, diluting classic masculinity—that which refers to a male-male relationship—into a shifting terrain of equivalences and equalities. In fact, what saves the theme of classic masculinity in the film is the macho sparring between Sonny and Rico, on the one hand, and Jose Yero, on the other, in one of the most powerful scenes of the movie, when the two detectives meet Yero in his bunker to discuss the drugs business. This scene leads Mann to construct a follow-up scene with Montoya and Isabella on one side, and Rico and Sonny on the other, in which Montoya delivers an authoritative lecture, along the lines of 'in this business I don't buy services, I buy results'.

What is even more fascinating is that Mann brings into the equation the global shadow banking economy of hedge funds and financial activities. The film tells us implicitly that not everything that is monetized and demonetized in the paper economy of global finance is legal. Montoya's transactions and operations are truly global and integrated with the banking system of Geneva, which Isabella visits often. This pre-dates the current discussion among international relations experts, who are seeing the current financial crisis, among other things, as a by-product of the shadow banking system and the extreme forms of liberalization and deregulation that capitalism assumed since the 1970s (Gowan 2009, Gökay 2009). However, Mann could not tell us more about it, as this would entail a different film. What Mann does with *Miami Vice* is, however, both provocative and instructive. He pushes us to see through the prism of the two parallel economies and bureaucracies exhibited—the legal and the illegal—the reality of their interaction and, in effect, integration.

'Can't do time, don't mess with crime', says Gina. Time, heat and the city are rehearsed and transcended here as stylish parameters of a faster film, either as dancing women in silver outfits gyrating to the superb piece of Linkin Park/ Jay-Z *Numb/Encore*, or as spy planes in the blue sky and fast boats in the sea, at times having Miami's skyline in the background. Class, masculinity and violence define the cultural settings of the film, while at the same time invoking the female presence as an equal to unfold unobstructed. The political and economic dimensions we attributed to the film's verisimilitude are more a tribute to its righteous instincts than an undercutting image of its veracity. On this count, our anti-war intellectual would have to reach out to a de-reified, neo-Marxian meta-narrative of the film to explain why lines such us 'this is the type of stuff CIA does, in Baghdad' invite pleasure.

CONCLUSION

We can now draw a few conclusions.

The fundamental task of this chapter was to explain how certain artistic commodities, such as movies, are so powerful as films and how they convey meanings that can transform an ugly and reified social reality into artistic pleasure. We have used three types of concepts, or variables, to explain this: violence, class and masculinity. We have also based this on the interpretation of three motion pictures: *Raging Bull, Heat* and *Miami Vice.*

The problem with the 1944 essay by Adorno and Horkheimer is that they viewed mass culture as the spin-off commodity of fascism, deprived of any autonomy, and as if it were fascism that created everything: mass production for mass consumption, a totalitarian regime based on a totalitarian ideology, and even a new mode of technological production affecting cultural commodities and creating a 'culture industry'.[13] Thus, reified cultural commodities appeared to them as being totalitarian entities solicited to inflict mass apathy and false identities upon the suffering working class. The real/social, as a historic form, was not de-reified through Marxist analysis. Instead, it was a direct reflection into their mode of thought. Parts of what Perry Anderson called 'origins of post-modernity', can indeed be traced back to the Frankfurt School, whose effort to break from the totalitarian ideology of fascism led them to present social reality as a fragmented entity ungraspable by grand narratives, or meta-narratives.[14] We have shown here that we need meta-narrative in order to demystify and de-reify, through critical, neo-Marxian analysis, cultural products of modernity. We have also shown through the meticulous examination of the three movies that themes such as 'class', 'masculinity' and 'violence' present both constant and changing features in cultural history. In this context, one could argue that cultural history runs along two philosophical and seemingly antithetical principles, that of *historicism* and of *structuralism.* To paraphrase Giuseppe Tomasi di Lampedusa (1896–1957) in *Il Gattopardo,* matters in cultural history appear as if 'everything changes [historicism] and everything remains the same [structuralism]'.[15]

A superficial look at *Raging Bull*, *Heat* and *Miami Vice* would attribute to them raw violence and extreme machismo. Obviously, appreciative film critics go further, reversing apparent flaws into a sublime expression of art, 'masterfully directed and performed by the actors'. However, such readings fail to explain why we like some things in the film which we definitely abhor in reality. We have shown that the answer to this crucial question is to make class, critical and holistic analysis the backbone of the investigation, thus coming to understand characters, societies and plots as de-reified moments of theoretical analysis.

Violence, class and masculinity in the movies, and in real life, have certain structural features but, above all, are historical categories. Violence is transformed through technology, and so is class and masculinity. However, a holistic 'Hegelian' perspective indicates that none of the meanings of these categories can be grasped separately. We cannot understand La Motta's attitude towards his wives if we fail to link this with his lack of education and the poor conditions in which he was raised. 'You can take the boy out of the Bronx', one film technician once famously said as the dubbing was going on, 'but you can't take the Bronx out of the boy' (Yule 1992, 296). Similarly, we will not be in a position to grasp the global logistics of Montoya's drugs money operations if we fail to comprehend his fatal dependence on Isabella, his financial adviser, whose business code of conduct draws from the philosophy of 'the diminished risk'. More to the point, Montoya's global operations show vividly the infusion of formal and shadow banking systems, both of which were in tatters in the wake of the financial crisis of 2008–09. In the final instance, an extended and holistic neo-Marxian analysis remains the key to de-reify signs of cultural products and expose power and capital as producers of social and political oppression.

NOTES

1 In writing this essay, I benefited enormously from Donald Sassoon's, *The Culture of the Europeans* (2006). Although I make no explicit references to this masterpiece, Sassoon's discussion of 'culture as a set of relationships', which moves from an elite phenomenon of the 17th century to a mass event of the 20th, helped me historicize my presentation here. I am also thankful to the following friends for help with the themes explored here: Carmen Rabalska (Kingston University) has provided extensive written comments on my text, for which I am immeasurably indebted; Steve Knapper (Kingston University) cautioned me about the origins of postmodernity, as I traced them back to the 'Frankfurt School'; Jairo Lugo, Jovo Ateljevic and Dejan Jovic (all at Stirling University); Bernadette Buckley (Goldsmiths College, University of London), Molly Greene and Carol Oberto (both at Princeton University) and Maritsa Poros (City College, CUNY), have all discussed parts of this text in the past and offered valuable critical observations. *Heat* was first 'discovered' by my old dear friend, Tolis Malakos, formerly at Queen Mary College, University of London, whom I thank enormously wherever he is. Thanks are also due to the Editor of this Routledge volume, Yoke-Lian Lee, for her overwhelming support and encouragement throughout a hectic university semester. Quotes from films in this chapter are the author's recollection of dialogue.

2 This should not be confused with any sort of Aristotelian catharsis, which could be an accepted element of a popular film, channelling displaced anxieties and aggression, which appeals to frustration experienced by the spectator.

3 By 'avant-garde masculinity', we mean the radicalization of the male-male relationship, the only acceptable form of equality in a macho context.

4 This definition of reification derives partly from Georgy Lukacs's early work, *History and Class Consciousness* (1922); see the English translation by Rodney Livingstone (London: Merlin Press, 1971), 83–222. However, the reader should also consult the first part of the first volume of *Das Kapital*, where Marx examines the same issue in a rather post-Hegelian manner, as opposed to young Lukacs's Hegelianism.

5 It should be mentioned, though, that despite the fact that the screenplay for *Raging Bull* was credited to Schrader and Martin (who earlier co-wrote *Mean Streets*), the finished script differed extensively from Schrader's original draft. It was re-written several times by various writers including Jay Cocks (who went on to co-script later Scorsese films *The Age of Innocence* and *Gangs of New York*). At any event, the final draft was largely written by Scorsese and De Niro.

6 A perceptive narrative on the issue of the 'American dream' in gangster movies is Edward Mitchell's classic 'Apes and essences; some sources of significance in the American gangster film', in Peter Lehman (ed.), *Wide Angle*, Vol.1, No.1, 1976.

7 Even a careful writer, such as John Baxter, contends that 'La Motta's book contained numerous scenes of sex, violence, verbal obscenity, and an overall tone of existential despair' (Baxter 2002, 186). The extreme form of internalization and subjectification of the character is more than obvious.

8 Scorsese himself is not immune to the esoteric fascination of the picture he produced, stating that '*Raging Bull* is about a man who loses everything and regains it spiritually' (Kelly 2004, 119).

9 For reviews of the film, see www.rottentomatoes.com.

10 'The performer is a complex transformer, a battery of metamorphosis machine', Perry Anderson states in his *The Origins of Post-modernity* (1998). The transformative capacity of *Heat*, unfortunately, had some negative aspects. The picture was cited as the model for a spate of robberies since 1995, including armoured car robberies in Colombia, Denmark, a bank robbery in North Hollywood, California, and Greece. See, among others, James 2002, 74.

11 However, it should be noted that the police duo is modelled after a type of 'the buddy movie black and white police duo', in which the black one is generally secondary to the white, and sexually made more neutral as he is in a stable relationship.

12 The extent to which these super-women can be taken as 'male fantasies' in order to provide spectacle is a possible, although a bit exaggerated, interpretation of the characters involved. This interpretation can only apply to a minority of cases, since the social stereotyping has the male in an active role and the female in a passive one.

13 See our discussion with Donald Sassoon in *Journal of Southern Europe and the Balkans*, Vol.8, No.3, 2006.

14 The best example here is Bauman's (1989), *Modernity and the Holocaust*.

15 *Il Gattopardo* (The Leopard) follows the family of its main character, Sicilian nobleman Don Fabrizio Corbera, Prince of Salina, through the events of the Risorgimento. The most memorable line in the book is spoken by Don Fabrizio's nephew, Tancredi, urging unsuccessfully that Don Fabrizio abandon his allegiance to the disintegrating Kingdom of the Two Sicilies in favour of Garibaldi and the House of Savoy: 'Unless we ourselves take a hand now, they'll foist a republic on us. If we want things to stay as they are, things will have to change'.

The gender politics of international relations

Jill Steans

INTRODUCTION

At the end of the 1990s Cynthia Enloe declared that this was a time to reflect on the many successes and achievements of feminist international relations (IR). In a little over a decade feminists had produced work on a wide range of themes, from the gendered nature of war and conflict to the reconceptualization of community and political space in IR.[1] Undoubtedly, feminists have been successful in carving out a space or creating a discourse on gender within IR over the past 20 years. However, while the achievements of feminist IR have been notable, since the late 1980s feminist scholars have encountered a particular antagonism towards the project of 'gendering' IR. Indeed, Enloe's comments on the health of feminist IR were in response to probing questions on whether 'you think this work is really doing anything on a significant scale to transform the way people think?' and whether the exchanges that did take place between specialists in gender and those of a traditional political science/IR persuasion, were not dialogues 'of the deaf'?'[2]

In so far as scholars located within the 'mainstream'[3] have engaged with feminism, this engagement has been rather selective. Mainstream IR scholars have expressed frustration with feminist IR because feminism had failed to produce a singular, coherent perspective or paradigm within IR. Ann Tickner has argued that in consequence while 'feminist perspectives on international relations have proliferated, they have remained marginal to the discipline as a whole' (Tickner 1997, 663). Resistance to feminist IR, particularly in its poststructuralist guise, has come from scholars who continue to insist that production of objective knowledge about the world (of international relations) is both possible and desirable. Feminists have retorted that in so far as mainstream scholars have engaged with gender, these 'engagements' have been motivated by a desire to appropriate gender and 'discipline' feminist scholarship in order to maintain the hegemony of positivism in IR (see Weber 1994; Tickner 1998; Tickner 1999; Weber 2001; Smith 2002; Carver 2003; Zalewski 2003; Kinsella 2003; Steans 2003).

If there were possibilities for meaningful debate between feminist scholars and 'the mainstream', they were closed-off by the binary oppositions set up in the initial starting points in the 'fourth debate' in IR:[4] I am *x* (positivist/rationalist), you are *y* (postpositivist/reflectivist), ergo we have nothing much to say to one another. However, the fourth debate in IR also opened up space for critical and constructivist approaches within the discipline broadly conceived and thus created space for the accommodation of feminist concerns and feminist modes of inquiry. As Marchand has argued, encounters between feminism and IR have been contingent upon and embedded in different realities, and attempts at making them more meaningful and substantive have to address these different contextual realities (Marchand 1998). Certainly, there is another story that could be told about the 'state of the discipline' of IR and the place of feminist scholarship within it that might be more upbeat about the possibilities for accommodating feminist concerns within an expanding body of critical and constructivist scholarship and generally, perhaps, be more celebratory about the impact of feminism in the field. Moreover, as Marchand noted, discussions about the marginalization of gender in IR are in danger of reproducing that marginality. However, Marchand also conceded that there has been a continual marginalization of feminism within IR to the extent that other 'critical theorists' have been more often engaged in discussion by the mainstream, whereas feminists have most often not been so engaged (1998).

Of course, attempts to map a field of study conceptually, or delimit the scope of inquiry, inevitably involve making decisions about what should be considered 'central' or 'marginal', but it is important to notice that such choices are never politically innocent. Academic disciplines are excellent examples of discursive communities that construct 'socially bounded' fields of knowledge. Within the academy certain forms of knowledge are institutionalized and valorized to the degree that they assume superiority to mere 'common sense' or opinion. Academic disciplines develop theoretical and analytical frameworks, generate concepts, construct categories and develop theories about the world and how it works. Whatever the chosen approach or method, the politics of 'gendering IR' cannot be avoided and all attempts to take gender seriously, sooner or later, necessitate confronting the gendered politics at work in the construction of IR. As V. Spike Peterson has noted, attempts to demarcate the sphere of IR as an academic discipline and establish its core concerns are the result of the ability of socially powerful groups to impose their definitions on others (Peterson 1992). Thus, it is unsurprising that feminist scholars have complained that there is a politics at work in the construction of the 'discipline' of IR that has marginalized and continues to marginalize feminist work.

This chapter[5] elaborates on the gender politics in IR through an assessment of the impact and status of feminist scholarship in IR adjudged by a selective review of the debates and conversations that have taken place between feminist IR scholars and those located in the 'mainstream' and 'middle ground' of IR. I will elaborate on what I mean by the 'mainstream' and 'middle ground' below: in short, in this context 'mainstream' means work that centres on an established IR research agenda of conflict and co-operation, and scholars who also demonstrate a certain scepticism, if not hostility, towards postpositivist approaches, while 'middle-ground' refers to scholars who have attempted to integrate ideas into a materialist-rationalist framework of analysis.

ENGAGING THE 'MAINSTREAM' OF IR

Feminist critiques of the 'orthodoxy' in IR

Despite the existence of competing approaches in IR and the sustained critique of mainstream approaches in IR since the late 1980s, it is nevertheless meaningful to speak of an orthodoxy in IR since positivism remains entrenched in IR scholarship in the United States particularly.[6] The term orthodoxy might be employed in a second, related sense, to mean that one approach is dominant within the field to such a degree that it serves as a 'common sense' view of the world against which all other perspectives should be judged; thus 'reality' is presented as given and known prior to 'theoretical understanding' (Tooze and Murphy 1993, 11).

To challenge orthodoxy in both usages of the term was to push back the limits of this universe of possible explanation, to give voice to that which could not be articulated or expressed because of the constraints of existing discourse. In the first wave of feminist scholarship in IR, feminists developed critiques of realist and neorealist approaches to IR particularly, since realism/neo-realism had achieved a dominant position in IR, such that it served as a 'common sense' view of the world. The critique was multidimensional, but centred on the following key points.

First, feminist scholars pointed to the deep gender bias embedded in its concepts and images of the world and challenged the core assumptions of realism and neo-realism. Second, specifically feminists argued that realism/neo-realism reified social structures and disguised the historical character of social and political relations. Third, feminists pointed to how attempts to draw rigid boundaries between the international (outside) and domestic (inside) and between private and public realms removed gender relations from the field of inquiry. Fourth, feminist work pointed to the complex ways in which gender was deeply implicated in the carving out of political spaces, the construction of identities and demarcating the boundaries of community in practices of 'state-making'. Political community has assumed gendered forms. For example, the linkage between the state, political loyalty and combat and citizenship historically had meant that in consequence women have been excluded from citizenship rights (see discussion in Steans 2006).

However, an 'orthodoxy' is not just a dominant world view 'but a particular mode of production of knowledge that specifies a particular relationship between objective and subjective and uses appropriate epistemological and ontological categories to support this relationship' (Tooze and Murphy 1993, 13). In this sense, positivist approaches generally can be described as the 'orthodoxy' or 'mainstream' in IR. While the term positivist can disguise subtle differences between ostensibly similar positions,[7] whatever their differences, neorealists and neoliberal institutionalists both assume that it is possible to understand the world objectively and, therefore, international relations is amenable to investigation using scientific methods. Thus, fundamentally feminist critiques of the orthodoxy or mainstream were centred on ontological, methodological but also epistemological issues. To simplify somewhat, feminist IR could be said to have identified a number of core tasks: first, to point to the exclusions and biases of neorealist IR, both in terms of the limitations of state-centric analyses and of positivism; second, to make women visible as social, economic and political subjects in international politics; third, to analyse how gender inequalities were embedded in the day-to-day practices of international relations; and fourth, and more controversially in terms of my implied consensus, to empower women as subjects of knowledge by building theoretical understanding of international relations from the position of women and their lived, embodied experiences.

Feminist scholars adopted the notion of 'feminist lenses' or 'gender lenses' which implied that there were other ways of 'seeing', 'knowing' and 'being' in the world, which could rise to different standpoints or perspectives. Peterson and Runyan, for example, argued that 'the knowledge claims we make, the jobs we work at and the power we have are all profoundly shaped by gender expectations. Through a gender sensitive lens not only the "what" of world politics but also the "how" we think about it looks different' (Peterson and Runyan 1993, 18). Runyan and Peterson pointed out that feminist IR was not exclusively about women, nor need it be only by women. The possibility of gender lenses was not premised on the existence of an unproblematic 'women's perspective'. It was possible for masculine perspectives to be held by women and feminist perspectives to be held by men, 'because those perspectives are politically not biologically grounded' (Ibid., 18).

Disciplining (feminist) IR

In reviewing the limited number of exchanges that have occurred between feminists in IR and mainstream scholars, one can identify a certain antagonism to the feminist project to 'gender IR'. At best, mainstream scholars have engaged selectively with feminist IR, ignoring or even disparaging the work of scholars who work with unsettled notions of gender and gendered subjectivities, while selectively engaging with those scholars who seemingly worked with stable and unproblematic gender categories. In turn, in embracing what appear to be settled and essentialist conceptions of gender, mainstream scholars have effectively attempted to reduce 'gender' to the status of one of many 'variables' that might be used to inform our theories on causality or to quantify 'impacts' in international politics. Thus 'legitimate' feminist inquiry has been confined within a rationalist/positivist approach to social science, which is seen as an exercise in 'making a conjecture about causality; formulating that conjecture as an hypothesis, consistent with established theory; specifying the observable implications of the hypothesis; testing for whether those implications obtain in the real world and overall ensuring that one's procedures are publicly known and replicable' (Keohane 1998, 196).

The acceptance of feminist scholarship in IR and judgements about whether feminism has contributed significantly to what is generally accepted as 'knowledge' in the field, have been profoundly coloured by consensual understandings of IR as a distinctive 'discipline' held by a community of scholars. These consensual understandings include agreement on the dominant preoccupations of IR and how our knowledge of this particular domain can best be advanced. The standing of a work or body of work within a discrete area of study is established by the judgement of one's peers, rather than being based on some absolute and objective measure of intellectual merit or academic worth. I do not wish to imply that scholars within the mainstream crudely and self-consciously perform the role of gatekeepers, rigidly defining and policing the discipline's subject matter and boundaries. The means by which Disciplines 'discipline' is certainly subtler than this. Nevertheless, engagements between feminist IR and the mainstream have been 'highly political' in so far as they have been read as attempts to 'discipline feminists as "goodies" and "baddies" in accordance with their perceived ontological, epistemological and methodological proclivities' (Carver, Cochran and Squires 1998, 29).

More so than any other branches of the social sciences, perhaps, debates within IR have centred on 'first order'-type questions concerning ontological and epistemological issues and/or appropriate methodological approaches. Given the nature of the debates that had preoccupied the IR community, one might have anticipated, perhaps, an initial openness to feminist questions and modes of inquiry and a willingness to engage with feminist critiques of the 'mainstream' (Grant and Newland 1991; Peterson 1992a, 1994). However, the implication of feminist critiques, as read by non-feminist IR scholars, was often taken to be that women occupied radically different life worlds and so 'women's experiences' could serve as a vantage point from which to construct knowledge of the world. Therefore, the mainstream of IR anticipated some kind of reconstructive project, a feminist paradigm or overarching feminist perspective on IR. However, feminism is not all of one kind and what it meant to be a feminist in IR could be quite differently understood. For example, it was perfectly possible for a liberal feminist to call for more attention to the position of women in IR, but to remain wedded to the possibility of objective analysis based upon sound empirical research and, as such, not to challenge fundamentally the basic epistemological premise or methodological preferences of positivism. On the other hand, scholars sympathetic to poststructuralism rejected the possibility of a neutral or objective stance, but also steered clear of epistemological claims made in the name of 'women' (Sylvester 1994a).

Some feminists were sympathetic to the idea of a reconstructive project, notably those who identified most closely with a 'standpoint' position. Standpoint theorists claim that the so-called 'objectivity' of positivism was based on the false premise that ontological considerations—who and what we think we are—have nothing to do with epistemology—the claims we made about the world. Drawing upon the work of feminist philosophers such as Sandra

Harding, standpoint feminists challenged the positivist distinction between the 'knower' (mind/subject) and the 'knowable' (nature/object), arguing that it was profoundly gendered (Harding 1991; see also Harding 1987). Hirschmann, for example, argued that in this way the perspective of a socially constructed 'masculine' experience is epistemologically validated, thus preserving male privilege and the social practices and structures that enable men to consider their own experiences the human experience (Hirschmann 1992).

Standpoint theorists argued that rather than accepting representations of 'human experiences' devised by men about the male world as seen by men as universal, one could start from the position that knowledge claims were rooted in concrete experiences. Women's experiences of child rearing and caring could serve as the starting point for rethinking key concepts like power. For example, if power were viewed as a capacity, energy and competence, rather than as the ability to prevail or dominate, one could both build up a critique of international relations as a realm characterized by anarchy, force and domination, and move beyond critique to articulate visions of a more interdependent and co-operative world (Tickner 1992).

Gender as a 'variable'

Perhaps it is because standpoint feminism potentially offered a coherent and singular feminist perspective that it appealed to scholars within mainstream IR. A feminist perspective or paradigm could be measured and judged against established world views. In embracing this particular strand of feminist IR, mainstream scholars assumed that standpoint theorists worked with settled and rather essentialist notions of gender, opening up the way for the introduction of a gender variable. Thus in various ways, a feminist standpoint could potentially contribute to the study of IR as a scientific enterprise.

Mainstream scholars could certainly point to examples of feminists who have seemingly acknowledged that this is part of their project. Enloe, for example, has been interpreted as proposing a new explanatory variable for the study of IR: the degree to which socially constructed gender hierarchies are important (Keohane 1998). Enloe claimed that in *Bananas, Beaches and Bases* (1989), and even more so in later books, her intention was not only to explore the effects of gender on IR, and on the lives of women particularly, but to sketch causalities (Enloe 2001). Enloe claimed that it was a mistake to portray feminist analysis as only interested in impacts. In both *Bananas* and *Maneuvers* (2000), Enloe was trying to show 'why states are so needful of ideas about masculinity and femininity' and, in so doing, 'make a theoretical argument about causality' (Enloe 2001, 656). She was also a self-proclaimed empiricist insofar as 'I want scholars to go out there and see which of two casual possibilities is at work in a given state at a particular time. To ask, "under what conditions do state officials invest state resources in the manipulation of masculinity?" The corollary question is "when do state officials try to manipulate women as people?" Neglecting to ask these questions produced "a very naïve understanding of how power worked and how inter-state relations worked"' (Enloe 2001, 657). Enloe also appeared to position herself within a standpoint tradition insofar as she not only wanted to take women's lives seriously, but also move women from the margins to the centre as subject of knowledge in IR.[8]

Whether or not Enloe would agree that what she had in mind was an enterprise in positivist social scientific analysis is a moot point. Moreover, this limited reading of feminist IR and the contribution of feminism to IR was problematic for a number of other reasons. First, while standpoint feminists made some claim in the name of women as knowers, even those who identify with a standpoint position differed as to whether this 'essentialism' could be justified on the grounds of biological/sexual differences, social experiences or whether it was merely a useful strategic ploy. Nor did standpoint feminists see an easy accommodation between their own epistemological claims and methodological approaches and the mainstream of IR. Indeed, Ann Tickner, a feminist IR scholar most often identified with a standpoint position, believed that the entire 'western philosophical tradition was too deeply implicated in masculinist assumptions to serve as a foundation for constructing a gender-sensitive IR' (Tickner 1998, 617).

While the mainstream was eager for a 'feminist perspective on Bosnia' (Zalewski 1995), feminist IR was perceived to be more troubling when it sought to expand the boundaries of IR and to ask the wrong kinds of questions (Weber 1994; Weber 2001, 82). Robert Keohane's thoughts on the 'contribution of a feminist standpoint' to IR theory provoked the first of a number of exchanges on the place and status of feminist theorizing within IR. Keohane's categorization of feminism as empiricist, standpoint and post-structuralist and his thoughts on the utility or potential contribution of each to advancing a mainstream IR agenda, prompted Weber to retort that this was an exercise in 'disciplining' feminist IR (Keohane 1989; Weber 1994). Weber objected that by categorizing feminists as standpoint, poststructuralist and empiricist—the 'good girls, bad girls and little girls' in IR, respectively—and explicitly rejecting poststructuralism as atheoretical and having no substantive research agenda, Keohane was seeking to confine feminist inquiry within the safety of the parameters of established discourse.

This was also a feature of what subsequently came to be seen as a significant contribution to the feminist/mainstream debate, an article by Adam Jones that questioned whether gender did indeed 'make the world go around' (Jones 1996). The Jones piece was significant for two reasons. First, the British International Studies Association deemed it to be of sufficient intellectual merit and its contribution substantial enough to be worthy of a prize. Second, and not unrelated, it prompted an exchange between Jones and his critics in the pages of a subsequent edition of the Review of International Studies (Jones 1996; Carver, Cochran and Squires 1998).

Jones began by acknowledging what he saw to be the key achievement of feminist IR, bringing a gender dimension to the study of IR by reclaiming women as subjects of history, politics and international relations. He then turned to the shortcomings of the feminist literature. Feminist IR, claimed Jones, limited its contribution to our understanding of the relevance of gender to IR by equating gender with women/femininity only and neglecting the study of men/masculinity (Jones 1996; Weber 2001). Jones later identified Cynthia Enloe as a prime culprit in perpetuating a partial and perverse view of gender in IR, by refusing to recognize men as victims and systematically marginalizing the male subject (Jones 1998, 303). Jones claimed that the concern with women only rendered the entire feminist IR project suspect since 'partisanship and scholarship do not easily mix' (Jones 1996). Jones later defended his position by arguing that it was not his intention to take issue with the normative feminist IR project, but only to point out that it had been 'one sided, selective and incomplete' (Jones 1998, 301). However, this limitation and distortion could be rectified if feminists (and others) asked broader questions about gender that incorporated both male and female experiences. Seeking a 'more balanced feminist IR' that addressed the position of men and masculinities was, in Jones's view, the first step in identifying a gender variable in IR.

It is questionable how much of the feminist IR literature Jones had read before embarking on his critique, since he made a number of claims that were without foundation. First, the central contention that feminist IR was concerned only with women, and so was unbalanced in its treatment of gender could not be substantiated by reference to the feminist IR literature (Zalewski 1999). As noted above, feminists recognized that 'women' and women's activities were constituted through the social relations in which they were situated, so the question immediately arose as to whether 'we should be concentrating on relocating/locating women within IR or should we concentrate instead on the functions of gender?' (Zalewski 1994, 428).

Nor were feminists guilty of ignoring men and masculinities; far from it. Enloe had used *Bananas* as a vehicle to illustrate how putting sustained effort into understanding the lives of women and asking questions about gender would lead to a deeper understanding of structures and processes which underpinned gender inequality and the complex way power worked in international politics. However, to understand the position of women and how gender relations worked, one had to look at 'when and where masculinity was politically wielded'. In turn, the ways in which masculinity worked to sustain inequalities in power could only be fully understood 'if we took women's lives seriously'. Thus, we learned a great deal about 'state anxieties about masculinity from paying attention to military wives' (Enloe 2001, 663).

Second, feminist IR scholars resisted the rather simplistic, essentialized categorizations of 'male' and 'female', 'masculine' and 'feminine' with which Jones appeared to be working. During the 1980s, academic feminism had grappled with issues of identity and differences among women, and had explicitly sought to problematize universal and stable categories, so it was unsurprising that feminists in IR also broached unsettling questions about 'who are "women", what is the difference and why does it matter?' (Zalewski 1994, 1). Enloe, often cited as a rather unsophisticated liberal empiricist or standpoint feminist, explored 'multiple masculinities' and how they 'got manipulated, the manipulators' motives and the consequences for international politics' (Enloe 2001, 663; see also Cooke and Woollacott 1993; Skjelsbaek and Smith 2001).

Jones called for more attention to men and masculinities, but was seemingly unreflective about how his approach rested on a conventional and essentialist conception of gender that was at odds with much of the contemporary social science literature on men and masculinities (see, for example, Connell 1995; Carver, Cochran and Squires 1998). His critics would later point out that to employ gender as a variable was to miss looking 'analytically and imaginatively at the who, how and why of power in the international context' (Carver, Cochran and Squires 1998, 297). Men and masculinity were generally treated and critiqued as privileged categories in society because women had been shown in feminist analysis to be a category of oppression (Ibid., 295).

Moreover, Jones was deeply critical of postpositivism, engaging in what was becoming a familiar attack on the 'bad girls' of feminist IR. Jones acknowledged as valid efforts to make knowledge claims in the name of women's experiences (the legitimacy of standpoint), because the epistemological assumptions of standpoint feminism could 'mesh with the classical tradition', the standard by which feminist contributions to IR should be judged (Carver, Cochran and Squires 1998; Jones 1998). Feminist empiricism was also to be welcomed, providing the research agenda made both men and women visible. However, post-positivism was explicitly rejected even though it was 'this form of feminist theorising that had arguably done most to address the tendency to collapse the categories "women" and "gender"' (Carver, Cochran and Squires 1998, 294).

Ultimately Jones's intervention into the gender/IR debate did little to advance understanding or encourage further dialogue between feminists and the mainstream. In charging feminists with partiality, selectivity and bias, Jones presented a selective, partial and rather distorted view of feminist IR, while his own project to 'gender IR' by identifying the gender variable in war and conflict did not go far beyond a crude measure of impacts or amounted to little more than 'stacking up dead male bodies against female bodies' (Carver, Cochran and Squires 1998, 296).

To feminist IR scholars, it was not immediately apparent how employing the 'gender variable', or focusing more on men and masculinities would necessarily advance a feminist agenda in international politics. Keohane, in an article published in 1998 and discussed at greater length below, argued that taking gender seriously opened up pertinent questions and potentially new avenues of investigation, such as: do countries with highly inegalitarian gender hierarchies behave differently from those with less inequality at home? As is seemingly the case with democracies, are states with more egalitarian gender relations less inclined to fight each other? Is gender, then, a variable that might be relevant to our understanding of the peaceful or hostile intentions of states or the likelihood of war? (Keohane 1998, 197). The problem was that 'taking gender seriously' in posing these kinds of questions, could work to engender a political backlash against feminism (Carver, Cochran and Squires 1998). Indeed, Keohane seemingly anticipated just such a possibility when in speculating further on the relevance of gender he suggested that 'perhaps states with less gender hierarchy would be less aggressive, but might be more easily bullied?' (Keohane 1998, 197). The fear that the breakdown of a patriarchal social order would result in weak and vulnerable 'feminized' states was the core theme of Francis Fukuyama's brief intervention into the gender/IR debate in his article on 'Women and the Evolution of World Politics' (Fukuyama 1998; see also Tickner 1999; Steans 2003).

Gender Trouble

Tickner has likened engagements between feminists and the mainstream of IR to the kinds of conversations that frequently occurred between men and women, fraught with the misunderstandings or the non-comprehension of people talking at cross-purposes (Tickner 1997). In asserting that these 'theoretical divides evidenced socially constructed gender differences', Tickner perhaps elided feminist IR too closely with a standpoint position (Tickner 1997, 613). However, misunderstandings have seemingly often been a consequence of the personal reactions that asking troubling gender questions tended to provoke. Keohane, for example, argued that IR scholars feared engaging seriously because they risked becoming targets for attacks on their motive (Keohane 1998). In Jones, the hostility to feminist IR was barely disguised in his charge that feminists chose to only recognize the male subject in international relations only 'where he could be brought in to explain female suffering and victimization' (Jones 1998, 303).

This accusation ignored the wealth of feminist scholarship concerned with diverse and multiple masculinities and the complex forms of power relationships that exist between men (Hooper 2001). Feminist analysis necessarily addressed issues of power in the construction of masculine identities, because it sought a deeper understanding of how the meaning of masculinity was constructed as a superiority of men (and the masculine) over women (and the feminine), but also involved the exaltation of hegemonic masculinity over other groups of men (Connell 1995). The world views of women and men are not so fundamentally at odds that we are incapable of understanding each other or engaging in meaningful discussion. Feminists were not incapable of empathizing with men as friends, fathers, brothers, partners, or sons, nor were feminists unaware that an inability to perform the masculinized role of breadwinner and provider could be a catastrophic and devastating experience for individual men. One could point to men within the discipline who had openly welcomed the contribution of feminism to IR, or had engaged with the feminist literature in a serious and reflective way, or had drawn upon feminist theories and concepts in their own work (see, for example, Halliday 1988; Neufeld 1995; Murphy 1996; Zalewski and Parpart 1998; Hoffman 2001).

Conversations between men and men, and women and women might be similarly 'troubled'.

However, conversations between the mainstream and feminists in IR have been rather terse and unaccommodating and, ultimately, have done nothing to further meaningful and productive dialogue. Tickner attributed the marginal impact of feminist scholarship, particularly in the US, and the continual questioning of whether feminists are doing theory at all to the different ontologies and epistemologies with which feminist and international relations scholars were working (Tickner 1997, 613). The ability to engage on a deeper level was impeded, in good part, by the continual insistence that it was possible to view and measure the 'gender variable' at work in IR, while continuing to be unreflective about how 'one is never outside of gender' (Weber 2001, 83). Feminists also rejected the notion of a neutrality of facts in favour of subjective epistemological positions, a position 'most unsettling to proponents of scientific methodologies' (Tickner 1997, 621). Consequently, feminist IR was likely 'to be dismissed as relativist and lacking objectivity' (Tickner 1997, 621).

CONSTRUCTIVE CONVERSATIONS?

Constructing a 'middle ground' in IR

While in the USA, at least, critical theory, poststructuralism and feminism have thus far expanded the margins of IR rather than 'transformed' the discipline in any significant way,[9] there have been more substantive engagements between the mainstream and those who identify with a social constructivist position (see, for example, Wendt 1999; Keohane 1998). 'Social constructivism is a "broad church" whose "followers" span from those merely integrating ideas into a materialist-rationalist framework to those who focus on the analysis of discourse and operate, both theoretically and methodologically closer to reflectivism, and especially post-structuralism' (Diez 2005, 188; see also Adler 1997). However, in this chapter the term 'social constructivist' is used specifically in relation to scholars who 'create theoretical and epistemological distance between themselves and critical theory' (Hopf 1998, 181; see also Christiansen, Jorgensen and Wiener 2001). Whereas critical theorists or 'critical' constructivists 'recognise their own participation in the reproduction, constitution and fixing of the social entities they observe, social (meaning here "conventional" or "moderate") constructivists are less concerned with epistemological questions, and thus "the observer never becomes the subject of the same reflective critical enquiry"' (Hopf 1998, 184). In this respect social constructivism is closer to rationalism/positivism than reflectivism/postpositivism. In summary, the middle ground position holds that in answering 'why' type questions, it is necessary to know something about identities and interests, about structure and agency, and the role of both material power and ideas in international politics. The 'middle ground' in IR is predicated on the assumption that it is possible to incorporate ideas as a 'variable' into explanatory theories and broaden systemic theorizing to account for the key role that is played by ideas. Basically, the construction of a 'middle ground' in IR comes down to a debate about the value ideas that a 'variable' takes on in specific instances.

Keohane has welcomed work that seeks to break down dichotomies between hermeneutic historically based traditions and positivist epistemologies modelled on the natural sciences, suggesting that these dichotomies should be replaced by continua with the dichotomies' characterizations at the poles (Keohane 1998, 194). Keohane envisages an accommodation between rationalist and constructivist approaches, arguing that 'choices are made on the basis of normative, descriptive and causal beliefs, all of which are deeply socially constructed. Choices are also made within structures of demography, material scarcity and power and within institutions that affect incentives and opportunities available to actors' (Keohane 1998, 195).

Keohane argues that an accommodation between rationalism and constructivism can be achieved if we recognize that knowledge is socially constructed and make efforts to widen inter-subjective agreement about important issues. However, while theoretical or methodological innovation might be tolerated, encouraged even, in Keohane's view there are clearly limits to how far this type of project can be allowed to go before it is categorized as unscientific, unverifiable and dismissed on the grounds that it lacks explanatory power. What is needed, he argues, are 'more cogent, contingent generalisations about international relations' that are 'scientific because they are based on publicly known methods and checked by a community of scholars, working both critically and cooperatively' (Keohane 1998, 196).

From the other side of the fence, so to speak, Fearon and Wendt have questioned 'whether progress in understanding international relations and improving human and plenary welfare is best served by the structuring of the field of IR as a battle of analytical paradigms', or whether 'important questions will be ignored if they are not amenable to the preferred paradigmatic fashion' (Fearon and Wendt 2002, 52). Or that 'we know so much about international life' that we should dismiss certain arguments or positions 'a priori on purely philosophical grounds' (Ibid., 52). Fearon and Wendt see the entire positivist/postpositivist debate as 'a dialogue of the deaf' in which each side tries to marginalize or subsume each other in the name of methodological purity. The challenge, they argue, is to 'combine insights, cross boundaries and if possible synthesise specific arguments in the hope of gaining more compelling answers' (Ibid., 68). IR needs to move away from debates on ontology in which 'a great deal rides on who wins', in favour of an 'ontological pluralism' which might tell us more about the conditions under which world politics would be more conflictual or co-operative (Ibid., 68).

They advocate a more pragmatic interpretation of rationalism and constructivism as lenses for looking at social reality or analytical tools with which to theorize about world politics. Keohane also favours a sophisticated view of science that overcomes any simple objectivist-subjectivist dichotomy. Feminists are not averse to the idea of adopting different analytical lenses or engaging 'empathetically' (Runyan and Peterson 1992; Sylvester 1994b). As is evident from the discussion above, feminist IR is not all of one kind, and undoubtedly some feminist scholars in the field might be interested to explore the possibilities for feminist engagements that are opened up by attempts to move beyond such dichotomies.

The co-option of gender in the 'middle ground' project

Just how far such endeavours might open up spaces for a deeper engagement with feminist work on the part of those who now claim to occupy the 'middle ground' in IR and where feminism(s) might sit within projects that attempt to bridge the gap between what appear, on the face of it, to be incompatible positions, would be to engage in a meta-theoretical debate that is beyond the scope of this chapter. Suffice to say that the effort to construct a 'middle ground' in IR has generated a substantive debate within IR, and among social constructivists (broadly defined) particularly about the intellectual and political credence of this project. At this juncture, I simply set out briefly what I understand this project to entail in so far as it is pertinent to my argument, since the main purpose here is to concur with the critics of the middle ground project that they do not take the politics of their endeavour seriously. In order to make this argument, I will begin with some reflections on the politics of 'bringing in', or 'co-opting', gender into a middle ground research agenda. However, before getting into the substance of my argument it is necessary to lay out in summary what the 'inclusion' or 'co-option' of gender entails.

Explanatory theories seek to go beyond the empirical description of an event (asking *what?* questions) and explain *why* an event takes place, or why some choices are made rather than others (and so aim to explain, for example, the reason for one policy outcome, rather than another). To speak of a 'variable' is to identify a factor that is pertinent to explaining certain events, actions or outcomes in international politics. So, for example, to identify a gender variable in international relations is to identify specific contexts in which gender makes a difference and is, therefore, relevant in explaining outcomes or actions in the 'real world' of international relations. Alternatively, it might be used to measure 'impacts', as above (Jones 1996).

Scholars as diverse as Goldstein and Keohane have argued that gender can be and should be employed in the study of IR, since it is clearly relevant to our understanding of 'real world issues of war in and between states' (Goldstein 2001, 2005; Keohane 1998). However, the only substantive engagement between feminism and social constructivism (as defined above) in IR has thus far proved to be similarly terse and unaccommodating (Carpenter 2003a, 2003b; Zalewski 2003; Carver 2003; Kinsella 2003). This is unfortunate, though not surprising, since the prospects for a meaningful dialogue were confounded at the outset by the proposed 'terms of engagement'. In her effort to embed gender studies in the middle ground of IR theory, Charlie Carpenter assumed that it was first necessary to uncouple the concept of gender, which she acknowledged as an important analytical tool, from feminism, defined as a normative commitment to the emancipation of women (Carpenter 2003a).

As with the feminist/mainstream debate, feminist scholars expressed concern that this was yet another project that sought to restrict the potentialities and possibilities of feminist scholarship, although Carpenter was at pains to point out that what she was advocating, in her view, complemented feminist analysis in the field (Carpenter 2003b; Carver 2003; Zalewski 2003; Kinsella 2003). It was not her intention to replace gender critique with the advocacy of gender as a mere variable, or an explanatory tool employed in the service of problem-solving in IR. Carpenter argued that her project was to further embed gender in IR by demonstrating that gender was relevant to the work of mainstream (and middle ground) scholars (Carpenter 2003b).

Of course, the relevance of a 'gender variable' in explaining actions, outcomes or impacts in international politics, must firstly be established by reference to an interesting empirical feature of the international realm that has not (yet) been noticed. The relevance of gender 'does not spring to one's eyes unless gender is actively used as an analytical tool' (Berg and Lie 1995, 344). In many, perhaps most, cases the relevance of gender will only become apparent if and when researchers go out and look for it. So, in my view, Carpenter was correct to argue that gender should be 'embedded' in the discipline, but was mistaken in thinking that gender and feminism can be easily and unproblematically uncoupled, since a certain feminist sensibility is a prerequisite to achieving this outcome.

Moreover, as feminists recognize, there are multiple meanings invested in 'gender'. Various appropriations of gender are reflected in the specific constitutive methodological criteria of research programmes. Specific statistical criteria indicate when a gender variable (difference) is 'significant' with respect to a given explanatory context, but to be deemed significant in the first place certain suppositions about the meaningfulness of that factor must already be in place; 'certain broader explanatory frameworks, specific understandings of what that factor is and certain kinds of expectations about how that factor will figure into the specific explanations at issue' (Rooney 1994, 110). So, to draw attention to the 'meaningfulness' of gender is to 'draw attention to the theoretical construction of gender itself' and so 'questions about the theoretical construction of gender cannot in turn be separated from broader gender issues in broader socio-cultural arenas' (Ibid., 110). How gender is rendered salient in such arenas is, in turn, 'a question of substantial social and political import, as feminist theorising shows' (Ibid., 110).

Carpenter hoped that scholars involved in both social constructivist analysis of gender and feminist scholars might engage in substantive dialogue. In actuality, the exchange between Carpenter and her detractors illustrated once again how even seemingly sincere efforts to engage across the boundaries of different intellectual traditions are apt to become bogged down in the politics of such engagements. It is not only contexts that are politically charged. The categories to be talked about—gender (or perhaps class or ethnicity)—are similarly infused with social and political meaning. No doubt, apprehension on the part of feminists about what is involved in 'uncoupling' gender and feminism can be (partially) explained by the knowledge that 'outside of academia, within development policy and activist arenas, the utility and relevance of gender has been highly contested' and 'in some neo-liberal policy applications, "gender" has come to lose its feminist political content' and become de-politicized. (Baden and Goetz 1997, 4).

The choices we make about our approach and/or our methods are political choices. Adler has argued that the engagement between rationalists and 'modernist' constructivists has proceeded from the assumption that science is not just one more hegemonic discourse and that science is compatible with the constructivist understanding of social reality (Adler 2002). Keohane, for example, has embraced social constructivists' insights, while remaining wedded to some notion of 'objectivity'. Both rationalists and 'middle grounders' hold that knowledge is best advanced through the testing of theories, or more particularly, hypotheses generated by scholars working consensually and co-operatively within an academic community (Keohane 1998). However, as noted above, 'consensual and co-operative' communities are not necessarily inclusive and, indeed, might be highly exclusive.

If one self-identifies as a constructivist, defending 'mature' theories on the grounds that they have 'proven successful in the world', as Wendt does, is a curious position to adopt (Wendt 1999, 59). Theories and concepts used to answer questions and the methods used in both scientific and social scientific inquiries are in themselves socially constructed. Even in the 'hard' sciences, early pioneers were not 'disembodied actors, detached minds passively contemplating the historical world', but people who 'carved out for themselves identities that allowed for a range of actions within particular social configurations' (Golinsky cited in Schiebinger 1998, 1,555). Phillips argues that 'disciplines, the content of which is handed down, ready formed, have laboured mightily over the generations to construct the content of the field and no doubt internal politics have played some role in determining the bodies of knowledge available' (Phillips 1995, 6). Indeed, it is the refusal to confront questions of social power and the political aspects of social constructivism that have led critics to argue that ultimately the 'middle ground' position is untenable.

It seems that projects that embrace ontological 'mashing' and give up on the possibility of discovering 'objective' facts about the world in favour of the inter-subjective negotiation of 'truths', would have to address the politics of knowledge claims, the politics of negotiating knowledge claims and indeed the conditions under which such engagements and conversations take place.

CONCLUSION

Feminist scholarship in IR has achieved much in the 15 years since a group of interested scholars met at the London School of Economics to ask questions about women and international relations. Enloe is perhaps right that this is a time to focus on the many successes and achievements of feminist IR. However, it is evident that feminist IR still faces considerable obstacles in gaining acceptance within a mainstream that has, historically, been 'crudely patriarchal' (Walker 1992). In Tickner's view, in the US positivism is no longer quite so securely embedded within the discipline, but it has yet to be displaced as the dominant approach to IR, and feminists' voices continue to be marginalized, ignored or appropriated in the interests of advancing a mainstream agenda (Tickner 1997). In the United Kingdom and elsewhere, where postpositivist approaches have been more widely embraced, the lack of attention paid to feminist perspectives by those within both rationalist and critical/constructivist schools has been disappointing (Marchand 1998). One continually finds examples of histories of IR that nod to feminism as a marginal voice in the postpositivist debate (Schmidt 2002).

Ultimately, the legitimacy of feminist work will only be recognized as part of 'the discipline' if 'the discipline' is rethought in ways that disturb the 'existing boundaries of both what we claim to be relevant in international politics and what we assume to be legitimate ways of constructing knowledge about the world' (Zalewski 1995, 352). Unsurprising, then that after a series of 'troubled conversations', feminist engagements with the mainstream have run their course and the debate has been characterized as one that is now 'exhausted' (Tickner 1997; Zalewski 2002). Feminism and IR is not yet an 'exhausted conversation' perhaps (Zalewski 2002), but as Baden and Goetz have argued (in a different context):

> It is important that we engage in dialogue with colleagues who work on gender issues from outside a feminist perspective, to attempt to broaden the scope of their studies and to see how their findings can inform our own work and campaigns.
>
> (Baden and Goetz 1997, 21)

I believe that the same argument applies in the context of IR. However, Baden and Goetz go on to qualify their statement thus: 'We also need to ensure that the pioneering contribution of feminist theorists and researchers is recognised' (Ibid.). Again, this is a sentiment that I echo. It is surely a precondition for any meaningful dialogue that one is aware of how one is situated or positioned in a debate or conversation, is reflective and open about one's identity and the identities of others, and cognizant of the complexities of our social and political relationships with others. Feminists would no doubt be sceptical that some kind of inter-subjective consensus on the substance of IR or on methods and approach is possible among feminists and the mainstream on the evidence of the conversations that have taken place thus far. Conversations between feminists and the mainstream are likely to be terse while feminists continue to be told that they will only be seriously engaged 'if feminists are willing to formulate their hypothesis in ways that are testable and falsifiable with evidence' by those who have for so long occupied the central territory in the discipline (Keohane 1998, 197). To limit or confine the discussion to the gender variable and to the scientific credentials of feminist theory is to refuse to engage in a serious discussion of either gender or feminism. To 'engage' feminism from the position of a constructed, yet increasingly privileged 'middle ground' in the discipline, and at the same time insist that this project is the *only* legitimate way to engage with gender in IR and, moreover, that it is an entirely separate enterprise from feminist IR, is to engage in the gender(ed) politics of IR: a politics that militates against meaningful and productive dialogue. To be genuinely open, dialogue requires 'the application of the principle of symmetry to the dialogue and mutual recognition of each others' strengths and weaknesses' (Berg and Lie 1995, 345). In this way, the possibilities of finding better solutions to the problems that beset both the theory and practice of IR might be better advanced.

NOTES

1 For example, on war and conflict (Elshtain 1987; Enloe 1993; Enloe 2000); security and community (Tickner 1992); policy-making in international institutions (Whitworth 1994); the gender dimension of globalization and global restructuring (Marchand and Runyan 2000); on ethics and human rights (Robinson 1999; Peterson and Parisi 1998; Hutchings 2000); on gendered identities and subjectivities in international politics (Zalewski and Parpart 1998; Hooper 2001); and the conception of the 'political' and political space in IR (Youngs 1999). More recently, postcolonial feminist scholarship with a focus on the interplay of race, gender and class in international relations has made a much needed and welcome contribution to the IR literature (Chowdhry and Nair 2002).

2 See 'Interview with Cynthia Enloe' 2001, 661. My thanks to Marysia Zalewski for drawing my attention to a number of issues that arise from this expectation or demand that feminism 'transform the discipline', including whether feminism is asked to do rather too much and hence likely to be presented as having 'failed'.

3 The term 'mainstream' in this context embraces both neorealism and neoliberal institutionalism since both are rationalist approaches to the study of IR.

4 The main divisions or positions in the fourth debate have been characterized in different ways. While the meanings and uses of these terms and categorizations might differ according to particular context, the former tend to 'see the world as something external to our theories, while the latter think our theories actually help construct the world' (Smith 1998, 226).

5 The argument developed in this chapter draws from an earlier journal article: 'Engaging from the Margins: Feminist Encounters with the "Mainstream" of International Relations', *British Journal of Politics and International Relations* 5:3, 2003, 428–54, and 'The Gendered Politics of IR', in J. Steans, *Gender and International Relations: Issues, Debates, Future Directions* (Cambridge: Polity Press, 2006). My thanks are extended to Marysia Zalewski, Peter Kerr, Daniel Wincott and Donna Lee for their helpful comments on an earlier draft of this article.

6 The neorealist view that the central concern of IR is the problems caused by an anarchic international environment is best exemplified by the work of Kenneth Waltz (1979), although the fusion of key concepts from traditional realism with a positivist, mechanistic view of the workings of the international system can be traced back to the earlier 'behaviouralist revolt' in IR (Vasquez 1983).

7 Since the late 1980s and 1990s, neoliberal institutionalists have contested the notion that international relations are 'fundamentally anarchic', the central contention of neorealism, and have pointed instead to the fundamentally institutionalized nature of international relations. Moreover, neoliberal institutionalists have contested the neorealist notion that international relations is a power struggle, or zero sum game, by highlighting the rationality or otherwise of co-operation given absolute rather than relative gains assumptions: the so-called neo-neo debate.
8 Enloe consistently refuses to allow herself to be pigeon-holed as any particular kind of feminist and, consequently perhaps, continues to be variously labelled as liberal/empiricist, standpoint and poststructuralist.
9 As Spike Peterson predicted over 15 years ago (see Peterson 1992a).

Plus ça (climate) change, plus c'est la même (masculinist) chose: gender politics and the discourses of climate change

Sherilyn MacGregor

'Even climate change cannot escape the gender wars'

(*New Scientist* 2007)

INTRODUCTION

'Are men to blame for global warming?' a recent article in the *New Scientist* asks before moving on to report, with thinly veiled scorn, that a Swedish Ministry of Environment study has claimed that men are more responsible than women for the current crisis. The author's incredulity is telling: it reflects the popular understandings that a) climate change is a scientific problem that affects all humans equally, and b) gender politics is an enduring yet increasingly irrelevant battle between men and women, not unlike a domestic dispute. His comment provides us with an apt starting point for a critical feminist interrogation of the contemporary climate change debate and the gendered discourses through which it has been framed. It is a good illustration that even in the face of potentially catastrophic socio-environmental change, the gender stereotypes, assumptions and lacunae that feminists have been analysing for decades remain much the same. It is tempting to suggest that the way we are going, patriarchy might be one of the last things standing when we lie gasping for breath on a dying planet.

Climate change presents serious challenges for politics: there will be a need to restructure institutions at all levels, to rearrange long-standing social, economic and spatial relations, and to devote resources towards regulation, mitigation and adaptation. What climate change presents for *the politics of gender* is less certain and much less discussed. The concept of gender is almost completely absent in policy documents and research reports on climate change at all levels. This comes as no surprise, since there is a long-standing blindness to gender within the broad field of environmental politics. However, what is both surprising and problematic is that feminists have largely ignored the issue of climate change. The work that has been done so far has focused almost exclusively on the material impacts of climate change on women in the Global South. My aim is to look beyond material impacts on empirical people to the dominant institutional and societal discourses that articulate and configure those impacts and formulate the agenda for their mitigation. I am interested in gender not as a synonym for women but in terms of hegemonic constructions of masculinity and femininity. I shall examine how four dominant and interconnected discourses effectively frame the climate change debate as a gender-neutral, scientific problem, while it is actually deeply gendered. An analysis of how these discourses emerge from and work to perpetuate prevailing gender roles and relations is useful in exposing the sets of unquestioned assumptions that inform the climate change debate.

In what follows, I look through the lens of gender politics to make three observations about climate change discourse, supported with examples drawn from academic, activist and popular literature. First, I argue that after several decades of women carving out a niche as advocates and exemplars of more sustainable ways of living on this planet, climate change has brought about a masculinization of environmentalism. Men dominate the issue at all levels, as spokespeople, experts, entrepreneurs and policy-makers, while women and their traditional environmental concerns have been shoved quietly to the side. Second, I argue that although climate change is a global problem with very public causes, responsibility for action has fallen disproportionately on the private sphere and, by traditional extension, onto the shoulders of women as mothers and consumers. Many women have embraced their eco-responsibilities with pride: here I give the examples of the ways in which the EcoMom Alliance in the USA and the Women's Institute in the United Kingdom have reinforced the gendering of environmental responsibility with equal measures of feminine righteousness and guilt. Third, and most disturbing, I suggest that climate change has led to a resurgence of neo-Malthusian population discourse. Feminists fought against populationism in the 1980s and 1990s as a sexist and racist discourse, a threat to women's reproductive freedom, and above all as a misdiagnosis of the environmental problem. But the same discourse has come back with a vengeance, and is epitomized by the work of the United Kingdom-based Optimum Population Trust. We now have the paradoxical situation where women are looked to for home-style solutions at the same time that the very thing that is thought to make them so caring towards people and the planet (i.e. their ability to give birth) is increasingly identified as the root cause of the global environmental crisis. All of this paints a rather gloomy picture of the state of gender politics today; in my view the climate change debate speaks volumes about the current *climate of gender politics* and the lack of change feminists have accomplished thus far.

GENDER AND CLIMATE CHANGE: BACKGROUND NOTES AND CRITICAL OBSERVATIONS

A good place to start this discussion is to explain that there is a lack of research on gender and climate change. This is not surprising, since there is a well documented case of gender blindness in the environmental social sciences in general (cf. Banerjee and Bell 2007). There has also been an almost total avoidance of environmental issues by feminist academics in recent years. If conference themes and journal articles are anything to go on, climate change is not on the academic feminist agenda. I conducted a CSA Illumina (CSAI) citation search of 20 feminist journals from 1990 to the present, using the keywords 'climate change' and 'global warming' (in the 'words anywhere' box). Surprisingly, my search resulted in a total of just 10 articles. This avoidance may be explained by the fact that feminists historically have been wary of the gender politics of the environment because its focus on 'nature' treads too close into biological determinist territory for feminist comfort (for a discussion see Alaimo 2000). The exception is the small and marginal branch of feminist theory called ecofeminism,[1] which has as its central concern the connections and intersections between the historic domination/exploitation of the environment by humans and the oppression/exploitation of women by men. Ecofeminist theorists have developed important, critical analyses of the gender politics of the environment (see, for example, Plumwood 1993, 2002; Haraway 1991, 1996; Warren 2000). However, unfortunately ecofeminism has been unfairly caricatured and associated with some essentialist ideas about women's unique relationship with nature that have led most feminists to avoid it altogether (Davion 1994).[2] As I discuss elsewhere (MacGregor 2009), this is bad news for feminism: by throwing the ecofeminist baby out with the essentialist bathwater, feminism has been left incapable of (uninterested in?) addressing the kinds of political issues that societies face in the wake of climate change. Ecofeminism is the only theoretical perspective that makes the exploitation of the non-human world—and the biosphere on which all human life depends—a feminist issue.

The small amount of research that exists on gender and climate change has been conducted by ecofeminists, Gender, Environment and Development (GED) scholars, and by feminist researchers working for government ministries, the UN and women's environmental organizations.[3] My CSAI citation search found that most of the articles on gender and climate change are in a special issue of the journal *Gender and Development* from July 2002.[4] A brief review of that special issue provides a good synthesis of the main themes, as well as some of the shortcomings, in the fledgling gender and climate change literature.

To demonstrate the lack of attention to gender in climate change research, Margaret Skutsch (2002, 30) conducted 'a scan of a number of prominent journals dedicated to the climate issue [which] reveals not a single article on the gender-differentiated implications of climate change in recent years'. She also notes that neither the Kyoto Protocol nor the United Nations Framework Convention on Climate Change (the two most important treaties on climate change) actually mention the words 'gender' or 'women'. There is a consensus among the contributors to the *Gender and Development* issue that the gender dimensions of climate change have been neglected by policy-makers. Not much appears to have changed since 2002. The Intergovernmental Panel on Climate Change (IPCC) is the main source of climate science and policy research upon which governments around the world rely when setting national targets and making policies to address climate change. I searched (using the 'Find' function in Adobe Acrobat) the IPCC's Fourth Assessment Report *Climate Change 2007* for the words 'gender' and 'women'. In the 52-page general synthesis report there is one mention of women (relating to the reduction of workload for rural women in developing countries that could be brought about by the switch to cleaner energy sources) and no mention of gender. In the three Working Group reports attached to the main Assessment, women are mentioned just once when listing the Millennium Development Goals, and gender is mentioned only a few times in reference to the articles in the climate change issue of *Gender and Development*. Finally, I found no hits at all in the 24-page synthesis report for policy-makers. At the very highest levels, then, climate change is cast as a human crisis in which gender has no relevance.

The second overarching theme in the special issue is that climate change is not gender neutral but has gender-differentiated causes and effects (Dankleman 2002, 24). The authors argue that, due to the scope of the impacts on ecological, economic, social and political life that climate change will bring, it is only reasonable to accept that men and women as gendered beings, and their gendered relations, will also be affected. All authors contend that climate change will affect men and women differently. Ecofeminists and GED scholars have been claiming for decades that women are more dramatically affected by environmental degradation than men due to their social roles as provisioners and carers and in their social location as the poorest and most vulnerable at the bottom of social hierarchy (alongside children) (Buckingham-Hatfield 2000). Climate change is no different, and the articles in the special issue give concrete evidence of gender-differentiated effects. For example, women are more likely to be hurt by natural disasters and extreme weather events than men; women's everyday provisioning work will be made more difficult due to climate change-related impacts (e.g. walking further for clean water and firewood); and economic and social breakdown caused by displacement will bring about a worsening of women's already low status and vulnerability in the face of that social breakdown. Put simply, the more socially and economically marginalized are more vulnerable to the detrimental effects of global warming. The IPCC accepts this equation (without explicitly mentioning 'women' or 'gender') by saying those with the fewest resources are the most vulnerable to the negative effects of climate change (Dankleman 2002, 22).

The GED approach to the issue also addresses gender differences in the *causes* of climate change, frequently pointing out that those who are most vulnerable to it are not the most responsible. There is a scientific consensus that climate change is in large part human-made (or anthropogenic) (Homer-Dixon 2006); feminists wish to point out that it has not been made by all humans equally. For example, Dankleman's contribution sets out the process whereby men have more money and power than women; in order to achieve this relative wealth and power men have exploited the environment, therefore men have caused and benefited from environmental destruction. This is similar to the approach taken by Gerd Johnsson-Latham (2007), author of the Swedish study that provoked the ire of the *New Scientist*.

In fact, the author cites a 2002 UN study that 'shows that the polluters are mainly men' (Johnsson-Latham 2007, 8).

Although the special issue presents important analyses and provides a much needed antidote to almost total gender-blindness on the issue of climate change, there are some inevitable shortcomings. One is that it focuses almost exclusively on climate change in relation to women in developing countries of the Global South. Seven out of 12 articles focus on the gendered impacts of and responses to climate change in countries such as Bangladesh, Peru and India. This is important and to be expected, given that the protagonists of the GED field are rural women in the Global South. Although the editor, Rachel Masika, wishes to steer clear of 'victim-talk', and there are a few examples of women's climate change activism, the dominant theme is women's vulnerability to harm. Another limitation is that there is little concern with men and, therefore, most analyses fall into the familiar trap that 'gender' means 'women'. Skutsch is one of the only authors in the issue to question the exclusive focus on women. While she agrees that the feminization of poverty is an important issue, she says we also need to disentangle the effects of poverty and gender inequality. She asks if women are more vulnerable because they are poor or because they are women (Skutsch 2002, 34), and responds by saying that poor men should get just as much attention as poor women because poverty is a more important variable than gender.

However, what of rich men and women in the affluent, overdeveloped North? Given the well-known links between affluence and climate change, it seems clear that research needs to consider explicitly the gender asymmetries within both the causes and impacts of climate change in the overdeveloped world, as well as how gender asymmetry may work in the solutions to climate change. Unfortunately, however, these issues are not addressed in the journal. Skutsch (2002, 34) argues, somewhat simplistically, that in 'better-off' societies there will be less gender differentiation in impacts of climate change. This highlights the rather narrow understanding of 'impacts' as only material and measurable effects (e.g. increased poverty, intensified burden of domestic labour) experienced by 'empirical' women that is deployed in much of the GED literature.

So while the analyses of gender and climate change in the articles discussed above are important, I want to examine the issue through a different lens. Thus far, little or no thought has been given to the cultural and symbolic (i.e. non-material) gender dimensions of climate change or to the ways in which gendered environmental discourses frame and shape dominant understandings of the issue. Gender is not just an empirical category (men/women), it is also a discursive construction that organizes the world. Gender analysis and the study of gender politics should involve the analysis of power relations between men and women and the discursive and cultural constructions of hegemonic masculinities and femininities that shape the way we interpret, debate, articulate and respond to social/natural/technological phenomena like war, economic crisis and climate change. I offer such an analysis in the remainder of this chapter.

CLIMATE CHANGE AS A GENDERED HUMAN DRAMA: THREE ISSUES FOR FEMINIST ANALYSIS

Gender is a concept that 'structurally organises […] virtually every aspect of social life in all cultures' (Peterson and Runyan 1999, 31). I want to look at climate change as a *gendered human drama* (Alaimo 2008, 300). All dramas do more than tell a story about what happened or what is happening to whom and why. Dramas have layers, players, plots and moral messages. They reflect their social and cultural context and provide insight into dominant values, the *zeitgeist.* This is not to treat climate change as a mere 'current event' or to downplay its seriousness. It is an approach that allows us to examine critically the contemporary climate change debate to see what the responses to climate change tell us about the current state of gender politics and what impact the unfolding crisis is having (or is not having) on the politics of gender. The lens of gender politics brings into acute focus the processes, norms and power relations through which we can recognize the workings of hegemonic masculinities and hegemonic femininities in all social phenomena (Peterson and Runyan 1999). Hegemonic masculinities and femininities are the particular discursive conceptualizations of maleness and femaleness that are 'dominant in a given set of gender relations at a particular time; and in relation to which other conceptualisations are seen as subordinated, marginalised or complicit' (Bretherton 2003, 104). It is useful to think of there being different 'scripts' in the human drama—ways of identifying, being, knowing—for males and females (Ibid., 31). I want to look at what scripts men and women are 'reading from' and what roles they are playing; what are the implications of gendered assumptions about men and women in and for the climate change debate?

In addition to taking this approach to gender analysis, I also want to apply a discourse lens to the climate change debate in the hitherto neglected context of the affluent world, my own geographical and cultural location. Environmental politics scholars have noted that in the affluent world (namely the USA and United Kingdom) climate change has been shaped by the three recognizable and hegemonic discourses of ecological modernization, environmental security and green consumerism (Schlosberg and Rinfret 2008). I would add population control discourse to this list. Explaining the use of a discourse lens in environmental politics, Charlotte Epstein (2008, 8) writes that 'a discourse is a cohesive ensemble of ideas, concepts and categorisations about a specific object that frame that object in a certain way, and therefore delimit the possibilities for acting in relation to it' (see also Dryzek 2005). Taking a critical look at discourses allows us to understand how people 'make sense of themselves, of their interests and their ways of behaving, and of the world around them' (Epstein 2008, 7). I argue that each of these dominant environmental discourses plays a central role in framing and shaping the issues I explore in this section: a) the masculinization of environmentalism; b) the feminization of environmental responsibility; and c) the 'greening of hate' (Hynes 1999).

The masculinization of environmentalism: the discourses of ecological modernization and environmental security

The rise of climate change to the top of the green agenda has brought about an apparent masculinization of environmental politics. There are several reasons for this assessment. At the most obvious level, it is clear that men are the dominant decision-makers and the prominent spokespeople on climate change. One could say they have simply *retained* a position of dominance throughout the development of environmental policy and politics (see Seager 1993), but this would unfairly paper over the times during the history of the modern environmental movement when the balance of influence seemed to have shifted. During the 1980s and 1990s, when the focus of the environmental movement was on the issues of health, anti-militarism/peace, right livelihood, and the conservation of wildlife habitats and natural resources, women were prominent activists. Key iconic figures included Lois Gibbs (fighting against toxic contamination of the Love Canal neighbourhood in the USA), the women activists of Greenham Common (protesting against the cruise missiles in the United Kingdom), Clayoquout Sound (protesting against the clear cutting of old growth forests on the west coast of Canada) and Chipko (hugging trees to save them from the logger's axe in India), Vandana Shiva (biodiversity activist), Anita Roddick (founder of The Body Shop) and Wangari Maathai (Nobel Peace Prize winner and founder of the Kenyan Greenbelt Movement). However, since the late 1990s or so, the shift of the dominant environmental focus to climate change has been accompanied by a relocating of the centre of environmental debate and action to within (rather than outside) the scientific and policy-making institutions. This has brought a shift in the key protagonists and has brought men to the fore as policy experts, scientists, political advocates, entrepreneurs, commentators and celebrities. Men far outnumber women in scientific and decision-making organizations that have responsibility for figuring out what to do about climate change. International environmental delegations are mostly made up of and led by men (Dankleman 2002). In the United Kingdom women are a small minority in fields that have influence over climate change policy-making, 'representing only 17% of FTSE boardroom appointments, 18% of Members of Parliament (MPs), 24% of Members of the European Parliament (MEPs) and 19% of scientists and engineers (Women's Environmental Network and the National Federation of Women's Institutes 2007, 11).

The IPCC is mostly made up of male scientists, with chairman Dr Rajendra Pachauri leading at the global level. The most prominent political climate change spokespeople in the USA are male: former Vice President Al Gore (co-winner with the IPCC of the 2007 Nobel Peace Prize), Governor of California Arnold Schwarzenegger, and Senator Robert Kennedy, Jr. At the time of writing, many greens are feeling optimistic about the eco-credentials of US President Barak Obama. In the United Kingdom political awareness about climate change has been aroused in various ways by such men as Sir Jonathon

Porritt (former chair of the Sustainable Development Commission), Zach Goldsmith (environmental adviser to the newly greened Conservative Party and ex-editor of *The Ecologist* magazine), Lord Nicholas Stern (economist and author of the Stern Review), and the Milliband brothers (David, the former Secretary of State for Environment and Ed, the new Secretary of State for Energy and Climate Change). His Royal Highness, the Prince of Wales won the title of Global Environmental Citizen in 2007 (Milmo 2007). Political columnists George Monbiot and Mark Lynas have become well known for their regular articles about the coming climate crisis. With precisely two exceptions, all of the above are white, wealthy and highly privileged men.

A 47-country internet survey of 'global consumers' conducted in 2007 by the Nielsen Company and the University of Oxford yields some interesting, yet unsettling, findings about who are the 'most influential spokespeople on climate change'. Perhaps not surprisingly, Al Gore was picked as the most influential, followed by Kofi Annan, Nelson Mandela and Bill Clinton. In the United Kingdom, the poll found that Richard Branson and Bob Geldof (with 23% and 18% of the vote, respectively) were the top 'influential spokespeople on climate change'. No women made it into the United Kingdom's top five. Globally, the top two women named in the poll were Oprah Winfrey (a talk show host who has invited environmental celebrities on to her programme to discuss climate change) and Angelina Jolie (an actress whose political work relates to refugees rather than global warming in any discernable way). There are 22 people on the list, of which five are women[5] and they are 'A-list' actor-models with highly dubious track records on being spokespeople for any political issue, let alone climate change. One can only surmise that these were the only women the respondents (i.e. 'global consumers') could think of at the time of the survey. Then again, I would be rather hard-pressed myself to name more than a small handful of women politicians, authors, scientists or entrepreneurs who are climate change spokespeople. Sheila Watt-Coulter (chair of the Inuit Circumpolar Conference) and Elizabeth May (leader of the Green Party of Canada) and Caroline Lucas (leader of the Green Party UK) are three of them. So why is it that women are so absent from the climate change debate? One possible answer is that the climate change debate has been shaped by stereotypically masculinist discourses that work to 'invisibilize' and alienate women and their concerns. This exclusionary process can operate in scientific institutions, policy circles, in the media and in society at large. Two masculinist discourses I have in mind are ecological modernization and environmental security.

Climate change is widely represented as a techno-scientific problem requiring technical solutions. This perspective and the discourse through which is it articulated (and made common sense) is called ecological modernization (EM). This discourse has become dominant in the past decade in the United Kingdom and Europe and is catching hold in the USA (Schlosberg and Rinfret 2008). Simply put, EM advocates the use of 'technological advancement to bring about [both] better environmental performance' and economic efficiency in a win-win situation (Schlosberg and Rinfret 2008, 256). Moving beyond the tired and overly politicized notion of 'sustainable development', EM has a supply side focus and depends on co-operation between government and business to solve environmental problems (cf. Hajer 1995). Climate change represents a welcome opportunity for this world view; it provides that a serious global problem can be solved by the wise partnering of techno-innovators and brave capitalists. Thus the crisis has brought about the development of all sorts of technical fixes that are economically lucrative and thus apparently win-win, such as carbon-trading and offsetting, carbon capture and storage, carbon sequestration, renewable energy (wind, solar, wave and geothermal power, biofuels), patented genetically modified crops, fake plastic carbon-absorbing trees, and the list could go on. Much of the response to climate change comes in the form of innovations and gadgets: 'lots of neat green stuff' (Schlosberg and Rinfret 2008, 268). While many of these innovations will be necessary for a sustainable future, EM advocates more technology and more development; a continuation of 'doing what we do' with more omnipotence rather than humility about both the benefits *and the costs* of human ingenuity to date.

The other (and related) dominant masculinist discourse that appears in the climate change debate is that of environmental security (ES). It is underpinned by Hobbesian predictions that climate change will inevitably lead to conflict over scarce resources between and within states (cf. Homer-Dixon 1999). Since the early 1990s, defence ministries (traditionally the domain of men) have been interpreting environmental 'insecurities' in ways that call for armed and militaristic readiness, alliances and responses (Elliott 2004). Some scholars have warned against this 'securitizing' move, arguing that it is contrary to the co-operation that is necessary on this shared planet and forgets the lessons of past wars and the environmental devastation caused by militarism. Some have said that environmentalism should stick with its core concerns for health, nature, and future generations rather than don the 'blood soaked garments of war' (Deudney 1990 in Schlosberg and Rinfret 2008, 262). However, such early warnings have not been heeded, and there has been growing interest in recent years in presenting climate change as a serious threat to national and global security. In 2004 the United Kingdom's Chief Scientist, Sir David King, made the connection clear, saying that climate change is a worse threat than terrorism (Connor 2004). Ironically, the environment was once considered a 'soft politics' issue in the field of international relations, far removed from the 'hard' political issues of security and military affairs (Peterson and Runyan 1999, 59–60). Now environment has become 'hardened' by the threats to national and international order that climate change is predicted to bring in its wake (e.g. climate refugees). By securitizing and militarizing it, the environmental crisis becomes a problem that requires technical, diplomatic and military solutions, entirely consistent with hegemonic (hyper)masculinity.

Climate change is the dominant issue for environmental politics today and, as I have suggested, it has been framed by two masculinist discourses and taken up by the institutions that are dominated by elite men. The worrying results of this are that women are excluded from decision-making and leadership roles and that the kinds of environmental issues about which women traditionally have felt most passionate have been pushed off the agenda, even though they have not gone away. Could women have become alienated from the climate change debate because they are less inclined than men to engage with science and technology? It is widely known that hegemonic femininity does not encourage women's aptitude for maths and science, and the under-representation of women in these fields provides convincing evidence. When it comes to a highly scientized and compartmentalized debate like climate change, it is quite possible that many women simply 'switch off'. According to survey data collected in the United Kingdom '... men are better informed about climate change science than women' (Hargreaves et al. 2003 quoted in Shackley, McLachlan and Gough 2004, 33). Reflecting on her experience of high-level climate meetings, Ulrike Rohr, director of the German gender and environment project Genanet, attributes the low participation of women to the exclusively scientific and technical approach to global warming. 'Women feel like they can't enter the discussions', she says (quoted in Stoparic 2006). The effect of feminine stereotypes and socialization may extend to both understanding and to action. Although I have no studies to support it, my instincts tell me that, given current gender socialization and division of domestic labour, far more men than women can get excited about insulating the loft, installing double glazing, generating electricity in the back garden, or retrofitting the car to run off used vegetable oil.

Another important consequence of climate change becoming the dominant issue is that most other environmental issues—those about which women care most—have been sidelined. Studies have found that women tend to be interested in the health and future prospects of children and about the environment-related issues that affect quality of life (cf. MacGregor 2006). For example, women have spearheaded campaigns on pesticide use, electromagnet fields, toxic contamination, hormones in cow's milk, multiple chemical sensitivity, asthma and sick building syndrome. Few of these now get much attention by climate change campaigners. As Bretherton (2003, 108) observes, 'environmental organisations such as Greenpeace and Friends of the Earth have witnessed the marginalisation of women, and of the issues most frequently raised by women around health, wildlife and local environment'. Where once women had some expertise and authority on environmental protection, they have now been alienated from the dominant issues and solutions.

In fact, climate change is now being used to 'trump' some of the issues about which women express concern. For example, nuclear energy has been put back on the agenda. Fears of the health risks of radiation from nuclear power and nuclear waste (significantly higher among women than men; cf. Solomon, Tomaskovic-Devey and Risman 1989; Kiljunen 2006; Freudenberg and Davidson 2007) have been put aside because it is allegedly a low-CO_2-emitting form of energy production. The potential health risks and the ethical uncertainties associated with genetically-modified organisms (expressed more by women than men; see, for example, Department of Trade and Industry 2003; Moon and Balasubramanian 2004) have been put to one side in the face of climate-related crop failures and the need for biofuels. Animal rights, a long-standing feminist ethical issue (cf. Adams 1999; Seager 2003), have become instrumentalized as a means to a climate saving end. The chair of the IPCC has recently told us to eat less meat because meat production causes more CO_2 emissions than the transport industry rather than because the factory farming and mass consumption of animals is both unnecessary and unethical.

Rather than look to technical fixes for environmental problems, women have tended to focus on the social dimension: sustainable lifestyles, ethical consumption and the precautionary principle (Johnsson-Latham 2007, 6). The Women's Environment and Development Organization have called this approach 'human security', possibly to mark it as distinct from environmental security discourse. They define it as the protection of 'the vital core of all human lives in ways that enhance human freedoms and fulfilment' (Ogata and Sen 2003); the security of individuals, their livelihoods, and human rights including economic security, food security, health security, environmental security, personal security, community security and political security (Dankleman et al. 2008, 2). It seems to reflect a stereotypically feminine set of concerns. From an ecofeminist perspective they are more ethically and socially relevant issues than those that seem to capture the attention of the masculinist people in power. This is the position taken by the contributors to the climate change issue of *Gender and Development*. Masika (2002, 3) writes: 'Predominant approaches and policy responses have focused on scientific and technological measures to tackle climate change problems. They have displayed scant regard for the social implications of climate change outcomes and the threats these pose for poor men and women, or for the ways in which people's political and economic environments influence their ability to respond to the challenges of climate change.' From this perspective, I argue, the twin discourses of ecological modernization and environmental security represent more of the same gender politics and threaten to intensify the underlying causes of climate change.

EcoMoms to the rescue? The discourse of green consumerism

It is a paradox of the climate change debate that women are absent and alienated from it at the same time as being increasingly implicated in its solution. While the masculinist EM discourse focuses on supply side, technical solutions and ES focuses on militaristic 'muscle-flexing' (Denton 2002, 18), there is a parallel *feminizing discourse* of green consumerism that places the onus on individual consumers to take responsibility for reducing the environmentally harmful impacts of contemporary life. Governments and environmentalists place emphasis on the role of individuals as consumers to tackle climate change by conserving energy, taking public transit, recycling waste, growing food and foregoing flights. One example is the United Kingdom Government's Act on CO_2 campaign, which gives helpful hints for reducing individuals' 'carbon footprint' when 'in the home', 'on the move' and 'out shopping' (campaigns.direct.gov.uk/actonco2/home). The cheerful website offers helpful tips on being green when planning meals, doing laundry, putting out the rubbish and buying clothes. Such policies and campaigns that focus on the duty of individual consumers are not new: environmentalism has been played out in the domestic sphere for decades. Climate change has helped to usher in a new agenda that adopts the same discourse with added urgency, not to mention added centralization of the message in the hands of government. The stakes are higher than ever.

The gender politics of this green consumerism discourse is largely unrecognized by everyone but ecofeminists. It operates without any thought to the workings of the private sphere; it takes for granted that the tasks will be accomplished by those whose consciousness have been sufficiently greened. However, in so far as consumption is a private sphere activity, and women tend to be principally responsible for household consumption, ecofeminists know that exhortations to 'live green' are directed at (and will be received primarily by) women. Men may hear them, but expect women to do the work. Studies have found that women are more likely than men to take on green housework such as precycling and recycling (MacGregor 2005; Schultz 1993). It is a key insight of ecofeminist scholars that women's socially ascribed roles as carers and provisioners bring them into direct contact with the environment in particular ways (for examples see Buckingham-Hatfield 2000). Because of these roles and this contact, because of the power of hegemonic femininity, women tend to feel more responsible for and more concerned about the quality of the environment (cf. Zelezny et al. 2000; Dietz et al. 2002; Hunter et al. 2004). Thomas Dietz, who specializes in social psychological research into environmental attitudes continues to report that 'generally, women are more concerned about environmental issues [than men]' (Dietz et al. 2007). So, when the state puts an emphasis on green household practices and green consumerism it downloads ecological duty to the private sphere and on to those most involved in unpaid domestic labour. Thus it effectively uses women's labour as a free subsidy for the common good. Ecofeminists have criticized this as a form of environmental privatization that is consistent with neoliberalism (cf. Sandilands 1993).

There is a longstanding ecofeminist critique of environmental politics that it pays insufficient attention to the politics of gender in general and the gender division of labour in particular. I have written about this lack of attention to 'life-sustaining labour' elsewhere (MacGregor 2005, 2006), as have other feminist environmental scholars who analyse the importance of social reproduction for environmental sustainability (cf. Shultz 1993; Sandilands 1993; Littig 2001). Giovanna Di Chiro (2008, 281) gives a useful definition of social reproduction as 'the intersecting complex of political-economic, socio-cultural, and material-environmental processes required to maintain everyday life and sustain human cultures and communities on a daily basis and intergenerationally'. She notes that social reproduction, a feminized sphere of activity, has been 'ignored or trivialised' in mainstream (i.e. non-feminist) scholarship, even though it has been affected in important ways by neoliberal capitalist globalization and ecological degradation. Because women's responsibility for social reproduction is *assumed yet ignored*, no one stops to raise questions of equity and fairness in a green agenda that depends on it.

There are differences between the way women's association with social reproduction plays out in North and South. In the Global South, according to GED researchers, gender politics have been manifested in development programmes that are explicitly designed to be carried out by unpaid women volunteers, based on the assumption that rural women are predisposed to taking an environmental care-tending role. This assumption, writes Masika (2002, 6) ' [...] continue[s] to translate into initiatives that place greater burdens on women's time and labour without rewards, and do[es] not provide them with the inputs (education, information, and land rights) they require'. In the affluent North, the assumption is subtler. Thus far there have not been any programmes that overtly target women as they do in the South. Perhaps there is no need for this, since women have internalized the sense of responsibility to 'do their bit' for the environment and have taken up the duties promoted by the 'green agenda' quite willingly and publicly. Invoking a Foucauldian analysis, some theorists have referred to this internalization as a form of 'environmentality' (Luke 1995; Agarwal 2005) whereby people are made into good green subjects by adopting the values of government. As I have analysed at length elsewhere (MacGregor 2004, 2005, 2006) many women have become environmental subjects (citizens?) who wear their green duty with feminine pride. Similarly, Sandilands (1999, xiii) describes what she calls 'motherhood environmentalism', arguing that women's concerns about nature tend to 'boil down to an obvious manifestation of natural protective instincts towards home and family'. She goes on to say: 'it is all about threats to children and self-sacrifice for the sake of future generations'. Women's maternal role is often used as a justification for their involvement in environmentalism. I have called this ecomaternalism (MacGregor 2006) in relation to women's 'quality of life activism'; I now recognize it in the climate change debate.

Two examples of a hyperfeminine, ecomaternalist approach to climate change are the EcoMom Alliance in the USA and the Women's Institute in the United Kingdom. The EcoMom Alliance is a non-profit organization that was established in 2006 with the aim of 'inspiring and empowering women to help reduce the climate crisis and create a sustainable future' (ecomomalliance.org). It has over 6,000 members and a trademarked motto: Sustain Your Home, Sustain Your Planet, Sustain Your Self™. The EcoMom Alliance pushes all the right buttons of hegemonic femininity in the effort to convince women that it is their duty as mothers to save the planet. Their website sums up the organization's *raison d'être* like this:

> Throughout history, during times of fever, flood, famine or flu, women step up and do what must be done […] As both role models and a market force, we believe mothers (and earth mothers alike), can help propel an environmentally, socially and economically vibrant and healthy future.
> (ecomomalliance.org/about)

The website also invites women to 'take the EcoMom challenge', which entails taking 10 well-known steps toward tackling climate change by changing household practices and making the right consumer choices. All of these steps are promoted by government and environmental campaigns; the difference is that whereas masculinist greens do not address who will take the steps, here there is an explicit acknowledgement that this is *women's work*, indeed work that most 'good moms' already do. For those who are not able to shoulder the burden of environmental responsibility, one of the 10 steps suggests ways to 'reduce mom guilt with carbon offsets, renewable energy credits or green tags'. Invoking a hyperfeminine metaphor, the site suggests that 'supporting renewable energy development to balance out your worst "eco-sins" is kind of like eating too many brownies one day and jogging extra the next' (ecomomalliance.org/take-the-ecomom-challenge). Rather than adopting a feminist position that critiques the traditional gender division of labour (where women take on all the unpaid housework while men are exempt from such domestic drudgery), the EcoMom Alliance accepts and affirms the gendered status quo. So internalized is this hegemonic hyperfemininity that many women find power in it and want to put their special power to good green use.

The same may be said for one of the very few women's organizations that campaign on climate change in the United Kingdom, the National Federation of Women's Institutes (or 'WI' for short). The WI has been around for 100 years and has traditionally been associated with stereotypically feminine concerns about domestic life (e.g. cooking, gardening, crafts). In recent years, the Institute has become involved in bringing women's perspective to such political issues as European agricultural policy, prostitution and the environment. In 2007 the WI paired up with the Women's Environmental Network (WEN) to conduct a survey of 500 women's opinions on climate change. Using the results of the survey, it published the Women's Manifesto on Climate Change, the preamble of which states:

> Women in the UK have a key role in tackling climate change as consumers, educators and 'change agents' in our homes, encouraging the adoption of lower carbon lifestyles and passing on green values to the next generation. We are also far more concerned about environmental issues than men.
> (Women's Environmental Network and the National Federation of Women's Institutes 2007, 2)

Here again we see the internalization of the green consumerism discourse: the acceptance of women's role in the private sphere amounts to a feminization of environmental responsibility. The WI celebrates women's 'power' to tackle climate change by making good decisions in the supermarket and in the household. 'As household managers, [women] are also key to controlling the 30% of UK carbon emissions that are produced in the home' (Ibid., 9). The Manifesto contains a list of 'what women want' the government to do about climate change, a list of demands that is justified by the argument that not only are women in a better position to take action on climate change than men, they are also more concerned about it than men. The survey data quoted lend support to the claim made by other researchers (e.g. Johnsson-Latham 2007) about masculine versus feminine environmental priorities: women in the survey were far more concerned about the threats of climate change to future generations (85%), animal life (81%), and food security (81%) (stereotypically feminine concerns) than about threats to the economy (39%) (a stereotypically masculine concern) (Women's Manifesto on Climate Change 2007, 14). While the women surveyed were found to have a high degree concern about climate change, the WI's Women and Climate Change Campaign seeks to spread the word to even more women so that they, too, will take up their responsibility as 'change agents' and thereby contribute to climate change mitigation.[6] The discourse of green consumerism has embraced the fundamental point that over-consumption by people in the affluent North is a key element of the problem. It has also been important in spreading the message that changes in consumption practices/choices have to be made by everyone if we want to halt the disastrous impacts that a Western lifestyle is having on the global environment. However, by targeting individuals in the private sphere, green consumerism effectively downloads responsibility to individual women whose 'gender script' tells them that they are better placed to solve the problem at home than powerful governments and industries run by men.

The discourse of demographic doom: the greening of hate?

> In 2007 the world finally woke up to climate change. It has not, however, woken up to one of its fundamental causes—human population growth.
>
> (Guillebaud 2007)

In the 1970s the modern environmental movement was dominated by the discourse of 'limits' driven by apocalyptic visions of an overpopulated planet; a 'population bomb' (Ehrlich 1968) with greater destructive power than nuclear weapons (for a critical discussion of this limits discourse see Dryzek 2005, 30). Invoking the 19th-century predictions of Thomas Malthus, environmentalists warned that the world's population was rapidly surpassing the planet's carrying capacity and that lack of resources would bring certain conflict, mass starvation and other miseries (cf. Hardin 1968). This Hobbesian view of humanity, in which conflict over scarce resources is inevitable, is a recurring theme in both environmental security and population discourses. Comparing population growth to a cancer, some early environmentalists advocated authoritarian government 'voluntary human extinction', and compulsory sterilization in developing countries as possible solutions (Ophuls 1996; Dryzek 2005). By the mid-1980s, this particular brand of environmentalism had been discredited; their extreme ideas were rejected on the grounds of questionable ethico-politics as well as faulty demographics and computer modelling.

Feminists, women's health and human rights activists had to work together throughout the 1990s to oppose the return of the population agenda at UN conferences: the Rio Earth Summit in 1992 and the International Conference on Population and Development in Cairo in 1994. They exposed the flaws in the diagnosis of the problem as well as the inherent sexism and racism in a discourse that located the problem largely in developing countries and in women's uncontrolled fertility. The dissatisfaction felt by women's groups participating in these UN conferences led to the establishment of the Committee on Women, Population and the Environment (CWPE) in 1991 (Silliman 1999). As explained in their 1999 anthology *Dangerous Intersections*, they reject the 'population paradigm' that assumes inherent conflict between humans and nature in favour of a more sophisticated analysis of the root causes of the 'ecological crisis': systems of technology, production, ownership and culture' (Ibid., xiv). The danger lies in the misdiagnosis of the problem in terms of sheer numbers of people rather than the levels of consumption of people in the overdeveloped minority world (Hynes 1999; see also Rees and Westra 2003). The CWPE, along with many other feminist critics, exposes the gender and race politics that are unavoidable in the population discourse as well as the kinds of policies advocated by populationists. As Joni Seager writes: 'Women's fertility is implicitly (sometimes explicitly) blamed for the global environmental crisis. Population control is a euphemism for the control of women' (Seager 1993, 216). Moreover, the disproportionate emphasis on controlling the fertility of women in the Global South (whose own ecological footprints, and that of their children, are miniscule compared to people in the affluent North), together with the support for draconian and misanthropic immigration controls, constitutes what has been called 'the greening of hate' (Hynes 1999; Urban 2007).

It seems that these important feminist critiques in the last decade may have had only a temporary effect in making the issue less acceptable for mainstream environmentalists. A growing number of environmentalists now lament the fact that population growth was allowed to 'fall off the agenda [...] because it is seen as too sensitive' [read: 'politically incorrect'] (Nicholson-Lord 2006). They argue that the current climate change crisis demands that we cast aside our liberal sensibilities and make a renewed effort to tackle population growth because 'technology and lifestyle changes by themselves will simply be incapable of delivering' the necessary reductions in CO_2 emissions (Nicholson-Lord 2006). The United Kingdom-based Optimum Population Trust, an organization led by some of the most respected environmentalists in the country (e.g. Sir Jonathon Porritt, Sir Crispin Tickell, Jane Goodall) makes this very case. Their website features a World Population Clock that counts in real time the number of children being born (6,798,144,999 and counting to the projected 9.2 billion by 2050), below which is the statement: 'We are rapidly destabilising our climate and destroying the natural world on which we depend for future life.' It also includes policy briefings that ask 'is there a right to have children?' and news releases with such titles as 'immigration threatens the environment', 'population growth is a security risk' and 'Britain [is] sleepwalking into an environmental nightmare' (www.optimumpopulation.org). Among their policy recommendations is a voluntary 'Stop at 2' campaign (applicable globally 'except in EU states with very low fertility rates'), increased access to family planning, and the empowerment and education of women to enable them to choose to have fewer children. They do not rule out coercive population policies, like China's one-child rule, but recommend them as a last resort (see Guillebaud 2007). Climate change, it is acknowledged, is a global crisis that calls for drastic measures. The neo-Malthusian discourse is back with a vengeance—and no apologies.

Climate change is not mentioned in the CWPE's important collection of essays critical of the underlying masculinism (and misogyny) of the populationist discourse. Much has changed in the decade since it was published. However, what has not changed is the dangerous intersections of the underlying sexism and racism with the wilful ignorance of the affluent elite that result in the kind of recommendations that are offered by the Optimum Population Trust. These recommendations continue to focus all of their attention on sheer numbers of people while offering no recommendations for how to end either overconsumption or poverty. What has also not changed is their failure to listen to and take feminist analyses of gender politics and unequal gender relations seriously. In the world beyond their narrow outlook, what has not changed are the complex reasons why women have more than the 'replacement number' of children (i.e. two), reasons which still include crippling poverty, the sexist preference for sons, male sexual aggression, rape, and the fact that the more women are devalued in society, the more their only social capital is the ability to bear children (Seager 2000). These are fundamental realities that all the 'education', 'empowerment' or contraceptive technologies in the world will not address.

CONCLUSION

Climate change arguably represents the largest challenge humankind has ever faced. The predicted impacts of climate change are frightening. Temperatures are projected gradually to rise to levels that will significantly alter ecosystems, bringing about sea-level rise, coastal erosion, water shortages and droughts, crop failures, increased flooding and extreme weather events. A scientific consensus about these projections, expressed by the IPCC among others, has shifted the debate away from the question of whether or not anthropogenic climate change is really happening, toward debates about what is to be done by whom, when and how. It seems that support for 'extreme' measures is growing in many quarters. We need carbon taxes, nuclear power, GMOs and population control because we have no time to wait for more socially complicated alternatives. As in most crisis situations (such as in times of war), critical reflection on the unjust human relationships that may have led to the crisis, and the intellectual tools that are used to interpret the impacts and devise potential solutions, is dismissed as a luxury that we can't afford. Understanding the gender politics of climate change is clearly not an urgent enough priority for it to be on the agenda.

I have argued that, contrary to popular perception, climate change is not simply a gender-neutral, scientific problem but is discursively constructed by four deeply gendered discourses. I have shown how the cultural constructions of masculinity and femininity emerge and are reproduced through these dominant discourses in different dimensions of the issue. From scientific expertise, to celebrity culture, to greening the household, the gendered discourses currently employed in the climate change debate give voice to a deeply-entrenched gender ideology that rests on exaggerated differences between men and women. They work to keep men and women in their ostensibly separate worlds of highly valued science, economics and defence, on the one hand, and devalued social reproduction and private domestic duty on the other. Masculinist discourses shape the issue in ways that effectively exclude women from leadership and make them responsible for solving it by donating their caring labour and curbing their fertility. For example, the shift to a low-carbon lifestyle is framed as an individualized and low-status duty for women in the private sphere, while techno-fixes are viewed as brilliant innovations, earning men more power, wealth and prestige. As the old saying goes, the more things change the more they stay the same. Perhaps they are not staying the same but getting worse; perhaps we are witnessing an intensification of dangerous and divisive patriarchal assumptions, justified by impending doom.

So what should feminists, and all those who care about the politics of gender, do in response to this dismal state of affairs? In short, much more than they are doing now. The lack of attention by feminists to climate change, and the regressive gender politics that it is ushering in, is worrying. It is telling that in an article considering the past and future of feminist thought (in which ecofeminism is absent), Rosemarie Tong does not mention it once. What does she cite as the most pressing issue for feminism's future? 'The ability to resolve the sameness-difference debate [among women]' (Tong 2007, 23). With reports about the seriousness of climate change everywhere we look, how has it come to pass that the inward-looking politics of identity has become a more important issue for feminism than the future of the planet? Feminists have been critical of environmental scholars for their blindness to gender. It is now time to be critical of feminist scholars who are blind to the environmental crisis.

Although the work done by a small number of Gender and Development scholars on the gender inequalities in the material impacts of climate change is important, I have argued that their analysis is too narrow. The analysis needs to move beyond seeing 'impacts' as only material and measurable effects (e.g. increased poverty, intensified burden of domestic labour) experienced by 'empirical' and 'vulnerable' women in the South. I have suggested that an important way of doing this is to look at the ways in which gendered discourses create and perpetuate gender asymmetries and make them seem inevitable and unchangeable. When rural women in the Global South believe it is their duty to eat last when food is scarce, we need to know both how this hurts them and how gendered discourses have worked to make this their feminine duty. When EcoMoms in the affluent North take the blame for 'eco-sins' and the responsibility for tackling climate change, we need to resist the urge to celebrate them as frontline 'change agents' and instead reflect critically on how they have come to internalize a gendered script that makes these more acceptable responses than demanding changes in governmental policy and industrial practices. It is not enough simply for feminists (like those in NGOs like WEDO and the WI) to argue for bringing more women into policy-making and expert positions. The lack of women sitting at the climate change policy table is not a cause of gender-blindness but a symptom of gender ideology and a framing of the issue by the exclusionary, masculinist discourses of ecological modernization and environmental security.

For several decades, an important yet seldom recognized contribution of ecofeminism has been to demonstrate the interconnectedness of social and environmental problems. Ironically, it seems that the climate change crisis has brought about a forgetting of the fundamental insights of both ecofeminism and environmentalism, that everything is connected to everything else and that it is separation, difference and compartmentalization that have got us into this ecological mess. Ecofeminists have insisted that our environmental problems are the product of gendered and culturally-situated discourses and practices, rather than being neutral or rational outcomes of human nature. More of this work needs to be done to understand climate change as a gendered human drama that involves men and women as actors, reading from masculine and feminine scripts, on our imperilled earthly stage.

NOTES

1 Space does not permit me to go into detail about the internal divisions within ecofeminism, or the complex reasons why some have eschewed the label but still do the work under the name of 'feminist environmentalism' (Seager 2003) or the more recent 'ecogender studies' (Banerjee and Bell 2007).

2 Although it is inaccurate and unfair to paint all ecofeminists with the same brush, several feminists have done just that: dismissed ecofeminism as 'frankly awful' and 'fluffy' (Sargisson 2001) and said that it should be allowed to 'die a quiet death' (Prentice 1988).

3 See, for example, the USA-based Women's Environment and Development Organization (WEDO), www.wedo.org/campaigns.aspx?mode=plantendorsements and the EU-based GenderCC—Women for Climate Justice, www.gendercc.net.

4 Since this chapter was written, a second climate change issue of Gender and Development was published in 2009.

5 They are: Oprah Winfrey, Angelina Jolie, Gwyneth Paltrow, Nicole Kidman, Salma Hayek (Nielson 2007).

6 In the great tradition of dystopian drama (and the great British tradition of self-referential irony), the campaign includes a short film called *A World without Jam*, which portrays the frightening future that awaits us if climate change proceeds as predicted. It is a future without the very comforts that women stereotypically hold dear: electric kettles, TV and, of course, jam.

The politics of gender: the case of the female trafficked migrant

SHARRON A. FITZGERALD

INTRODUCTION

In the 21st century the perceived threat posed by 'irregular migration' to 'Fortress Europe' has led to the increased politicization of European national identity politics. States parties have always used the regulation of national identity, the ability to govern specific territory and the movement of people over that territory in processes of state-formation. Yet what sets contemporary national identity politics apart is 'the growing normative incongruities between international human rights norms, particularly as they relate to the "rights of others" [...] and assertions of territorial sovereignty' (Benhabib 2004, 6–7). Changes in the global political economy mean that, in the first decade of the 21st century, female migrants from the former Soviet Bloc and the developing world constitute over half of the world's migrant population (Zlotnik 2003). In this chapter, I want to use these issues as a backdrop in relation to some arguments concerning the female trafficked migrant and the attendant issue of transborder sex work, and, thereby, ask some questions about what challenges 'the rights of others' may pose for the politics of gender and feminist identity politics in a globalizing world.[1] Feminist empirical and theoretical work has provided critical advances and insights into how processes of globalization and the politics of neoliberalism intersect with racialized class and gender (Acker 1998; Sassen 2002; Eisenstein 2004). Yet, in relation to human trafficking and transborder sex work, feminists remain divided on political and philosophical grounds (Barry 1979, 1995; Doezema 2005). Inasmuch as the pro- and anti-sex work lobbies, repectively the Global Alliance Against Traffic in Women (GAATW) and the Coalition Against Trafficking in Women (CATW), adopt different positions on prostitution, both organizations present the non-Western sex worker as the 'victim subject' (Brown 1995, 27). Such an approach has, in turn, shaped how the state understands and attempts to regulate the phenomenon across domestic and international jurisdictions (Munro 2005).[2] States parties invoke feminist rhetoric that claims that the female 'victim' of human trafficking is vulnerable to the vagaries of Western capitalism and organized crime gangs. They also cite the state's responsibility to uphold these women's human rights to legitimize their exclusion for 'their own good'.

Engaging in this dialectic extends other work in which I have argued that the feminist production of knowledge about the female trafficked migrant not only influences the state production of knowledge of sex workers, but also influences Western nation-states' attempts to legitimize stricter immigration law and border control and the *de facto* exclusion of geospecific populations from Western nation-states (FitzGerald 2008). Challenging such understanding for its negative impact on non-Western women's mobility and their right to autonomy enshrined in international law requires an understanding of how these discourses operate and from where they get their legitimacy within feminism as well as within domestic and international legal jurisdictions.[3]

From the outset, I wish to make it clear that it is not my intention to deny that human trafficking is exploitative. The question of *why* this is the case has been analysed by skilled researchers and needs no further elaboration here (Hughes 2000; Kelly 2005). However, I do not accept that discursive strategies of female 'victimhood' and trafficking as 'modern-day sex slavery' explain fully the complexity of this phenomenon. Thus, in this chapter, I want to consider how unreflexive feminist identity politics can have a disempowering rather than an empowering impact on the lives of non-Western women. In order to do this, I examine aspects of feminist politics of gender by focusing on its technologies of regulation that define the parameters of 'appropriate' expressions of female sexuality and subjectivity. This approach raises some key questions with regard to transnational feminist identity politics and its 'right' to speak for the 'sisterhood'. In this I am not suggesting that we abandon feminist methodologies, only that we use its concepts more carefully, self-reflexively and always with an eye to challenging its implicit inclusions and exclusions. It is this critical position that structures the form of the discussion that follows.

First, I turn to some key philosophical contexts in which feminists have theorized the representation and politics of the gendered subject. As part of this discussion, I draw on key but often diametrically opposed thinkers who interrogate the issues of subjectivity, representational politics, agency, sex work and the non-Western female subject. My rationale for doing this is two-fold. Primarily, I want to demonstrate some problematics in feminist identity politics. Second, it is my concern that while partisan politics between pro- and anti-prostitution remains, states parties will continue to manipulate the schism and use it to justify the *de facto* exclusion of non-Western women from the west. In order to discuss this issue, I have selected Judith Butler's work on gender, agency and subjectivity, Martha Nussbaum's work on prostitution and Ratna Kapur's critique of what she perceives as liberal feminism's (especially Nussbaum's) failure to tackle the politics of inequality and exclusion that define women's lives in post-colonial contexts in the developing world. In addition, at a more applied level, I interrogate the discourses of the female trafficked migrant as a 'victim' of 'sex slavery' that dominate legal and political debates in feminist activism on human trafficking. This is important because I want to illustrate why feminists need to be mindful of the broader and real consequences that flow from our theorizations.

UNDERSTANDING THE POLITICS OF GENDER

Feminist analyses of subjectivity have produced critical insights into the intersection of racialized class and gender categories, and how these modalities interact with normative identity politics in a variety of contexts (Irigarary 1988; hooks 1990; Spivak 1992). The concept of normative identity politics introduces some interesting questions with regard to how the construction of the female 'subject' within feminist theory and politics has become problematic. With this in mind, I turn to Butler's critique of feminist representational politics because she suggests that the proliferation of feminist research on woman as 'the subject' of feminism has generated a challenge to contemporary representational politics from within feminist discourse.

Representational politics

It is a key lesson of Butler's analysis of power that gender and agency are 'processes of repetition' that are linked to a 'forced iteration of norms' that impel and sustain the construction of gender identity (Butler 1993, 94). For Butler, the 'performative' reiteration of gender norms facilitates agency because it 'permits the stabilization of a subject who is capable of resisting these norms' (Butler 1993, 10). In this she notes that processes of resistance to gender norms occur at the boundaries of 'excluded and de-legitimated sex' (Butler 1993, 16). Therefore, the female trafficked migrant who is a divisive figure in feminist representational politics is a good example to illustrate why feminists need to be mindful of gender norms that 'idealise certain expressions of gender, that in turn produce new forms of hierarchy and exclusion' (Butler 1999, viii). It is 'those regimes of truth' writes Butler, that 'stipulated that certain kinds of gendered expressions were found to be false or derivative, and others, true and original' (Ibid., viii). Other scholars such as Wendy Brown have engaged with this problematic, remarking that 'politicised identities are based on exclusion for their identity' (Brown 1995a, 65). Such ambivalence operates in certain feminist quarters to reject the selling of female sexuality, but it also depends on the idiom of the exploited sex worker who is a victim of patriarchy, gendered inequality and misogyny to re-enforce specific political agendas such as women's universal human rights. Ambivalent power relations sustain a hierarchy of womanhood where certain excluded women, such as the female trafficked migrant, need the 'sisterhood' to advocate on their behalf. While I agree in principle that it is right and good that feminists must concern themselves with the plight of the disenfranchized, I am less convinced that using poor women's experiences to further specific political issues is beyond moral reproach. Butler observes that intrinsic to this hierarchy of womanhood is the dominant tradition of feminist analysis where a certain 'normativity' is intrinsic to the social and political formation of gender (Butler 1999, xxii). Inasmuch as normativity underpins the politics of representation in aspects of feminist theory, it takes a concrete form, writes

Butler, when feminists use it to 'foster the political visibility of women' (Ibid., 4). For, as Butler recognizes, it is impossible to resist normativity without simultaneously privileging an ideal of gender.

Fundamental to this political vision is that 'the presumptions about normative gender and sexuality determine in advance what will qualify as "human" and the "liveable"' (Butler 1999, xxii). In light of these concerns, Butler's critique of normative representational politics requires us to consider the following question: When do certain political agendas actually codify and entrench extant social relations and inequalities? Therefore, if we follow Butler's argumentation, feminist identity politics tends to construct the female trafficked migrant and the attendant issue of transborder sex work through a series of discursive strategies inbuilt into the political system (feminism) that is supposed in fact to facilitate her empowerment. It is this paradox, then, that suggests that universalizing theories of female subjectivity and women's common oppression have denied the possibility of multiple 'cultural, social and political intersections in which the concrete array of "women" are constructed' and try to live (Butler 1999, 20).

Butler's theorization of the 'subject' of feminist identity politics and her critique of the concept of 'woman' has been debated widely. Lois McNay suggests that Butler's dependence on Michel Foucault's analytics of power has led to an underdeveloped concept of agency that alludes to symbolic and material social relations 'but does not disaggregate them analytically' (McNay 2001, 61). This failure to clarify her interpretation of agency, Penelope Deutscher (1997) argues, has led to a series of misinterpretations around the category of 'woman'. Allison Weir contends that, despite Butler's attempt to interrogate subjectivity and agency beyond dominant meta-narratives, she 'subverts her own call for a subversion of identity by rendering identity so omnipotent [...] that subversion becomes impossible' (Weir 1996, 113).

Still, I do not reject Butler's theorization of representational politics. It seems to me to continue to be a salient political question in relation to human trafficking and the attendant issue of transborder sex work. Cutting across these issues the question of whether the possibility that some women can choose of their own volition to engage in sex work goes to the heart of many feminist issues on bodily integrity and agency. Prostitution signifies, in certain quarters, women taking ownership of their lives and bodies and living their lives as best they can in their particular situation. Conversely, prostitution represents the persistent sexism and violence in all sectors of patriarchal society. Yet, these reactions are not isolated from the contexts in which they occur. They also constitute normative understanding of female identity and sexuality. In light of this, I want now to place Butler's theorization of the politics of representation of the 'subject' of feminism alongside a consideration of a more explicit theorization of prostitution as a problematic in feminism and the politics of gender. To do this, I now turn to the work of one of Butler's critics, Nussbaum, and her engagement with the interconnection between law, sexuality and sex work from a liberal perspective. Nussbaum has been vociferous in her criticism of Butler because of what she perceives as Butler's lack of concern for 'the material conditions of real women' (Nussbaum 1999b, 37). Indeed, I want to suggest that some interesting points of convergence between the two scholars exist. These spaces of overlap may assist us in moving beyond the current stalemate in feminist anti-trafficking initiatives.

Sex work

Nussbaum examines the options available to poor women. Consequently, she promotes what she calls a 'capabilities approach' that advocates providing poor women with sustainable conditions and resources to improve their lives. Nussbaum positions herself as a scholar who is concerned with 'those real bodies and those real struggles' (Nussbaum 1999b, 37). An important aspect of this framework is the interrelated questions of human dignity and difference as it relates to gender and sexuality, and how legal and political regimes should respond to these issues (Nussbaum 1999b, 2001). One such issue is the question of sex work and the sex worker's subjectivity. Thus, in *Sex and Social Justice,* Nussbaum questions cultural beliefs and practices with regard to some of the moral and legal debates concerning women taking pay for sex. From this perspective, then, Nussbaum asks whether or not our current judgements and political responses to sex work are accurate. More profoundly, she asks even if these responses are accurate 'what does this tell us?' (Nussbaum 1999a, 277).

Radical feminists have generated a number of arguments against the legalization of prostitution. Generally, these issues comprise: a) prostitution involves health risks and violence; b) it allegedly deprives the sex workers of their autonomy; c) it invades intimate bodily space; d) it commodifies sexuality; e) male clients shape the prostitute's activities and this perpetuates relations of male domination of women; and finally, f) prostitution is not a trade people enter by choice. It is useful here to consider these objections in the context of the female trafficked migrant. What is troubling here is that the divisive form of this dispute suggests a persistence of political and cultural hierarchies of womanhood that privilege a normative gender politics which denies that prostitution can be in certain contexts deemed a 'liveable life' for certain groups of women (Butler 1999, xxii). The stigmatization of sex work in this way does not provide good legal reasons or sound moral arguments for the continued criminalization of the occupation.[4]

Responding to the suggestion that prostitution involves health risks and violence, Nussbaum states that so far as these claims are true, 'the problem is made much worse by the illegality of prostitution' (Nussbaum 1999a, 288). Recent feminist scholarship reveals that law and social policy attempts across different legal jurisdictions to criminalize sex work often make trafficked women more vulnerable to health risks, violence and coercion (Munro 2005; Askola 2007; FitzGerald 2008). For example, initiatives to curb on-street prostitution in the long run have little bearing on the 'harms' experienced by trafficked women. Trafficked migrants tend to be located in off-street premises. In fact, the emphasis on repressive law enforcement in relation to prostitution feeds into exploitative practices such as detaining sex workers in dangerous environments against their will. Nussbaum asserts that in situations where law enforcement advocates for rather than against women, risks such as these are reduced significantly (Nussbaum 1999a, 288).

This links to the second objection, which is to say that the prostitute lacks autonomy. Female autonomy is an important principle of feminist identity politics. Nussbaum asserts that the right to self-determination about one's body is a fundamental aspect of women's human rights, and that feminists must always be vigilant for attempts to roll back legislative protections. The right to bodily integrity, Nussbaum claims, 'does not seem to support legal regulation of prostitution, provided that [...] it is consensual' (Nussbaum 1999a, 288). Whatever the basis for this approach it is clear that in relation to human trafficking much of the feminist debate revolves around this issue of consent and coercion to transborder sex work. The framing of the debate in this dichotomy ignores the finer scales at which women choose to move or be trafficked. As Vanessa Munro argues, in this context, stakeholders 'have interpreted the [legal] provisions in line with their own (polarised and politicised) agendas, privileging victims who can prove their coercion and discrediting those who cannot' (Munro 2008, 11).

For Nussbaum, the question of the commodification of female sexuality has entered the anti-prostitution feminist lexicon in two main ways. Feminists have deployed the Marxist theory of capitalist labour relations to argue their case (Sullivan 2003). First, they argue that the prostitute alienates her sexuality on the market. Second, they assert that she converts her body and her sexuality into a product to be bought and sold. Yet, in a world where thousands of women from the former Soviet Bloc and the developing world make the decision, of their own volition, to be trafficked against a backdrop of variable economic opportunity and gender inequality, then, according to Nussbaum's rubric the political question for contemporary feminist politics must be: are the above-mentioned assertions true? If we take the dual concepts of self-determination and human agency as our starting point, then, as Nussbaum observes, cutting across all the difference 'the prostitute still has her sexuality; she can use it on her own, apart from her relationship with a client' (Nussbaum 1999a, 291). This excludes instances of kidnapping or coercion, but it also provides for instances where individuals change their minds. This argument is pertinent to human trafficking. It is often the case that a woman has agreed to be trafficked, but somewhere along the way she changes her mind and is unable to leave her circumstances because her 'employers' took her documents and money (FitzGerald 2008). Furthermore, she can cease to be a prostitute, and her sexuality will remain her own. It is in this regard, then, that we must ask ourselves: does self-determination or the lack of it describe a situation that is unique to the sex worker? The evidence suggests that this is not the case. For most of us, except the privileged few, work in circumstances not entirely within our control. We work in contexts where we commodify our labour, and by extension alienate our bodies on the open market for a fee.

Furthermore, this argument becomes even more problematic when we consider that most anti-trafficking policy focuses on non-Western women and transborder sex work. This understanding of the problem eschews a plethora of other forms of labour exploitation sourced by international human trafficking rings such as agricultural labourers. All of these 'irregular' migrants experience to some degree or other a level of exploitation. Fundamentally, privileging the female trafficked migrant and sex work illustrates the intersection of race, class and gender that continue to permeate anti-trafficking initiatives in many quarters. If, then, the political sticking point for feminism is that the taking of money for sex is problematic, two questions suggest themselves. First, feminists must ask: why is prostitution undesirable in circumstances where women have exercised agency and have chosen this option? A second question flows directly from the former. If, in a world where globalization and the policies of neoliberalism continue to exacerbate gender inequalities especially for the poor, and where poor women have responded to global restructuring and their inherent inequalities by developing transborder sex work in what Saskia Sassen refers to as 'strategies of survival' (Sassen 2002, 212), why, then, are feminists more divided about women engaging in transactional sex work than the lack of choice that may contribute to their decision to move in the first instance?

How to distinguish between the subjectivity projected by Nussbaum and the liberal agenda of equality and human dignity she represents, and that of the female trafficked migrant who has consented to be trafficked but ultimately has become entrapped in circumstances that deny her autonomy over the acts she performs becomes the key question at this point. Clearly, we are talking about two very different situations. Any effective response to this question must separate out the difference. While Nussbaum offers critical insights into the production of female sexuality in feminist identity politics, her particular brand of transactional politics of gender is not without its critics. Therefore, I now wish to turn to the work of the Indian feminist scholar, Kapur.

Non-Western female subjects

Postcolonial feminism is critical of what Kapur observes as the liberal tradition's 'inability to transcend assumptions about the Other on which legal reasoning and the liberal project are based' (Kapur 2005, 17). Kapur points to the European Enlightenment thinking and the history of colonial discourse and practice with its intrinsic denial of the personhood of millions of colonial subjects (Mehta 1999). In this context, the liberal notion of stewardship legitimized the superiority of European colonizers and their 'God-given right' and obligation to dominate territorially and to protect the colonized subject 'from themselves' (McClintock 1995). Kapur argues that in the postcolonial moment the idea that we continue to make progress because we endeavour to include more voices within the discourse of rights is not a reality for the vast majority of the world's population. For, while Kapur agrees with Nussbaum's assertion that stigmatization of 'Others' is endemic in the postcolonial world, she challenges Nussbaum's claim that 'liberal individualism' (Kapur 2005, 18) has the universal potential to improve the lives of women world-wide.

Inasmuch as Nussbaum is critical of the ways in which liberalism has been misappropriated, particularly in relation to the advancement of women's rights, Kapur is not content to critique its misapplication. The concept of responsibility appears in Kapur's work not simply to apportion blame, but to make visible the consequences of the feminist liberal rights discourse in the postcolonial moment (Eisenstein 2004). Therefore, rather than deploying universal standards to tackle human rights issues, Kapur argues that 'the starting point of an empowering politics should commence from a critical position' (Kapur 2005, 20). This requires us to understand that politics of reform or reparations cannot correct the ongoing consequences of colonial subordination and exploitation. In tracing the contradictions within the liberal feminist agenda, Kapur turns to the legal treatment of the transnational migrant and, in particular, the female trafficked migrant. It is anti-trafficking initiatives, she argues, that have 'failed to distinguish between consensual migration, albeit illegal, and coerced movements' (Ibid., 141). This error has reinforced assumptions about women from the Global South and East. Essentialist assumptions about the female trafficked migrant maintain colonial stereotypes of weak, vulnerable and unsophisticated women 'incapable of choosing to cross borders' (Ibid., 142).

The rhetoric of representation and the subsequent politics of gender that surround assumptions about women conform to an approach 'that views women in need to protection, naturally weak and incapable of exercising agency' (Kapur 2005, 147). In turning to this, I want to consider in concrete terms the problematic of gender essentialism that eschews differences between geographically dispersed women through the idiom of the 'global sisterhood'. Two central concerns emerge. To do this, in the following section I will focus on some of the policy initiatives and identity politics of the pro- and anti-prostitution lobby.

DEFINING THE FEMALE 'VICTIM' SUBJECT

The previous sections illustrate how methodological debates on female subjectivity, agency and sexuality have created a schism among feminists. This division is evident in the policy interventions taken by the pro- and anti-prostitution lobbies, respectively. As mentioned earlier, each camp is opposed diametrically on the question of whether sex work can be defined as legitimate labour. Inasmuch as it is patriarchy and issues of gender equality as well as the perennial question of male domination of female bodies and behaviours that dominate the landscape of feminist politics, I want to suggest that, in the case of the female trafficked migrant, the politics of difference from a feminist perspective takes on paradoxical form. This paradox eschews the material differences in the lives of geographically dispersed women, their subjectivity and their attitudes, often by necessity, to their sexuality. My ultimate concern here is to highlight the potential for circular argumentation to facilitate how Western states parties use the idiom of the female victim of trafficking to legitimize their *de facto* exclusion of geospecific groups from the west. In fact, I would suggest that the ways in which law and social policy understand and attempt to regulate the female trafficked migrant rely upon the feminist concern with non-Western women's sexual exploitation, insofar as it must locate its concerns in humanitarian questions to achieve legitimacy. This seems to me to be precisely *the* political question—indeed the question of gender politics—today, and one that explains, in part, the surge of international interest in 'irregular' migration including human trafficking, that I began by noting. Thus, at the risk of what may seem a digression from my theoretical discussion at this point, I want to suggest that we might approach this 'question' through a discourse analysis of feminist anti-trafficking policy.

The global sisterhood

An analysis of the impact of feminist gender politics as a mechanism of knowledge and power centring on the female trafficked migrant and prostitution necessitates shifting through what Butler terms the normative discourses that determine what gender presumptions are 'true and original' (Butler 1999, viii). It is interesting to note that anti-prostitution CATW policy initiatives demonstrate how the organization uses the notion of female vulnerability and the exploitation of geospecific groups of women to speak for the global sisterhood. For example, it suggests that 'factors that impel women to take the risk of illegal migration are [...] the resurgence of traditional discriminatory practices against women' (Healy and O'Connor 2006, 6). The reconceptualization of trafficking and sex work as violence against the entire 'sisterhood' is problematic because it erases contextual differences between women. In response to this critique the CATW contends, 'we cannot dissociate prostitution from other forms of male violence against women [...] in all countries around the world' because 'sexual exploitation preys on women made vulnerable by poverty' (Healy and O'Connor 2006, 27). Consequently, the CATW's position is that sexual exploitation erotizes women's inequality and is a vehicle for racism and Western patriarchal domination (Barry 1995; Ekberg 2002). As Butler has shown, the feminist normative discourses of 'appropriate' subjectivity operate to differentiate modes of sexuality and agency through definitions of what is 'a liveable life' for all women (Butler 1999, xxii). This attempt to 'foster the political visibility of women' (Ibid., 4) works on the principle of universality, and does not permit any consideration of the will or consent of the women involved (Wijers 1998).

Despite its different political agenda, a discourse analysis of the pro-sex-worker GAATW's policy on trafficking reveals that it mirrors the abolitionist tendency to represent trafficking as an historical 'injury' against all women (Brown 1995). Official GAATW documents reveal a deep-seated racialized problematic in its objectives. This takes material form in debates on 'forced' and 'voluntary' prostitution. Consider the following: 'the gendered nature of trafficking derives from the universal and historical presence of laws, policies and customs [...] that justify [...] the discriminatory treatment of women and girls' (GAATW 1999, 3). Activists symbolically deploy the image of non-Western female trafficked migrants as follows: 'the majority of the persons trafficked [...] are women and girls due to their inferior and vulnerable status at home' (GAATW 1999, 3).

Concepts such as the female victim of trafficking perpetuate an image of non-Western women as naïve and ignorant and a 'passive object of others' actions' (Agustín 2005, 107). Here, surveillance, through the enforcement of normative sexual identity politics, corresponds to another deployment of the 'gaze' among and between the bodies and behaviours of 'sisterhood'. The preceding paragraphs provide the elements for a critique of the idiom of the female 'victim' of trafficking as a problematic in feminist identity politics. I argue that this subjectivity arises from the insistence, in certain quarters, in viewing all female trafficked migrants as hapless and vulnerable. The representation of female trafficked migrants as vulnerable to Western capitalism invites increased state surveillance and regulation of non-Western women's mobility and sexual identity (Kapur 2005). In my view, this representation continues as a blind spot in our policy interventions on human trafficking. It overlooks how some women of different backgrounds can be driven by different assumptions and values, but can be also motivated by different commitments and react to their circumstances in a variety of ways. Attention to the politics of difference permits us to 'uncover the very ways in which the very thinking of what is possible in gendered life is foreclosed by certain habitual and violent presumptions' (Butler 1999, viii). Another of these habitual presumptions occurs in discursive strategies around the female trafficked migrant as the modern-day sex slave. It is this idiom to which I now turn.

Discourses of slavery

In the first decade of the 21st century, the anti-prostitution lobby has recast trafficking as a form of modern-day sex slavery against all women (Barry 1995). Abolitionists contend that prostitution 'is always [...] degrading and damaging to women' (Anderson and O'Connell Davidson 2002, 8). Activists argue that, in the final analysis, sexuality in prostitution is ultimately the preserve of male sexual titillation (Ekberg 2002). Fundamentally, the CATW positions itself as a force against those 'who promote the alleged "right to prostitute" and romanticise prostitution as "sex work", in essence defending the vested interests of the sex industry' (Healy and O'Connor 2006, 3).

Similarly, the GAATW's objectives on trafficking uncover entrenched normative discourses on the female trafficked migrant as a victim 'of this contemporary form of slavery' (GAATW 2004, 1). This position gains purchase when lobbyists situate it in debates on 'forced' and 'voluntary' prostitution where 'victims of trafficking are treated as objects or commodities' (GAATW 1999, 3). What is interesting here is how the pro-sex-worker strategies focus on the conditions of coercion, in which trafficking violates women's human rights. The organization asserts: 'The core elements of trafficking are the presence of deception, coercion or debt bondage [...] the victim may have agreed to work in the sex industry but not to be in slavery-like conditions' (GAATW 1999, 5).

Yet the pro-sex-worker lobby is careful to define the scale of sex work as legitimate labour when Western sex workers engage in it. These workers are autonomous individuals who are fully in charge of their bodies and understand the ways of the world (Murray 1998). However, a calculated rescaling of the same labour retrieves a colonial paradigm of non-Western women as naïve and passive victims of their circumstances, whose ignorance makes them especially prone to ending up 'in a situation analogous to slavery' (GAATW 2007, 12). This description has interesting links to Kapur's assertion that anti-trafficking initiatives 'fail to distinguish between consensual migration and forced migration' (Kapur 2005, 141). Indeed, research shows that most trafficked migrants have not entered a third country illegally and often have done so independently—for example, on student or holiday visas (Munro 2008).

In many ways, the politics of gender that surrounds human trafficking has produced a 'politics of connectivity' (Deforges et al. 2005, 444) among feminists evident in their sense of social responsibility for the female 'Other'. While this development offers the potential for an alternative politics and 'social justice' for all women, I suggest that feminist politics on trafficking and prostitution indicates some pitfalls with this approach. While I accept that the idea of global connectivity is important, I am concerned about how feminists deploy this conceptual framework to reinforce their international 'humanitarian' interventions from the platform of universalism. Furthermore, it reinforces a hierarchy of womanhood buttressed through colonial stereotypes of non-Western women. This neo-colonialism places Western feminists at the centre of the spaces of authority to advocate for all 'Others'. Therefore, I argue that despite their claims to internationalism and the politics of differences, certain feminist activists work within a regional framework that is geospecific in character.

CONCLUDING REMARKS

There is not space here to develop this argument further, but by way of some final remarks, I want to suggest that the politics of gender is no idle matter. Butler's conception of 'normative identity politics' requires some serious reconsideration in the context of the female trafficked migrant. What kind of normative identity politics is this? What would be the 'political' form of non-Western women's understanding of their subjectivity, when as Kapur puts it the West views 'all consensual migrant females as trafficked and thus rendering women's cross-border movement as illegitimate' (Kapur 2005, 145). Certainly Butler is clear that 'the violence of gender norms' (Butler 1999, xix) underpins contemporary representational politics. If, then, feminism is not to fall into the problems associated with false dichotomies that determine what model of identity has the right to exist, the right to be equally visible, then the political question must be how do we do more than pay lip service to the 'rights of others'? Perhaps one way forward is that we, as Nussbaum asserts, locate our responses to the female trafficked migrant in context-sensitive politics.

NOTES

1 In this essay I use the terms 'sex worker' and 'prostitute' interchangeably.

2 In December 2001, the UN released for ratification its two optional Protocols to supplement its *Convention Against Transnational Organized Crime.* The first of these is the *Palermo Protocol.* The second is the *Protocol Against the Smuggling of Migrants by Land, Sea and Air.* Article 3(a) of the Protocol to prevent and suppress trafficking states: '"Trafficking in persons" shall mean the recruitment, transportation, transfer, harbouring or receipt of persons, by means of the threat or use of force or other forms of coercion, of abduction, of fraud of deception, of the abuse of power or of a position of vulnerability or of the giving or receiving of payments or benefits to achieve the consent of a person having control over another person, for the purpose of exploitation. Exploitation shall include, at a minimum, the exploitation of the prostitution of others or other forms of sexual exploitation, forced labour or services, slavery or practices similar to slavery, servitude or the removal of organs'.

3 This is a difficult issue under international law. The 1948 United Nations Declaration of Human Rights, Article 13, recognizes the individual's right to the freedom of moment (to emigrate)—but not their right to enter another country (to immigrate). The Declaration upholds the sovereignty of the state and its right to regulate its borders.

4 It should be noted that in certain EU domestic jurisdictions the state has decriminalized prostitution. I speak specifically about the Federal Republic of Germany and The Netherlands. While prostitution is not illegal in the United Kingdom, the procurement of sexual services is. In the main, this is how many legal regimes tackle prostitution. Sweden has a unique approach to prostitution where a zero tolerance to prostitution operates. The Swedish Government argues that this is a women-centred law that punishes clients rather than suppliers of sexual services.

Feminist imperialism: travel writing and journalism past and present

CORINNE FOWLER

'I and my guide (it is not grammar, but racial pride is above grammar any day—and the guide was black) arrived at the hospital mission station.'

(Beatrice Grimshaw, *From Fiji to the Cannibal Islands*, 1907)

'[We need to] drag Afghanistan's brutalised men and invisible, downtrodden women out of the dark ages.'

(Jonathan Miller, Channel Four website, 2004)

INTRODUCTION

The epigraphs to this essay may seem unlikely bedfellows. After all, the traveller Beatrice Grimshaw journeyed to Fiji in 1907, while Jonathan Miller's journalistic statement about 'downtrodden' Afghan women was published nearly a century later. In many respects, Grimshaw's nakedly racist declaration is of a different order to Miller's expression of concern for women's rights. Nevertheless, this essay contends that there are important continuities between protofeminist travel writing and the feminist content of reports by journalists who are committed to the ideal of secular democracy. In order to explore this connection, the following essay interweaves two apparently distinctive strands of debate. The first pertains to the colonial content (or not) of British women's travel writing from the period of high empire.[1] The second concerns the 'integrationist feminist' practice of advocating Western-style secularity as a remedy for gender inequality. As Liz Fekete (2006, 2) and Arun Kundnani (2007, 24–44) observe, the latter phenomenon has particular political consequences. They point out that prominent feminists in Western Europe and the United States are today increasingly inclined to oppose liberal immigration policies on the grounds that multiculturalists have been indifferent to patriarchal customs such as polygamy, clitoridectomy, forced marriages and honour killings. Given this trend, the essay also attends to the feminist content of current news media coverage.

PROTO-FEMINISM AND IMPERIALISM

Before embarking on such a discussion, it is first necessary to provide a brief background to the study of colonial women's travel writing. The 1980s saw a sustained focus on the politics of literary taste, which heightened awareness of the disproportionate amount of scholarly attention to the literary canon. During this period, many researchers began to study middlebrow and low-brow writing in earnest (Kerrigan 2003; Baldick 1983; Eagleton 1996; Lovell 1987). This collective foray into popular narratives and low-status genres coincided with a sea-change in academics' readings of journeys in general and travel writing in particular.[2] By the 1990s, travel accounts had long-since ceased to be read mimetically. Researchers became increasingly interested in travel writing's rhetorical function and, more often than not, its role in producing and disseminating 'knowledge' about (post)colonial subjects (Ghose 1998a, 2). It is important to add, however, that, around the close of this decade, influential scholars in the field began to tire of what Reinhold Schiffer (1999, 1) uncharitably describes as the 'fervent ideological loyalties' associated with this particular mode of enquiry. Many critics in the field turned to widespread evidence of past experimentation in travel writing, particularly from the modernist period.[3] However, others believed that anthropological theory and postcolonial studies had left an important critical legacy and that, despite the recent increase in scholarly activity regarding travel writing and the accompanying proliferation of publications relating to the form, its ethical dimensions had yet to be theorized with sufficient rigour.[4] Nevertheless, travel writing's porous generic boundaries have been a consistent focus of investigation since the late 1980s, most notably its kinship with novels and the antecedent genre of ethnography (Pratt 1986). Given that it is also akin to historiography and autobiography, an important principle here is that the genre of travel writing is exceptionally amorphous. From Edward Said (1978) to Mary Louise Pratt (1982, 1986), Sara Mills (1992) and Tim Youngs (1994), therefore, many influential scholars have investigated travel writing's generic entanglements as well as focusing on the ideological implications of its intertexual dimensions.

Studies of colonial and proto-feminist women's travel writing coincided with the growth of Women's Studies as a discipline.[5] Initial anthologies and scholarly monographs tended to be written in a celebratory vein and have subsequently been criticized for concentrating on travellers' gender to the exclusion of other significant factors, such as class and race. A related criticism of this early work has pertained to the way in which colonial women travellers have been valorized as less colonial than men, or even as anti-colonial.[6] The problem here, as Shirley Foster and Sara Mills (Foster and Mills 2002, 3) have pointed out, is that, for critics such as Birkett and Wheeler (1998), Morris and O'Connor (2004) and Schriber (1995), it was 'self-evident' that travel writing by Western women is distinctively different from that of its male counterpart. Together with scholars such as Lisa Lowe (1991), Billie Melman (1992), Jenny Sharpe (1993), Reina Lewis (2004), Alison Blunt (1994) and Indira Ghose (1998a, 1998b), Foster and Mills have warned against operationalizing essentialist notions of gender. Instead, and without dismissing the possibility of individual women's agency, they have focused on the range of discursive pressures and contextual influences that shape women's travelogues.

Most of the controversy about the colonial content of women's travel writing has centred on the stereotypical figure of the plucky woman traveller who defies social convention, a figure who has, as Foster and Mills suggest, continued to influence modes and discourses of travel. They argue that this eccentric figure has 'set the discursive boundaries for women writing about their travels', even to the present day (Foster and Mills 2002, 3). Moreover, the brave woman traveller ideal has assisted a particular conceptual leap, namely, that colonial women who trespassed into the male domain of travel writing also, by definition, rejected the basic tenets on which empire was founded. The contention that colonial women's tendency to breach prescribed gender boundaries predisposed those same woman to transgress racial boundaries is implicit in many studies of colonial women travellers. By way of countering this tendency, Foster and Mills advise that critics should ideally attend to the writing of a far wider range of female travellers, including those who identify with inconvenient forms of femininity or who make baldly imperialist statements. While they argue that gender does indeed make a difference, they suggest that this difference is due to women travellers being differently positioned than their male counterparts rather than because of any innate psychological difference derived from gender *per se*. Consequently, they counsel against making 'global statements' about women's travel writing, instead observing that factors such as race, age, class, historical period and political ideals all interact with gender in significant ways (Foster and Mills 2002, 1).

FEMINISM AND WESTERN JOURNALISM

The relevance of gendered positionality has provided a similar focus for debates about the relationship between gender and present-day news media coverage. Prominent women journalists are occasionally prompted to make statements about the significance of gender to their news reports. Some of these statements embrace notions of gendered interiority. An example of this is the contention of successful reporters, such as Christina Lamb and Anne Sebba (Allen 2005, 77 in Fowler 2007, 19) that they are less inclined than male journalists to focus on the machinery of war. Other assertions pertain to gendered positionality. A claim with some justification is that women journalists have greater access than men to private female spaces. This is nothing new of course, since it is reminiscent of women's unique access to the harem in the 18th and 19th centuries. A related assertion is that women journalists attain a 'bi-gendered', or 'honorary male' status on account of their Western affiliations. However, though it is rarely stated explicitly, this status is presumably also partly conferred because of their access to mainstream news networks.[7] Again, this circumstance also has an historical precedent, since colonial women's status in the colonies was often conferred by their perceived closeness to centres of power. As Sidonie Smith (in Fowler 2007, 432) has noted, such silences serve as a reminder of the important principle that 'centres and margins shift against "various horizons of power"'. The problem here is that research into gender and the news suggests that the difference in content between news media coverage by men and women is negligible. There are a wide variety of reasons for this, but explanations tend to cluster around the assimilationist politics of the newsroom. My own research into the coverage of Afghanistan in 2001 certainly bears this out. During the military operation in Afghanistan known as Operation Enduring Freedom, both male and female journalists tended to deny Afghan women access to media spaces (Fowler 2006a, 4–19). Despite this tendency and in line with comparisons of colonial male and female travel writing, women's journalism is often placed in opposition to a putatively 'male' brand of journalism that is supposedly qualitatively different. After all, if it is advisable to avoid making 'global statements' about women's travel writing and journalism, researchers might also allow that men's journalism may be more self-reflexive than such critiques generally imagine. For this reason, it remains important to adopt a comparative approach to journalism by men and women in order to avoid assumptions of difference on the basis of gender alone.

PROTO-FEMINIST TRAVEL WRITING FROM THE PERIOD OF HIGH IMPERIALISM

The central claim of many critics working on colonial women's travel writing has been that it is qualitatively different from that of its male counterpart. This is a reasonable, if exaggerated, assertion. Investigations of 'the female gaze' have tended to focus on the various factors that explain such differences.[8] There is now widespread agreement that they are due not so much to women's gendered interiorities as to their uncertain status within the patriarchal colonial order. Colonial women's writing, as Elleke Boehmer (1998, xxxi) has argued, necessarily reflects the permissible discourses and rhetorical stances of the times. Additionally, particular patriarchal expectations have placed pressure on the narratives of women travellers. As Foster and Mills (2002, 9) put it, colonial women negotiated the 'arena' of travel in distinctive ways partly because 'their departure from the interiority of the domestic sphere involved greater justification and management, both ideologically and practically'. Moreover, their accounts were also being written with an eye on the prejudicial reception to which they would undoubtedly be subject (Mills 1992). The force of critical opinion in the 19th century prescribed the topics and tone of female-authored travelogues in ways that are likely to have structured the narratives of women who wanted to get published (Foster and Mills 2002, 11). Where selected passages in particular travelogues can be found to subvert the dominant order, it is not always possible to make straightforward assertions about a writer's political intent. While women colonial travellers' encounters with other women often led them to reflect on their own position in society, this did not automatically lead to a revolution in their thinking about race.[9] As Mills argues, rather than pursuing autobiographical readings of women's travel writing, it is important to identify the separate strands of discourse that are woven into the narrative tapestry.[10] This is partly because readings that celebrate the bravery and unconventionality of individual women travellers risk 'colluding' both with the self-image of the pioneering colonial explorer advancing civilization but also, as Ghose argues, with the humanistic pursuit of unified personal identity (Ghose 1998a, 4–5). Moreover, in their quest to valorize women travellers, such readings often turn a blind eye to narratives' colonial content.

Instead of simply drawing conclusions about an individual writer's personal political stance, therefore, it is worth attending to the unstable and contradictory nature of colonial travel writing *per se*, which reflects the textual ambiguity of *all* colonial narratives, irrespective of the author's gender (Mills 1992, 9). This ambiguity is not the preserve of writing by colonial women alone. It derives from the insecurities associated with colonial power: the fear of power's usurpation, anxiety about unknowable colonial subjects, the contradictions housed within colonial ideology, and internal disagreement among its advocates. An excellent example of the latter phenomenon can be found in the travel narrative of Beatrice Grimshaw, the author of the epigraph to this essay, where Grimshaw can be found subjugating grammatical rules to perceived racial hierarchies. In the incident I am about to discuss, Grimshaw's Fijian guide leads her to a picnic spot. Upon first surveying the view, Grimshaw appears to replicate a familiar colonial cliché as she peers down at a conquered landscape: 'Tanna l[ay] like a bright green map under our feet' (Grimshaw 1907, 222). The imagery of the map renders the land as reassuringly readable as charted territory. However, this moment of colonial dominance is quickly shown to be illusory when her guide informs her that the picnic site was once the scene of a cannibalistic feast. The guide explains that he considered it a 'good joke' to direct her to the very spot where a girl was killed and eaten. The most notable effect of this disclosure is that it hampers Grimshaw's ability to maintain a colonial view. Tanna is no longer a prone landscape that acquiesces to her colonial gaze but instead assumes a more threatening form, one that cannot be enclosed within the folds of a colonial map. As this passage shows, Grimshaw's narrative has a complex relationship to the discourse of colonial dominance since her guide's revelation raises the spectre of that which is uncolonizable. This narrative deviation from the colonial stance could be attributed to her gender. As a woman, her skewed relationship to colonial power perhaps leads to a disavowal of the master-of-all-I-survey posture. However, this cannot be concluded for certain, since bouts of colonial anxiety erupt elsewhere in the narrative, most notably in relation to the suspicion (by no means peculiar to this travelogue about the South Pacific) that colonization is only 'varnish deep' (Ibid., 80). It is equally possible that the passage plays out a familiar imperial anxiety about the proximity and prevalence of uncolonizable spaces. Such anxiety can be found in many men's narratives. In this example, as with countless others, gender interacts with a range of factors in the way that Foster and Mills suggest.

As with many travelogues by women from the period of high imperialism, Grimshaw's travel narrative is packed with illustrative examples to this effect. In its protofeminist passages, the travelogue frequently subverts notions of gendered interiority without transgressing colonial boundaries. On one occasion her horse, Somo-somo, gets stuck in a muddy ditch, causing her to lose patience with her guides: 'I would have given £20 for the liberty to sit down on the bank and cry [...] But my men were three children of Nature, which meant three useless babies in trouble of any kind, and Somo-somo's life hung on me'. As the animal starts sinking in the mud, she stirs them into action:

> It does not matter what I said. There is a kind of English that every Fijian understands and obeys. I gave them that English, reproducing it phonographically from my recollections of the sort of thing that Island mates used to say to the cargo-workers on the quays.
>
> (Grimshaw 1907, 99)

Regardless of the sense that her gender mismatches such language, both statements imply that masculine behaviour is not so much innate as learned and 'reproduce[d]'. She adopts a masculine persona in order to maintain her supremacy as a colonial woman over three colonized subjects.[11] Such examples serve as a reminder of Antoinette Burton's point that (proto)feminist 'battles for rights' are to some degree also 'quests for power' (in Ghose 1998a, 31).

COLONIAL TRAVEL WRITING AND NEWSPAPER READERS BACK 'HOME'

Historically, women's letters and travel writing have had an important impact on journalistic coverage of current affairs. At times of colonial crisis, the letters and diaries of women travellers were published in British newspapers or cited by feature writers, both in London and British India. Indeed, such accounts have shaped ideas about the cultural orientations of former colonies right up to the present. A prominent example occurred after the disastrous 1842 British retreat from Kabul in Afghanistan, when around 16,000 British and Indian soldiers were killed. As one of the few surviving British witnesses of the events leading up to the 1842 massacre, Lady Florentia Sale was an important figure. Not only did her *Journal of the First Afghan War* (1841–42) bring her to public attention, but it subsequently proved an important source for definitive historical accounts of the invasion and retreat, including that of the famous colonial historian Sir John Kaye, whose 1851 volume, *History of the War in Afghanistan* also informed Rudyard Kipling's fiction about Afghanistan. Sale's jingoistic yet damning account details the military and tactical errors leading up to the retreat and describes the events of the massacre in florid and graphic language. Her *Journal* undoubtedly helped to fix notions of Afghan brutality in the British popular imagination, notions that have haunted British reporting about Afghanistan right up to the present.

Sale's *Journal* was reprinted in 1969 and re-issued in 2002 following Operation Enduring Freedom, which began in October 2001.[12] Her account of the first Anglo-Afghan War was read by a British public back 'home' that was familiar with the series of events that had led to the British invasion of Afghanistan and eager to know more about the massacre that put an end to it. As Patrick Macrory points out, the narrative presents a *memsahib*-like courage in the face of hardship and terror. He also observes that it was almost certainly written with future publication in mind. The journal was published the year following the War by John Murray. It quickly became a bestseller and was reprinted four times with a total of 7,500 copies sold (Macrory 2002, 12). Married to the second in command at Kabul, Lady Sale received the accolade of 'Soldier's wife *par excellence*' by *The Times* and she became a minor celebrity. It was the printing of excerpts in the newspaper that enabled her account to have such wide exposure and an enduring influence on travel writing about Afghanistan right up to the present.

An even better-known illustration of the force and influence of women's letters in the press relates to the 1857 'Cawnpore massacre' during the First War of Indian Independence (1857–59), which English journalists, MPs and periodical writers preferred to call 'The Mutiny'.[13] The Cawnpore massacre was one of a chain of events set off by 'Native' soldiers of the Bengal army shooting their officers on 10 May 1857 and, on the following day, claiming the king of Delhi as their new leader. The resultant fighting led to the temporary loss of large swathes of northern India for the British, who were trapped or expelled from numerous towns and cities across the region. The Cawnpore massacre was a key incident at the time. The besieged British negotiated an evacuation from their stronghold, but most were killed or captured during a battle as they attempted to leave. Around 200 British women and children were killed and their bodies thrown down a well. Letters by women who escaped death either on that day or in similar circumstances elsewhere were enlisted to what Albert Pionke describes as 'the Mutiny's wider fictionalisation' (Pionke 2008).[14] In his survey of Victorian journalism, Pionke observes that articles in the press during this period were generally written in a patriotic, if sometimes critical, tenor and attempted to portray the conduct of the British in a favourable light. However, reports of raped British women led to hysterical and sensationalist images of the Indian rebels as savage rapists even though none of these reports was verified by an independent enquiry (Ghose 1998a, 92). Meanwhile, as Jenny Sharpe points out in relation to British reprisals, Indian women were not 'rapeable' (Sharpe 1993, 232). The relentless focus on sexual violation diverted attention from the violent excesses of the British retaliation and avoided dwelling on the actual forms of humiliation to which colonial women were subject since, in a telling reversal of power, they were required to perform menial chores and manual labour (Ghose 1998a, 92).

The letters of the women caught up in the Cawnpore incident were published in *The Times*. They provide a convincing example of divergence from male accounts of the incident. As Ghose observes, the letters have several common features. First, they often deviated from the official story of events by revealing embarrassing tactical mismanagement or poor military strategy. Second, though they were dramatic in tone, they stopped short of the 'near pornographic' reports of sexual violence to be found in many men's accounts (Ghose 1998a, 23). This may be explained in part by the rules of propriety that governed their writing. For women writers, there were social embargoes against dwelling on events in lurid detail. There were book-length accounts of the First War of Indian Independence, too. In an incisive close reading of Frances Isabella Duberly's 1859 *Comprising Experiences in Rajpootnana and Central India, During the Suppression of the Mutiny, 1857–1858*, Ghose identifies a passage that appears to convey a sense of sympathetic identification with colonized women suffering all forms of abuse during violent British reprisals, since it toys with the possibility that 'the Indian woman is [...] rapeable too' (Ghose 1998a, 104). However, despite these subtleties newspapers in London and Bombay (now Mumbai) bombarded their readers with graphic stories of Indian men violating British women. The effect, as Susan Kingsley Kent points out, was to 'electrif[y] the British public', who continued to be transfixed by images of 'scalped and dismembered white women' for decades to come (Kingsley Kent 1999, 218). As Sharpe observes, the newspaper-fuelled hysteria surrounding the incident revealed the extent to which colonial authority rested on the ability to control the sexuality and conduct of British women (Sharpe 1993, 232). The crucial point here is that such expressions of concern over women's welfare helped to fortify claims about the savagery of Indian men even while women's suffrage was being vigorously opposed back in Britain.[15]

AGENCY, GENRE AND INTERTEXTUALITY

There is plentiful evidence that colonial women's travel narratives commonly subvert, as well as reinforce, oppressive and gendered interpretative practices. However, as I have argued, such subversions by no means guarantee any wholesale renunciation of colonial authority. Despite the huge temporal and contextual variation in circumstances, I would argue that the same principle applies to women's war reporting. Before elaborating on this point, however, it is necessary to detail some common assertions about differences in reporting style and content by female journalists. The comparative study of men's and women's reporting is hampered by the over-representation of men in fields such as war reporting. This said, there are clear assertions of difference that do bear investigation. As I mentioned earlier, female war correspondents have commonly claimed that they are not as preoccupied as their male counterparts with the machinery of war (Allan 2005, 77). However, the blame for this preoccupation does not always rest on the shoulders of male correspondents themselves. The *Telegraph* reporter Christina Lamb, for example, argues that it is male editors who demand details of military conflict rather than what she terms 'behind the scenes' coverage.[16] Another prominent assertion is that women's gender provides a passport to inherently less accessible female news sources. According to this premise, the bi-gendered status of female correspondents allows reports by women to be more inclusive than those by men. However, as I have argued elsewhere (Fowler 2007a, 18), the privileges conferred by honorary manhood status are predicated on women reporters' perceived access to the power of the mass media. Moreover, any benefits conferred by this status are tempered by wider conditions of reporting. As with men, the agency of women reporters is restricted by such factors as the established conventions of reporting, the cultural norms of newsrooms, time constraints, stringent editorial processes, market forces and political restrictions on scope and content (Fowler 2007a, 18). The homogenizing force of these processes makes it difficult to produce clear-cut statements about the agency of female war correspondents. Moreover, as with studies of women's travel writing, it is important not to dismiss the possibility that men's reporting is similarly subject to a kind of internal critique inspired by feminist concerns.

In her bestselling book about Afghanistan entitled, *The Sewing Circles of Herat* (2002), Lamb remembers her early career as a war reporter and makes an explicit claim about her gendered positionality:

> There was an American club [in Peshawar] where one could drink Budweisers, eat Oreo Cookie ice-cream and listen to middle-aged male correspondents in US army jackets with bloodstains and charred bullet holes [...] hold court with stories of conflict and 'skirt' from Vietnam to El Salvador. [T]hey knew the sound of every weapon ever invented [...] and all of them went to the Philippines for R and R [rest and relaxation].
>
> It was different for me. I was a young girl in a place where women were regarded as property along with gold and land—the three zs of the Pashtuns, *zan*, *zar* and *zamin* [gold, women and land]—and kept hidden away behind curtained doors.
>
> (Lamb 2002, 36)

This feminist dual condemnation of male reporters'—as well as Pashtun—sexism is complicated by a number of factors. The first concerns the way in which her disassociation with masculine norms of reporting aggregates power to her own putatively feminized mode of reporting. By identifying with abandoned lovers in Viet Nam, prostitutes in the Philippines and Afghan women behind Pashtun doors, she implies that her reports are imbued with a more sensitive and inclusive moral vision. However, having distanced her own reporting from that which she characterizes, even caricatures, as essentially masculine, she is faced with the perpetual dilemma of any feminist reporter whose news organization is aligned with political centres of power. The dilemma poses itself as follows: having defended the reporting style of female correspondents such as herself, might any subsequent form of de-authorization on account of her association with Western military aggression amount to a self-defeating form of renunciation? Although such questions get to the ideological crux of the matter, they remain difficult to answer satisfactorily. However, the passage does yield two particular examples of the kind of restrictions that curtail any autonomous vision on her part. The first example relates to the way in which her sympathy with Pashtun women is harnessed to an assumption of male brutality and female victimhood. In this sense, the passage complies with a long representative tradition by the West, which under-historicizes Afghan women's active role in the country's political and military affairs, either as guerrillas during the Soviet occupation or as participants in key organizations or in oppositional politics more generally. The second example relates to the co-presence in Lamb's narrative of a founding text by Louis Dupree, the father of 20th-century anthropology about Afghanistan (Dupree 2002). On this occasion at least, the ideological orientation of this particular intertext compromises Lamb's interpretative autonomy, since she inherits Dupree's distinctive conception of Pashtun female victimhood. This notion is encapsulated in her phraseology, which by definition reinforces, rather than contests, Dupree's particular understanding of uncontested male ownership ('*kept* hidden away').[17] This minor example illustrates a few reasons why, despite women's critiques of *machismo* male reporting styles, anthropologists of journalism such as A.C. Delzotto and Alf Hannerz argue that there is little variation in the tone and content of news reports produced by men and women overall (Delzotto 2002; Hannertz 2004).

FEMINIST ALIGNMENTS WITH CONSERVATIVE POLITICS

Since the 1980s, anthropologists have mounted an effective challenge to the widely held assumption that women's oppression is more or less universal outside Australia, Western Europe, Canada and the United States (Cornwall and Lindisfarne 1994, 20; Moore 1998). As Henrietta Moore points out, feminist anthropologists have emphasized the importance of variations in family structure, national legislation and political context across the globe. Even so, as I have argued elsewhere (Fowler 2006a, 13), there are clear ideological advantages in subscribing to the view that patriarchy more or less rules outside the West, since it serves the self-image of Western men and women as liberators and liberated.[18] Paradoxically though, pervasive images of malevolent and patriarchal Third World, often Muslim, men tend to deny the agency of women world-wide. This point was made in a statement, issued in 2007, by four Iranian scholars to protest against Western public debate about Iran. In the statement, Niki Akhavan, Golbarg Bashi, Mana Kia and Sima Shakhsari argue that Iranian women have been 'at the forefront of literacy, educational, artistic, journalistic, and legal advancements'. Despite frequent public declarations of concern over women's rights in Iran, the group points out that these same rights would deteriorate drastically in the event of a future military attack on Iran (Akhavan et al. 2007). However, what is significant here is that the scholars' target is not news media coverage, but the 'lucrative industry of Iranian and Muslim women's memoirs', which 'appropriate the legitimate cause of women's rights and place it [...] in the service of Empire building projects'. Ignoring antecedent critical feminist scholarship in Iran and the Middle East, the writers of such memoirs embrace 'an idealised middle-class norm of Euro-American consumption' and are heralded as 'experts' on Islam (Akhavan et al. 2007). However, the statement also points out that critical engagements with these memoirs are quickly dismissed as misogynistic. A notable example of this is a critique of the memoirs in the same vein by Hamid Dabashi (2006), who was lambasted by the press and dismissed as a sexist Middle Eastern man. Akhavan, Bashi, Kia and Shakhsari point out that the systematic failure to register critical feminist scholarship, combined with the singling out and distortion of work by thinkers such as Dabashi, exposes the essentialist, gendered assumptions that inform the memoirs' reception and the select band of Iranian women, such as Azar Nafisi, who are granted the right to speak for their silenced compatriots 'courtesy of international publishing houses'. They argue that it is imperative to read such memoirs in the context of the politics of reception and with an eye on the ideological justification for the current wars in Afghanistan and Iraq. Here, the memoir genre is identified as part of an 'industry of "knowledge-production" that juxtaposes "freedom" and "progress" [...] to "backwardness" and "barbarism" in Iran and in the rest of the Muslim world'. This critique of the genre centres on the apparent authenticity of the inexorable 'silenced to vocal' trajectory of what are termed 'salvation narratives', which trace the journey from 'lost' (Muslim) woman to 'redeemed' woman with a newfound subjectivity. That this subjectivity is generally comfortable with a 'secular' environment, I would add, reveals much about the self-image of 'secular feminism', which places itself in opposition to the tyranny of religion.[19] It is relatively rare to find any mention in Western feminist scholarship of the wealth of work by Muslim scholars who have sought to remove questions of women's agency from the 'secular' arena. Among the most important contributions to this parallel debate are Leila Ahmed's seminal study of feminism and Islamic interpretation (Ahmed 1992) and Saba Mahmood's ethnography of women's piety movements (Mahmood 2005). The latter offers a sound critique of the 'secular-liberal' basis for condemning Islamic revivalism. However, given the secular foundations of much feminist argumentation, these separate lines of debate seem destined never to meet.

Recent years have seen the rise of what Arun Kundnani (2007, 26) terms 'integrationist feminism' in Australia, Europe, Canada and the United States. One prominent feminist integrationist was the late Susan Moller Okin, the Harvard professor of ethics and political science. Liz Fekete notes that, just as a particular interpretation of Samuel P. Huntington's 'clash of civilizations' thesis was being popularized, Moller Okin published an article in the *Boston Review* arguing that multiculturalism was 'bad for women'. Blaming the Western liberal tradition of embracing cultural diversity at the cost of hard-won feminist gains, Okin proposed a form of 'culturocide' as the safest means of integrating immigrants into US society (Moller Okin 1999, 12–13). As Kundnani (2007, 26–27) points out, integrationist feminism promotes a radical redefinition of 'integration'. Quoting Roy Jenkins's 1966 definition of the concept as 'equal opportunity accompanied by cultural diversity, in an atmosphere of mutual tolerance', Kundnani shows that, in Britain at least, integration has come to mean 'assimilation to British values'. Nowhere is this phenomenon more strikingly evident than in recent public debates about sexual equality. In an effort to present the British as naturally liberal and Muslims as naturally sexist, he argues, integrationist feminism overlooks Britain's own problematic engagements with gender. Throughout Europe too, as Liz Fekete (2006, 9) points out, integrationist feminists participate in a form of monoculturalization, whereby national culture is increasingly reified as local, bounded and hereditary. Such processes, she argues, collude in the process of shearing the national culture of its 'seamier side' and 'steering race relations policy away from multiculturalism towards monoculturalism'. Parodying Europe-wide discourse about the failure of multiculturalism, Fekete writes:

> If there is anything amiss in this, our European homeland, it is the consequence [...] of too much goodness. Over-tolerance towards people from different cultures is our Achilles heel. We must preserve our cultures at all costs and not let them be contaminated.
>
> (Fekete 2006, 10)

This issue is thrown into sharp relief by 'honour killings' and forced marriage. As Kundnani (2007, 11) points out, although two British women are murdered every week by their male partners, disproportionate attention to Muslim 'honour killings' reinforces the notion that Muslims are the sole bearers of patriarchy. While violence against white women is considered a straightforward matter of human rights, he argues, combating violence against Muslim women is couched in terms of the urgent need to oppose a culture (Ibid., 41–42). Of course, Kundnani is not the first to point out this contradiction. Amrit Wilson (2007, 31–32) notes that the Southall Black Sisters have long called for the mainstreaming of issues such as forced marriage and honour killings by incorporating it in the national strategy for violence against women and children. However, this call has been consistently ignored. As a consequence, it is difficult to implement effective, integrated responses to violence against women. As both Fekete (2006, 13) and Wilson (2007, 35) observe, the increasing tendency to conflate forced marriage with honour killings and the explicit linking of these practices with terrorist threats to the state serve to justify tighter immigration controls.

It is against these discursive backdrops that several ironies are being enacted across Europe, revealing a disjunction between feminist rhetoric by government officials and actual policy decisions. In an astonishing reversal of long-standing feminist resistance to coercion, women and girls are legally forced to unveil, women are deported without regard to the threat of violence that awaits them, and South Asian women's organizations continue to undergo the threat of closure due to underfunding.[20] The paradox here is that punitive reactions to public concern, such as tightening immigration controls, can actually exacerbate violence against women, forcing them to remain in violent relationships to avoid deportation to their countries of origin, where their marriage breakdown may provoke further violent reprisals.[21] Wilson (2007, 29) provides a powerful example of this, noting that international expressions of concern over women's rights under the Taliban were undermined by the British Home Office policy of deporting Pakistani women to areas of the North West Frontier Province that remained under Taliban control. She provides a further example of the British Government's support for many South Asian organizations belonging to the religious right. This encouragement was quickly forgotten when the same government began to 'posture as confronting [… "Muslim" patriarchy], reacting with shock and horror as though patriarchy was a monster unknown to it, which had suddenly appeared from […] alien and backward land[s]' (Wilson 2007, 31).

CONCLUSION

As Fekete (2006, 2) points out, the views and policies associated with far-right parties are 'no longer on the political fringe' in Europe. It seems ironic, therefore, that the growing feminist alignment with an increasingly conservative politics has coincided with the scholarly recognition that colonial women's travel writing cannot be uncritically celebrated as straightforwardly anti-imperialist in tone. While the scholarly assessment of colonial women's collusion in the colonial project is being thoroughly revised, in the political sphere, controversial figures such as Ayaan Hirsi Ali[22] have been causing a stir across Europe. Meanwhile, autobiographical accounts of Western-style emancipation by Afghan, Iranian and Iraqi women continue to top bestseller lists. Moreover, critical feminist responses to the way in which women's rights have been placed in the service of domestic anti-immigration campaigns and aggressive military incursions have yet to register on the radars of war reporters. The critical point here is that feminist-leaning news media coverage of the wars in Afghanistan and Iraq has its precursor in protofeminist travel writing. There are, of course, countless points of continuity and discontinuity between these traditions. Even so, there has yet to be a sustained comparison between them.

Despite the pervasive culture of the newsroom, contemporary news media coverage by men and women *is* undoubtedly subject to a form of internal feminist critique. However, for centuries now, travellers and journalists with (proto)feminist identifications have faced a similar predicament. Now, as then, feminist battles for rights remain vulnerable to being co-opted in pursuit of discrimination at home and military invasion abroad.

NOTES

1 The period known as high empire denotes a phase of colonial British expansionism from around 1870 to the period leading up to the First World War.
2 See Susan Bassnett in Indira Ghose 1998a, 2.
3 There is a convincing argument to suggest that travel writing from this period reflects, as Andrew Hammond argues, a broader 'loss of confidence in the Western gaze' (2003, 185). I have discussed (and to some extent challenged) this phenomenon in Fowler 2007a, 214–19. Key figures in this current of thought include Andrew Hammond (2003) and T. Youngs (2004a). Correspondingly, postcolonial critics of travel writing have argued that it would be wrong to level the charge of imperial certainty at colonial travel writing since such writing is, in any case, riddled with anxiety. For a helpful summary of this last point, see the introduction to Boehmer 1998.
4 A key publication in the field of travel writing and ethics was Islam 1996. See also the forthcoming volume by Forsdick, Fowler and Kostova 2010.
5 Although it is important to mention that Women's Studies departments in the United Kingdom and USA are gradually being amalgamated with Sociology and Cultural Studies departments.
6 See Kabbani 1986; Mills 1992; Youngs 1994b; Ghose 1998a.
7 Again, this issue is discussed at greater length in Fowler 2007a, 17–24.
8 See Kabbani 1986; Mills 1992; Lowe 1994; Lewis 1995; Melman 1995; Lewis 2004; Smith 2001; Ghose 1998a; Blunt 1994.
9 For a detailed discussion on this, see Foster and Mills 2002, 9, in the section of their book entitled 'Women writing about women', which starts on 12.
10 For a seminal discussion of the tendency to produce 'partial and coherent readings' of women's travel writing, see Mills 1992, 5.
11 A related discussion of these sections of Grimshaw's narrative can be found in Fowler 2006b, 40–41.
12 This was also the case with many out-of-print ethnographies of Afghanistan, including the work of anthropologists such as Louis Dupree (1973), Nazif Shahrani (1979) and Whitney Azoy (1982), whose works were all republished or reissued after 2001.
13 See Pionke 2008.
14 This process typically took the form of fabulated letters—often based on actual originals—to an imaginary English addressee. Pionke draws attention to a series of letters from John Company to John Bull in *Blackwood's Edinburgh Magazine* and *The Perils of Certain English Prisoners*, which was co-written by Charles Dickens and Wilkie Collins for *Household Words* and published at Christmas of 1957.
15 See Ahmed 1992.
16 Interview with Christina Lamb by the author, www.studiesintravelwriting.com/publications.php?id=366.
17 Italics are mine.
18 Indeed, the 2008 World Economic Forum Gender Gap Index places Great Britain below Sri Lanka in its league table of favourable conditions for women. In the United Kingdom, female full-time workers are paid 17% less than their male counterparts, while women in part-time work are paid 37% less than part-time male workers.
19 Scholars such as Talal Asad (2003) and Timothy Fitzgerald (2007a) have written about the need to engage critically with formulations such as 'secular democracy', which depend on unexamined assumptions about the essentialized category of 'religion'. This is because 'religion' has been placed in opposition to 'secularity' since the 17th century.
20 For an intelligent discussion of this topic, see Wilson 2007, 25–38.

21 Wilson notes that, before 2003, those coming to the United Kingdom to join spouses were denied permanent residency in their own right for a probationary period of a year following their arrival. In 2003 this period was extended to two years. If the marriage breaks down due to violence during that period, a woman faces deportation to her country of origin pending an appeal, which can take any time from several months to a number of years. Since she is denied access to any public funds during this period, the rule actually strengthens the hand of the violent party. See Amrit Wilson 2007, 28–29.

22 Dutch politician Ayaan Hirsi Ali is author of *De Maagdenkooi* (published 2005), known internationally as *The Caged Virgin, An Emancipation Proclamation for Women and Islam* (Hirsi Ali 2006).

Gender and/as genocide

TAWIA ANSAH[1]

INTRODUCTION

This chapter examines gender in relation to the crime of genocide, specifically gender as it relates to the legal discourse on genocide. I am interested in how gender, as well as sex and sexuality, enter the international law on this particular form of mass exterminatory violence. In this relation, gender hinges on sexual violence, primarily against women, although at least one commentator has deployed a gender analysis or 'lens' to describe the primary targeting of men during genocide.[2]

The discourse on sexual violence against women during genocide tends to fall into three categories: in the first, there is a conflation between sexual violence and genocidal violence. The shorthand for this gender modality is 'rape as genocide'. In the second, sexual violence and genocide are seen as related but separate, the former instrumental to the latter: 'rape and genocide'. The gender analysis here sees an intersection between the two forms of violence. In the third category, the issue of sex and sexual violence in relation to genocide is analysed outside a framework of gender, or the 'gender lens', as such. That is, the commentators within this modality attempt a critique of the gender analysis as it affects the legal discourse and, thereby, the political activism that results.

My analysis of the three gender modalities is an attempt to see what aesthetic of gender and/as genocide ensues: what idea of genocide and of gender each produces in relation to law. The aesthetic approach attempts to see how we see and do not see the nexus of genocide (race violence) and gender (sex violence). I attempt through this approach to encounter the ambivalent investments pursuant to sex and race as juridical constructs and ways of seeing state-sponsored and exterminatory violence. In the first instance, this means discovering what subjectivity is produced through each modality of gender in relation to genocide.

Gender viewed under this aesthetic prism sees both the conflation or 'as' model and the critical or 'and' model as related in their desire for fixed juridical forms. In this sense, 'rape' and 'genocide' are projects of desire in relation to their contingent objects: rape fixes gender, just as genocide fixes ethnic (or other) identity. Their conflation or disaggregation therefore produces different but dyadically related ideas or images of gender and of race. Put otherwise, gender analysis sees genocide through the lens of sex, whilst the race model sees genocide through the lens of blood. The development of an aesthetic lens focused on each sees them in their figurative and constructive element—for instance, the assimilation of sexual violence to race figures the former as a metaphor—the nation is 'raped' by the enemy—rather than as set categories of analysis or vehicles to the 'truth' of sexual and exterminatory violence. The aim of this chapter, then, is to examine gender in relation to genocide as an aesthetic project in order to think through them as juridical and immanent categories.

PART 1: GENDER → GENOCIDE

The emphasis in this chapter is on the relationship between gender and genocide: as juridical discourses, each has had an impact on the other inasmuch as recent instances of genocide in the 1990s (but not exclusively by any means) have been marked by an increased deployment of sexual violence by the perpetrator in the course of the exterminatory violence characterized within legal and political discourse as genocide. As a result, there has been a sharper awareness within the public perception of genocide of the instances of systematic rape and sexual assault. This awareness is the result of the concerted efforts of human rights advocates, mainly feminists.

One may characterize the lens through which to 'see' the genocide and, with it, the significance of the sexual violence that is visited upon women during the course of it, as a 'gendered' lens. One definition of such a lens comes from Laura Sjoberg and Caron E. Gentry, wherein they state that, 'the authors are committed to the view that genocide is a very gendered phenomenon: that is, it affects men and women differently, and it does so because of its situation in a world in which femininities are subordinated to masculinities, and women to men'.[3] Those who see rape *as* genocide are committed to the view that once we see the different forms of violence experienced by men and women we cannot but see the violence as a whole *as* gendered. With that view predominating, we are charged with eradicating the unequal treatment of men and women before the bar of justice. It is pursuant to a gendered perspective that rape, for instance, enters the juridical lexicon of genocide.

However, to begin with, how might one characterize a *non*-gendered view of genocide? In his two-volume study of the 'meaning' of genocide through an historical retrospective, Mark Levene notes that gender, along with culture, may function symbolically for the protagonist targeting 'a particular ethnic or social group' (Levene 2005, 140). However, gender may also be instrumental to the perpetrator's aims, i.e. instrumental to the destruction of that which represents a threat to the group's sense of identity: 'It [i.e. the cause] may [...] also be very functionally grounded. The perceived need for land, for resources, for popular empowerment, for the unfettered mobilization of national assets, or the opening up of a frontier region—all of which a real or imagined communal group may resist or deny—is certainly intrinsic to the majority of genocides' (Ibid., 140). Thus, when Levene relates stories of sexual violence, they read more as exceptions to the rule: of the specifically gendered violation of Jewish and Romani women as perceived by the Nazis, Levene asks:

'To what extent the dangerous sexual allure of female Roma and Jewess, or their alleged skills, especially in the Roma case, as clairvoyants and sorceresses, are reflections of a broader culture of European misogynism or, rather, are logical gender extensions of archetypal Jewish and Romani male "construction" is not quite the issue here. This is partly because any special female potency, bar of course their all-important reproductive power to bear more Jewish and Romani children, was excluded from the Nazis' genocide equation' (Ibid., 139).

Levene sees gender as one of a number of justifications, for example, a rationale for the Sudanese imam's 'fatwa' endorsing the 'jihad' against the Nuba people: 'The rebels in South Kordofan and southern Sudan started their rebellion against the state and declared war against the Muslims. Their main aims are: killing the Muslims, desecrating mosques, burning and defiling the Koran, and raping Muslim women [….]' (Ibid., 128). However, those who adopt a gendered perspective or framework of analysis see the same thing quite differently.

For Catharine A. MacKinnon, who wrote extensively on the sexual violence that took place during the Bosnian conflict in the 1990s, gender is central to an understanding 'as such' of the violence of genocide. Also, gender and sexuality are of a piece; Janet Halley quotes MacKinnon in 1989: 'Since sexuality largely defines gender, discrimination based on sexuality is discrimination based on gender'.[4] MacKinnon links the sexual violence that takes place during genocide to the language and, per the Holocaust, the paradigmatic 'situs' of genocide: 'Sexual atrocities thus give a distinctive content to the term "as such" in genocide's definition [….] When Jewish women were sexually used in brothels in concentration camps during the Holocaust, they were used as Jewish women "as such", no matter who else was used with them' (MacKinnon 2006, 229). Furthermore, 'Women not being considered a people, there is as yet no international law against destroying the group women as such. "Sex" is not on the list of legal grounds on the basis of which destruction of peoples as such is prohibited. For women as such, there is no legal equivalent to genocide—the destruction of women as such that Andrea Dworkin proposed calling gynocide—presumably because it is commonplace, built into the relative status of the sexes in everyday life' (Ibid., 230). Here, rape and genocide are fused: rape is genocide, genocide is rape.

Part of the enterprise of bringing to the law the view of genocide as essentially a gendered phenomenon is to counter a history of silence around the role gender has played in instances of genocide. As MacKinnon notes, 'Genocidal rape did to race, ethnicity, religion, and nationality what rape outside genocides does to sex. When and to the degree that it works, rape destroys national and ethnic groups in genocides as it destroys women as a group under sex inequality'. As noted, she applies the lens historically: 'In the Holocaust, in particular, the genocidal role of sexual atrocities has been largely denied, the women as such being as relatively invisible as their sexual abuse has been' (Ibid., 232).

Thus, where Levene sees the gender story in parts or fragments, MacKinnon sees it as constitutive with respect to genocide. In a real sense, MacKinnon's view of sexual assault has become paradigmatic for feminism.[5] At one level, sexual abuse during genocide is itself genocidal: describing the ways in which women in Rwanda and in Gujarat were raped and murdered, MacKinnon notes that, 'In such spectacles, sexual abuse performs, and in so doing enacts, the destruction of the target peoples. It performs genocide' (Ibid., 231). At another register, though, MacKinnon seems to suggest that when a woman is raped it is as such an act of genocide; it is genocidal *tout court.*

MacKinnon's influence on the international apprehension of genocide after Bosnia and, later, after Rwanda, each conflict marked by systemic sexual violence, was quite profound. A sampling of the literature on how rape should be adjudicated under international law reveals agreement with MacKinnon's 'rape as genocide' view on the one hand, and 'rape and genocide', or rape as torture or a grave breach, on the other. I want to look at some of these commentators to see how gender entered international discourse as it pertains to genocide.

Mary Deutsch Schneider examines the mounting attention to gender violence under international law. She notes that, 'The very creation of the ICC [International Criminal Court] may be seen as a global response to the extreme atrocities perpetrated in the last century, including gender violence' (Schneider 2007, 916). Schneider observes that sexual violence during armed conflict 'takes many forms, such as: forced sexual intercourse or other sexual acts with family members, forced impregnation, forced pregnancy', and other equally brutal forms. Schneider highlights the fact that conflict's horrors may affect both men and women, but '[t]here is an overlay of additional brutality [...] that occurs with greater violence and frequency to women. Male civilians are killed; female civilians are typically raped, then killed. In interrogations, males are savagely beaten; females are savagely beaten and raped' (Ibid., 915).

In relation to sexual violence and genocide, Schneider suggests that '[s]exual torture of women is also used to cause terror sufficient to drive whole populations out of an area, or to deter male activists from revolutionary activity. In this respect, the gendered division of violence acts to keep repressive regimes in power' (Ibid., 917[6]). In Rwanda, '[i]n the hundred-day period when 600,000 or more people were slaughtered [...], [r]ape was the rule and its absence the exception' (Ibid., 932).[7] In Darfur, 'Gender violence is a prominent feature of the current conflict [...] which has complex roots' (Ibid., 960). Indeed, 'Rape and sexual assault have been used by government forces and government-backed Janjaweed militia as a deliberate strategy with the aim of terrorizing the population, ensuring its movement and perpetuating its displacement. While there are many pressures for women not to report victimization, the numbers of sexual assaults are thought to be in the tens of thousands' (Ibid., 964).[8]

Schneider's argument is that with the increase in gender-based violence during these recent conflicts, more needs to be done through law to redress the problem. I read Schneider as committed to a gendered analysis of genocide in order for law to effect justice for the victimized women.

The conflicts in Bosnia and Rwanda in the 1990s in particular led to inclusion of gender violence as justiciable within international legal instruments and decisions. The law developed in this regard both procedural and substantive rules to address and incorporate rape within the crimes adjudicated in relation to genocide. Schneider celebrates the procedural rules developed by the International Criminal Tribunal for the former Yugoslavia (ICTY) designed to protect witnesses that come before it to provide testimony (Schneider 2007, 950), and she lauds the decisions of both ad hoc international tribunals (the other being the International Criminal Tribunal for Rwanda, or ICTR) for the substantive law developed within its decisions addressed specifically to gender violence. In the *Akayesu* judgment the ICTR 'recognized sexual violence as causing "extensive harm and that it is intentionally used during periods of mass violence to subjugate and devastate a collective enemy group"' (Ibid., 947).[9] Schneider notes that, 'The decision found rape part of the genocidal regime carried out by Hutus, "an integral part of the process of destruction". Sexual violence was a step in the process of destruction of the Tutsi group—destruction of the spirit, of the will to live, and of life itself. The Tribunal noted the collective nature in targeting of the broader group' (Ibid. 947).[10] *Akayesu* 'presented the first ever conviction for sexual violence based on genocide' (Ibid., 947).

The next logical step would have been for the Rome Statute of the ICC, ratified in 1998, to have adopted the *Akayesu* interpretation, which comes closest to defining rape as genocide. As Schneider notes, 'The Rome conference produced a final draft of the ICC Statute incorporating gender-based crimes in progressive and previously uncharted ways. War crimes, crimes against humanity, genocide, and aggression are the general categories of crimes which fall under ICC jurisdiction, and include gender violence'. The implication is that rape could be prosecuted as genocide or an 'act' of genocide under the Rome statute. However, 'The Rome Statute did not define rape' (Schneider 2007, 934),[11] leaving the link between rape and genocide to decisional interpretation: the *Akayesu*, *Delalic*, and *Furundzija* cases[12] represent the pre-Rome Statute legal jurisdiction over gender violence within the context of genocide.

Schneider, like MacKinnon, sees the potential for international law to transform the domestic legal understanding of categories of violence, especially sexual violence. She writes that the ICC has 'the power to make enormous inroads into the elimination of violence against women. It must be made clear that the world considers rape and sexual violence to be militarily, legally, and personally reprehensible, whether the purpose is to harm women, to message men, or to excise [sic] masculinity [….] Efforts must put blame for the destruction of lives and community where it belongs—on those who commit and permit sexual violence' (Schneider 2007, 994–95).

The ICC Rome Statute has the power to ratify a gendered lens in relation to genocide thereby marking, for Schneider, a certain clear visibility for the victims of sexual violence both within conflict and more broadly. However, the Statute does not conflate rape and genocide, or sex and race; instead, it relates sexual violence to crimes against humanity under Article 7(1)(g),[13] and war crimes under Article 8(2)(b)(xxii).[14] If the assimilation of rape to genocide as such would grant maximum visibility to the issue of gender, then Rome falls short. However, others, who also adopt a gender perspective on genocide, disagree with this, as I discuss in Part 2.

PART 2: GENDER / GENOCIDE

In the *Commentary on the Rome Statute of the International Criminal Court*, 'gender' is carefully defined to include a constructed and ambulatory element. The authors of the text note that, 'In United Nations usage, "gender" refers to the socially constructed roles played by women and men that are ascribed to them based on their sex. The word "sex" is used to refer to physical and biological characteristics of women and men, while "gender" is used to refer to the explanations for observed differences between women and men based on socially assigned roles. The phrase "in the context of society" seems to imply the notion of these socially constructed roles and differences'.[15]

The definition permits for the adoption of a view of gender from the perpetrator's perspective. Indeed, this is Katie C. Richey's critique of the law and its tendency to conflate or assimilate rape with genocide and with crimes against humanity. Richey notes that, 'legal discourses on this issue [of rape as an act of genocide] equate women's bodies with national territory, integrity, and honor—a metaphorical slippage that reinforces perpetrator motivations for war rape' (Richey 2007, 111).[16] In other words, rape *as* genocide can only be understood metaphorically and, as such, erases the actual bodies and lives harmed by the acts of rape during war or genocide. For Richey, the focus should not be on 'genocide' as gendered but rather on the rape itself as 'a deeply gendered act that carries symbolic meaning for the victim, the community, observers outside the community, and the perpetrators themselves'. Rape is *like* genocide, in that it 'reduce[s] the female body to an object [...] The rape is an offensive assault on the integrity of the body, but the literal invasion is committed with metaphorical ends in mind'. As such, 'One way women's bodies are objectified by war rape is their symbolic association with territory, nation, and group. When soldiers rape women, "they emasculate the enemy by rendering him helpless to protect his women, whose bodies sign in for the nation or group"' (Ibid., 122–23).[17]

Richey's criticism of the conflation of rape and genocide is that it plays into the metaphorical and normative obliteration of women as subjects of international law: 'The international community has been more attentive to this symbolic aspect of war rape as genocide and less responsive to its literal and corporeal implications. The international legal discourse surrounding war rape accepts the rapes on the same metaphorical terms as its perpetrators, equating the rape of women's bodies with the invasion of territory and the integrity of women's bodies with the integrity of the ethno national [sic] groups to which they belong' (Ibid., 123). A gendered lens, therefore, is required to 'see' the woman through the violence of rape itself, on her body, and to 'see' the woman beyond the conflation with race, nation, or group, whether as metaphor or as intersectional axis of redress. The gendered lens does not deny the intersection with race but requires their disaggregation.

Richey proposes, not unlike MacKinnon herself, as it happens (MacKinnon 2006, 238, 246), that rape be assimilated to torture rather than to genocide: 'Conceiving of war rape as a grave breach of the Geneva Conventions is the best way to ensure women's rights are recognized fully as human rights under international law' (Richey 2007, 129). This gets around the problem of having to prove that a single act of rape can only be prosecuted under international law if it is part of government policy, i.e. 'widespread, systematic and discriminatory', or 'committed under orders' (Ibid., 115). In effect, she wants to achieve the same result as the conflationists: for the latter, if an act of rape is 'as such' an act of genocide, the international community is obligated under law to 'prevent and punish'. For Richey and the assimilation of rape to torture, a single sexual assault will no longer be elided under widely-held presumptions: 'International law begins with some expectation of the inevitability of rape, and rape as a war crime is unique in this context. The legal discourse on torture does not reflect the same "boys will be boys" mentality. There is no similar expectation of the inevitability of minimal levels of torture during war; a single instance of torture is considered a grave breach of the Geneva Conventions and there is no attempt to delineate normal torture from egregious torture' (Ibid., 115–16).

For both Richey and the conflationists, law is the key to repair the damage and reinforce a prohibition against sexual violence during genocide, but only by adoption of a gendered perspective that excavates an unequal treatment of men and women both during the conflict and at the bar of justice. For both, the emphasis is on what at law is called the *actus reus*, the act itself divorced from the mental element of intent, the *mens rea*: the latter is required for a finding of criminal liability in most domestic criminal jurisdictions, and the intent requirement is also built into the convention prohibiting genocide. However, Richey fears that too much attention is focused on the latter, cautioning that, 'The rape is an offensive assault on the integrity of the body, but the literal invasion is committed with metaphorical ends in mind' (Richey 2007, 122). By this rigorous exclusion, Richey's critique of 'rape as genocide' by assimilating the *actus reus* to torture has the paradoxical effect of reinforcing, at an aesthetic level, the 'literal' view that led to the conflation model itself. That is, within the gendered lens deployed by Richey, the *actus reus* is sufficient to 'see' the harm, and must be seen in order to achieve justice for the violence visited upon the women regardless of the context (war, genocide, etc.). In both, there is a tendency toward essentialism.

A third gender modality also looks at the act but seeks to place it within its context and, thereby, avoid the problem of essentialism or reduction: this modality is often described as the 'intersectional' model. Intersectionality avoids the conflation of rape with genocide but largely agrees that rape, for international adjudication, is best assimilated to torture or a grave breach of the Geneva Conventions. The intersectional model shares with other gender perspectives a sense of the immediacy of the harm, but includes in its analysis the mediation of other factors within the sexually violent act itself: the woman's race or other communal and identity ascription.

The intersectional model avoids elements of the other gender models seen thus far: the conflationist's collapse of all identitarian indexes within the 'gender' of the woman on the one hand, or Richey's purification of contingent identity markers within the assimilation of rape to torture, the 'literal' view of the act as such. The intersectional model treads a fine line between them and, in the result, falls prey to a certain element of distortion.

The intersectional model is one of the more pervasive within genocide discourse and takes the form of arguing for apprehending the nexus of gender and race or other community-based identity, i.e. to see the victim of rape during genocide as a target because she is a woman and because she is a member of the group slated for extermination. The model seeks to complicate each category, 'rape' and 'genocide', especially as it affects female victims.

Justin Wagner, in a detailed examination of the incidence of sexual violence in Darfur, aims in his article 'to forcefully make the case for the inclusion of gender-related crimes in war crimes [and genocide] prosecutions' (Wagner 2005, 193)[18] because the link between gender crimes and local 'tradition' is 'often overlooked' (Ibid.). In this sense, Wagner's approach is intersectional: precisely because race violence is front and centre, and because sexual violence, particularly rape, is suppressed and denied in the aftermath of conflict because of its stigma,[19] prosecutors and judges must recognize that the forms of violence are not separate: 'violence against civilian women appears to be just another tactic of terror perpetrated by government and Janjaweed forces' (Wagner 2005, 203, citing an Amnesty International report). An intersectional analysis of rape within war or genocide would 'serve to delegitimize the cultural stigma that rape carries with it' (Ibid., 242).

In her call for an intersectional analysis of the propaganda leading up to the genocide in Rwanda, Llezlie L. Green notes that the Tutsi women 'were targeted by propagandists based on a dual animus—on the basis of their ethnicity and their gender. As such, an "intersectional" legal analysis by the tribunal in the Media Trial would provide a more effective and equitable result for the victims of sexual violence. Furthermore,' she continues, 'such an analysis would ensure that a focus on the broader context does not obscure Tutsi women's experiences' (Green 2002, 735).

Green's focus is on the prevalence of a gender element within the genocidal propaganda: 'Gender hate propaganda was perhaps the most virulent component of the propaganda campaign' (Ibid., 748). The propagandists both demonized and exalted Tutsi women, making them on the one hand 'prostitutes who used their sexual charms to seduce the Western forces stationed in Rwanda', and on the other 'more beautiful and desirable, but "inaccessible" to Hutu men whom they allegedly looked down upon and were "too good for" them' (Ibid., 747[20]). Green notes that four of the infamous 'Ten Commandments' of Hutu Power, a set of norms that fuelled the Hutu revolution that precipitated Rwanda's independence from the Belgian colonists (and the Tutsi rulers) in the 1950s, were devoted to a prohibition against Tutsi women, either as wives or as employees.

Yet, notwithstanding its devastating effects in leading to the mass rape and sexual violation of Tutsi women during the genocide itself, there is something pathetic and plaintive about the gender-baiting within the propaganda. One of the excerpts Green points to suggests a profound sense of insecurity of the Hutu 'revolutionary' leadership begging its male members to come to their senses: 'Every Hutu should know that our Hutu daughters are more suitable and conscientious in their role as woman, wife, and mother of the family. Are they not beautiful, good secretaries and more honest?' This is followed by the next commandment: 'Hutu woman, be vigilant and try to bring your husbands, brothers, and sons back to reason' (Green 2002, 748–49, citing 'Shattered Lives', 17–18). What is necessarily elided from Green's gendered analysis is a very real history of decades of oppression and subjugation of the Hutu at the hands of an imperial Tutsi hegemony only made worse under the Belgian colonialists.

Her gaze firmly fixed on the gender issue, Green is able to extrapolate that 'These commandments clearly demonstrate a fixation with demonizing Tutsi women', thereby 'increasing their vulnerability to sexual violence throughout the genocide' (Green 2002, 749). As such, 'The link between gender-based hate propaganda and sexual violence is clear. Propagandists used sexualized images of Tutsi women to instigate ethnic hate and conflict. These images incited hatred of these women and of their sexuality. Thus, both ethnic and gender stereotypes, functioning individually and jointly, fueled the sexual violence committed against Tutsi women' (Ibid., 749–50). In short, 'It is not difficult to see a link between the sexual, often pornographic, nature of the propaganda and the resulting sexual violence. Thus, the hate propaganda campaign was often gendered in specific ways that merit attention in the Media Trial [at the ICTR]' (Ibid., 771).

Green quotes a young female victim of racial and sexual violence: 'Rape is a crime worse than others. There's no death worse than that' (Green 2002, 750, citing 'Shattered Lives', 59, quoting 'Jeanne, a young woman who was abducted and "forcibly married"'). This echoes MacKinnon's view, and there is something within the intersectional argument here that corroborates the subordination feminism model. As such, although intersectional, the trend is toward seeing the sexualized images as a distillation of race and gender in its attempt to see the victim within her context, her gender and her ethnicity, and thereby to see the violation as worse than gender violence or racial violence as separate harms: 'Just as state statutes on group defamation in the United States increase the criminal penalties for those convicted of crimes motivated by racial, ethnic, or religious hatred, crimes based on both a gender and a race/ethnicity animus demand a more severe penalty' (Green 2002, 774–75). Intersectionality here both corroborates the conflation model and extends its implications: because one form of violence is 'compounded' by another it is worse for the victim, and so the law should recognize and punish the perpetrator more severely.

Citing Green among others, Fiona de Londras wishes to access the truth of the experience of sexual violence and genocide. Under the subtitle 'The Notion of Truth', she writes: 'Survivors have noted their desire for the truth to out in the tribunals [….] It is not enough to simply deal with genocidal sexual violence in a vacuum—to see it as an incident. Such an approach does not acknowledge or appreciate the real truth of that experience' (de Londras 2007, 117). De Londras also employs an intersectional analysis to see the nexus of sexual- and group-identity-based violence: 'appreciating that a woman suffers genocidal sexual violence not only because of her sex (i.e., her biological make up [sic] and capacity to reproduce) and her gender (i.e., her position within a particular society, or at least her constructed position within that society), but also because of her position within one of the protected groups. It is the intersection of all of those characteristics that make women and girl children perfect targets in genocides that occur within "traditional" societies, i.e., within societies such as those targeted in Yugoslavia and Rwanda' (Ibid., 117).

De Londras draws on the work of Kimberlé Crenshaw, referencing the latter's 'three elements of the subordination of women: "the structural dimensions of domination (structural intersectionality), the politics engendered by a particular system of domination (political intersectionality), and the representations of the dominated (representational intersectionality)" […] stressing that the traditional analysis of these experiences in the context of gender or race marginalizes women of colour within both categories' (Ibid., 119).[21] As such, 'To provide an effective remedy to women victims of genocidal sexual violence international criminal law must acknowledge that women experience genocide in a very particular way *because* of the intersection between their gender and their identity within one of the protected groups, and tailor its response to sexual violence as genocide in order to fully appreciate and respond to these experiences' (Ibid., 120, emphasis in original).[22] Here also, intersectionality moves toward either conflation or the coincidence, as intersectional, of race and sex.[23]

As de Londras points out at the conclusion of her essay, notwithstanding the benefits of obtaining the truth of the woman's story, caught within the dual harms of race and gender violence, there are weaknesses to the intersectional analysis, not least in 'the potential for intersectional theory to result in essentialism' (de Londras 2007, 123). The three weaknesses de Londras summarizes, following Davina Cooper, are, in brief: a) the production of 'only two, bipolar positions: powerful and powerless', even as many persons in reality occupy 'contradictory positions'; b) 'Intersectional analysis sees only the intersections of axes and not the axes themselves, which are also formed by entanglements of various natures as a result of their existence within society itself'; and c) 'Cross-intersectional comparison is impossible: an understanding of the relative positioning of people with vastly different intersectional experiences cannot be engaged in as a result of their differences', an inherent weakness due to 'the concentration on the intersection of axes' rather than on the formation of the axes themselves (Ibid., 123).[24] De Londras suggests that intersectional analysis should, none the less, be used by judicial institutions, but with an awareness of these weaknesses as 'warning signs of essentialist and reductionist thinking' (Ibid., 124) in the application of the theory.

De Londras suggests, in the result, that 'in order to be complete, intersectional analysis must be personal, in-depth, expansive and non-reductionist: a heavy task for already over-burdened judicial institutions'. However, the benefit, which far outweighs the burden, is, for the victims, 'the chance to tell their story and have that story heard in its complete, personal, intersectional and societal truth' (de Londras 2007, 124). De Londras's view, in effect, highlights two aspects of the juridical discourse on the forms of violence pursuant to the crime of genocide: the difficulty of apprehending the intersection of forms and, even when this is attained, the difficulty of avoiding essentialist and reductionist representations of those forms.

A third problem, intimated in between the lines, is the incommensurability, on the one hand, of the need for justice and, in pursuit thereof, the need to tell the 'truth' of one's story before the law; and, on the other hand, the complex, even contradictory reality of the experience and of the (intersected) 'identity' that lives within and, through the conflict, that is harmed by that story. In other words, for de Londras, the very benefits of an intersectional analysis come with both its weaknesses (tendency toward essentialism and reductionism) and its inherent suppressions and pre-juridical constructedness in the pursuit of justice. In effect, each victim's story, however much it purports to represent a 'complete, personal, intersectional and societal truth', is already, by virtue of its being a 'law story', something of a distortion.

The adoption of a gendered lens either by those who see a conflation between rape and genocide, or those who see them as intersected and coincident with each other, requires the suspension of the idea that the lens itself is a distortion or a construction of the event. The strong sense of the truth of perception inheres in the factual reality—Richey's 'literalism'—that this happened to this body, this body is female, this did not happen to that other, male body. None of this is necessarily excepted by the critics of the conflation and intersectional view of sexual violence and genocide as a gendered phenomenon, but the critics, examined in Part 3 below, point to two things in relation to the gendered lens: first, that deploying a gendered view of genocide may have an impact on how we come to view both gender (as socio-sexual construct) and genocide (as juridico-racial construct); and second, the gendered view may elide aspects of the experience of genocide that alternative, or non-gendered, perceptions or modalities may submit. Each of the critical perspectives discussed below suggests the extent to which gender, as a lens or approach or modality, is both powerful and to some extent denies its power to shape the meaning of genocide.

PART 3: GENDER | GENOCIDE

One of the salient points Karen Engle raises in her critique of the 'growing consensus' around 'rape as genocide' is the extent to which the gender frame elides its own constitutive power as a form of governance (Engle 2007, 191). This elision is in tandem with an avoidance of the violence of international law, an elision necessitated by the desire to elicit its salvific operation ('calling in the troops') to save the victims of rape during conflict. Engle calls, therefore, for a 'break' from the consensus in order to halt the projection of gender into this form of violence (of military intervention).

Engle notes that the readiness with which human rights advocates are given over to military intervention as the solution to conflicts, and the increasing dependence upon the rhetoric of genocide and, with it, its conflation with rape, has created a 'crisis mentality' within the community. She writes: 'In the feminist context, the proliferation of calls for intervention based on existing or impending genocides suggests a new motivation for claiming that rape is genocide. Moreover, as some feminists have succeeded in equating rape and genocide, rape itself has become a significant justification for intervention. Documentation of rape helped drum up support, for example, for sending troops into Darfur' (Engle 2007, 191).[25]

Engle critiques the underwriting conflation model leading to this call: 'The conflation of rape and genocide suggests that what is unique about rapes in question is that they are based on ethnic hatred. But that distinction, whether made in the context of Bosnia and Herzegovina or Darfur, or made by feminists or nationalists or both, often distorts the understanding of rape that many feminists have long promoted' (Ibid., 224). Such a view, she maintains, highlights a crisis mentality that inevitably acquiesces in the extreme of violence as the solution: 'If crisis is the point at which harm is attended to, every harm must be made into a crisis to receive attention' (Ibid., 224).

Furthermore, Engle notes that, 'This focus on crisis both displaces and distorts attention to "the everyday", whether it be "everyday" killing, rape, hunger, or gross wealth disparity. It also reinforces a pre-realist understanding of intervention: imagining a world in which not acting militarily is "not acting", and refusing to see the ways in which many of the same powers that ultimately send in the troops often have played a significant role in creating conditions ripe for a crisis' (Ibid., 224). Citing the work of Marti Koskenniemi, she notes the importance of this elision and its distortive effects: 'What about the violence of a global system [...] in which "more than 30,000 children die every day of malnutrition, and the combined wealth of the 200 richest families in the world was eight times as much as the combined wealth of the 582 million people in all the least developed countries"' (Engle 2007, 224–25).[26]

Engle concludes by following Janet Halley and David Kennedy in their calls to 'take a break' from feminism and humanitarian intervention, respectively.[27] The call to take a break attempts to recognize and recalibrate the extent to which the ways of seeing conflict, through a gendered lens that, in its escalation, creates a crisis mentality around injury, implicates the observer in power deployments and forms of governance. Engle's critique of the gender modality is a story of the modality's elision of the law's violence, elided in the interests of making visible the woman harmed, but, she cautions, at cost.

The gender lens, whether as conflation or intersection, requires in the first instance a focus on the harm to the woman in the course of genocidal violence. This focus is described by Halley as a 'politics of injury' or the 'Injury Triad' (Halley 2006, 319–20). It is this politics, and the feminism that expresses it in the forms of gender analysis described above, from which Halley calls for taking a break. Halley's critique of the gender lens, similar to Engle's, is a story of that modality's elision of its own, inherent violence.

Injury politics, according to Halley, has three basic elements, a 'triad of descriptive stakes: women are injured, they do not cause any social harm, and men, who injure women, are immune from harm—female injury + female innocence + male immunity' (Halley 2006, 320). Halley goes on to note that 'Feminists often produce this triad as if it were feminism; and as if the three stakes were tied so tightly together that each requires assertion of the others. This is the crux of the contemporary politics of injury' (Ibid., 320). In the discourse on rape within the context of genocide, Engle sees a growing feminist 'consensus' around the link between rape as genocide and the call for military intervention. In effect, with the systemic nature of rape in the conflict in Bosnia, Rwanda and, more recently, in Darfur and the Democratic Republic of Congo (DRC), the bona fides of the Injury Triad appear normative.

Halley, however, suggests through a 'thought experiment' what might happen were one to depart from this gender model. Halley asks, in effect: what would happen if all the categories within the Triad were complicated, i.e. if gender were suspended, freeing sex and sexuality from its normative purchase on how we see an instance of abuse suffered by 'a woman' at the hands of 'a man'?

In the case of *Twyman v. Twyman*, Halley reads 'the "facts" of the case "against the elements of the cause of action for intentional infliction of emotional distress"—and then reread[s] them as if they were our best examples of non-feminist theories about morals, power, and sex that I derive from Nietzsche's On the Genealogy of Morals: A Polemic and Foucault's [History of Sexuality] Volume One' (Halley 2006, 348–63). The point of the exercise is to read the facts against the judgment, which corroborates the intuitive and 'normative' (Ibid., 23) sense of a powerless and innocent woman, an abusive man who, since he is the husband has 'immunity', and a rule of law that 'saves' the woman from the man. Nietzsche provides the moral template, and Foucault the political framework, for seeing a different side to the story.

In brief: Sheila, the wife, sued for divorce after an eight-year marriage in which initially she gave in to her husband's demands for sado-masochistic (S/M) sex but ultimately refused, saying that since she 'had been raped' before their marriage she could not endure the torture. William, the husband, then sought S/M sex elsewhere and, when he was caught, said that he wouldn't have strayed but for his wife's resistance. She relented one more time, and the incident was apparently bloody. She prevailed at court alleging that William 'intentionally and cruelly' imposed 'deviate sexual acts' on her (Halley 2006, 349).

Halley offers a detailed analysis of the different elements of the 'facts' the different judges focused on, some evincing a feminist cast of mind—for instance, Halley notes that the judges used 'a strange temporal locution—"the experience of having been raped"; "the trauma of having been raped"—that locates the moment of injury in a perpetual present'. Halley suggests that '[i]n much feminist rape discourse, this is exactly right. Once raped, always raped' (Ibid., 354). Then she asks whether feminism can 'accommodate a completely reversed image of the Twymans' marriage'. In the new schema, William is the one subordinated and Sheila has power, both moral and legal. 'Can feminism read the case as male subordination and female domination—and *still* as bad for women?' (Ibid., 356). Using Nietzsche, Halley charts the extent to which the moral power that Sheila is presumed to have under the actual ruling and the feminism that supports it is one that also causes her to suffer, over and over again: Sheila 'suffers terribly with every new access of subordinated sensibility', and feminism 'might be responsible not only for her power but also for the terrible suffering that grounds it' (Ibid., 359).

This form of top-down power is contrasted with that of Foucault's conception of modern sovereignty: Halley charts four central points about Foucault as she applies his theory to *Twyman*: first, that power is not 'a relation of dominance and subordination', but rather is a 'highly fragmented and temporally mobile "field of force *relations* [...]"'. Second, that 'power was not necessarily bad. It might be *pouvoir* (the capacity to create effects) rather than *puissance* (the capacity to dominate or coerce)'. Third, 'sexuality emerged historically as a discourse and produced as its effect the people we are—people who think their lives crucially involve knowledge of their deepest sexual selves'. Fourth, the principal paradigm of power was constitutive (Ibid., 360).

Reading *Twyman* through Foucault, Halley writes: 'One of the things that then immediately emerges is the intense, and formally almost identical, sexual pathos of both Sheila *and* William. Both are committed to the idea that they have deep, inner, injured sexual selves beyond which they cannot move one micron, and which they must enact with near-fatal completeness. William *must* live out the affliction of a perverse implantation, a deeply resisted fetishistic desire. He is a classic subject of the psychiatrization of perversions. Sheila *must* live out the affliction of rape trauma. Rape trauma is her deep inner truth, and her experiential life must make it manifest. In a terrible way, William and Sheila are perfectly matched to provoke the complete manifestation of their diametrically opposite desires; but oddly, this is because they are basically the same.' Hence, Halley's reading moves from a dominance/subordination framework to the struggle for power: 'There is both strength and danger in framing the possibilities described by these readings. Only if we articulate and explore them will we ever look into the world and see whether it matches them' (Ibid., 362–63, emphases in original).

Halley's thought experiment involves a domestic dispute between husband and wife, one that suspends the dominance–submission, victimizer–victim paradigm.

Can this be applied, with the same urgency as is the gender lens, to the systemic sexual violence that plays out not in a marriage but in a whole community in furtherance of a genocidal aim? Considering the intimate nature of the violence of genocide, where neighbour kills neighbour, the question gives pause. However, Marie Fleming, for one, thinks not; and with the same fervour of a rejection of law that she attributes to Foucault, she refuses Foucault's sexuality and biopower model at least as it applies to genocide. For Fleming, genocide is a special case that 'returns' the idea of the sovereign, however much within the modern era 'it' may be biopolitical (all about sexuality and the 'power of life'), to law, and returns law to blood. In the result, what Fleming's take on genocide, an ostensibly non-gendered view that none the less vindicates the gender analysis, shows is that beneath the gender lens lies a strong desire to embrace a mode of apprehension that, at least since the advent of modernity, has been in question.

Fleming is concerned to understand how it is that genocide comes about: 'I believe it's an urgent task facing all of us to try to figure out just what this repeatability of genocide turns on, and how unthinkable acts of genocide became possible in modernity' (Fleming 2003, 98). She examines various thinkers on genocide, from Habermas to Horkheimer, and essentially resolves the case according to a bipolarity: on the one hand, we have a Hobbesian conception of the sovereign and, on the other, a Foucauldian conception. Hobbes has articulated a theory of the juridical sovereign that is not tied to a 'King's head', i.e. top-down, but instead to the 'body politic' as the repository of the true sovereign. This understanding allows us to see genocide as 'a crime in modernity [that] reflects the status of the "body" of the people. As such, it stands in contrast to, yet extracts from, the traditional crime of regicide in which the body of the king was sacrosanct' (Ibid., 107).

For Foucault, on the other hand, the sovereign, represented by the state, involves a top-down theory. Foucault's theory 'turns away from the idea of sovereignty altogether' (Ibid., 104). Foucault sees power in 'relations, a more-or-less organized, hierarchical, co-ordinated cluster of relations' (Ibid., 105, citing Foucault 1980). In effect, 'The old power of death now gives way to the "administration of bodies and the calculated management of life"' (Fleming 2003, 105, citing Foucault 1976, Vol.1). Sex 'enters into history at the juncture of the "body" and the "population" [….] In the normalization processes of modernity, individuals get a sex and they are made visible […] but only in their abnormality, in their deviance from the norm'. As such, 'We have moved, according to Foucault, from a society of blood to a society of "sex". We are a society "with a sexuality"' (Fleming 2003, 105, citing Foucault 1976, Vol. 1).

For Fleming, the 'gender' of this sexualized, top-down analysis is fraught: 'Whereas [Horkheimer and Adorno] see Nazi anti-Semitism and other racial projects as crises of civilization, Foucault understands the great processes of modernity as a (homo)sexual event.' As such, she concludes that for Foucault, 'race and genocide, like everything else in modernity, should be fitted into the discourse of sexuality' (Fleming 2003, 106). However, sexuality and genocide just don't seem to fit: 'Genocide, he concedes, might well be the "dream" of modern states, but the reason for this has nothing to do with blood'. Rather, 'it is because power is situated and exercised at the level of life, the species, the race, and the large-scale phenomenon of population' (Fleming 2003, 106, citing Foucault 1980). With this, Fleming suggests that Foucault, in his 'utopian wish for a liberation of "bodies and pleasures"', attempts to repudiate every sort of law. This wish he generalized into a rejection of any attempt to understand power in terms of sovereignty. The law, he claims, 'always refers to the sword'. Foucault, therefore, advocated that we reject the law, i.e. 'cut off the King's head' (Fleming 2003, 107).

CONCLUSION

Fleming poses an interesting juxtaposition to the conflationists, the intersectionists and the critics of the gender model. In a sense, by returning the idea of genocide to the will of the sovereign as the body politic, her view is ostensibly non-gendered. Genocide is returned to 'race' or blood. Sovereign will calls up the law and, thereby, its exception, the latter inherent to a genocide wherein the body politic—the sovereign—gives itself over to the power of death over which the law rules but, in normal times, constrains. Genocide in this story has sharply delineated perpetrators and victims; it is saved from the ambivalence of the rule of sexuality—the power of life—in the midst of death. However, this story of the revived Hobbesian sovereign omits the extent to which blood is symbolic, constructed, an effect of power.

Therein, it seems, lies the fear and the investment inherent to any rigorously stabilized prism through which to see a large-scale, complex and often self-contradictory event such as genocide: the fear that it is the lack of sharp lines and clear delineations that underwrites its repeatability: 'The scandal of the Foucaultian discourse of sexuality is that the victims of the Nazi and other genocides become unintended consequences of a politics that issues from nowhere in particular and whose target and object is always sex, never race and ethnicity, not even in the case of the Jews murdered in the name of "Aryan" blood' (Fleming 2003, 113).

A similar conception underwrites the gender prism through which both the conflationists and the intersectionists see clearly before them the woman, injured. In that story of genocide, the assimilation of sexuality to gender is the first move in the return of gender to law as the means to 'save' the woman from the injury. However, the cost of juridical salvation (by the sword) is the denial of life—sexuality as life, in its complex and difficult variegation—to gender and, therefore, to law, whose power of death continues to enshrine its legitimacy.

NOTES

1 My thanks to Dr Yoke-Lian Lee for inviting me to participate in this project. My thanks also to Ms Kirsten Resnick for her wonderful research assistance and helpful conversations.

2 See Adam Jones, 'Genocide and Humanitarian Intervention: Incorporating the Gender Variable', at www.jha.ac/articles/a080.htm (accessed 9 November 2008); noting that the 'gender variable [...] one of the least-analyzed and most misunderstood elements of genocidal killing' requires a new framing as 'gendercide', i.e. gender-sensitive killing, with an emphasis on 'the genocidal or proto-genocidal targeting of males, especially "battle-age" men', and includes 'the demonization of "out-group" women' via 'abuses including rape and sexual assault' (Abstract).

3 Sjoberg and Gentry 2007, 145 (emphasis in original). In the article 'Gendered Genocide', which concerns the view of female genocidaires through a gendered lens, the authors note: 'If men and masculinities dominate and narrate genocide, then the hybrid roles that women and femininity play need to be taken account of *through gendered lenses* which recognize both where women are and the role that a gendered global political and social context played in getting them there.' It is 'about using gendered lenses to analyse not only female perpetrators but the genocidal war as a whole' (173, emphasis in original).

4 Halley 2006, 57. See also: 'By far the most brilliant and forceful thinker about sexuality in U.S. feminist legal theory for the last twenty-five years has been Catharine A. MacKinnon. Her formulation—which for short-hand I will call power feminism—has become the paradigmatic understanding of sexuality in sexual-subordination feminism in the United States' (27).

5 Halley 2006, 45, quoting MacKinnon in the journal *Signs*: 'If the literature on sex roles and the investigations of particular issues are read in light of each other, each element of the female *gender* stereotype is revealed as, in fact, *sexual*'. Furthermore, 'Women experience the sexual events these issues codify as a cohesive whole [...] The defining theme of that whole is the male pursuit of control over women's sexuality' (530–32).

6 Schneider cites Peterson and Runyan 1993.

7 Schneider cited Kelly D. Askin, 'Prosecuting Wartime Rape and Other Gender-Related Crimes under International Law: Extraordinary Advances, Enduring Obstacles', *Berkeley Journal of International Law* 21 (2003) 288, 296, at 305 nn.94–95 (citing Human Rights Commission Resolution S-3/1.L, 3d.sp. Sess., UN Doc. E/Cn.4/S-3/1.L (1994); S.C. Res. 935, UN SCOR, 49th Session, 3400th mtg at 11–12, UN Doc. S/935/1994 (1 July 1994)).

8 Schneider cited: Amnesty International, 'Darfur: Rape as a Weapon of War: Sexual Violence and its Consequences', AI Index: AFR 54/076/2004, 3.1 (2004), 94; Amnesty International, 'Lives Blown Apart: Crimes Against Women, in Times of Conflict', AI Index: ACT 77/075/2004 3 (2004), 16 (describing the stigma of rape and how survivors and their children 'are likely to be ostracized by their community').

9 Schneider cited Askin, 'Prosecuting Wartime Rape', 320.

10 Schneider cited Prosecutor v. Akayesu, Case No. ICTR-96-4, Trial Chamber Judgment, 731, 732, 692–94 (2 September 1998).

11 Schneider continued: 'The Rome Statute did not define rape and the landmark *Akayesu* Trial Chamber Judgment was the first of either Tribunal to do so.' Schneider reports that the indictment with 12 counts of international crimes excluded gender-related violence, but a woman judge with 'extensive expertise in international human rights law and gender-related crimes, was on the bench'. Furthermore, 'The compelling witness testimony supplemented by international pleas to include sexual violence in the charges against Akayesu, resulted in the trial being adjourned so that the office of the Prosecutor could investigate charges of sexual violence and consider amending the indictment to include such charges if warranted' (934, cited Prosecutor v. Akayesu, Case No. ICTR 96–4-T, Judgment, 416 (2 September 1998)); see also Prosecutor v. Akayesu, Amicus Brief Respecting the Amendment of the Indictment and Supplementation of the Evidence to Ensure

the Prosecution of Rape and Sexual Violence within the Competence of the ICTR, May 1997 (prepared by Joanna Birenbaum, Lisa Wyndel, Rhonda Copelon and Jennifer Green).

12 Prosecutor v. Delalic, Case No. IT-96-21-A, (Judgment 20 February 2001); Prosecutor v. Furundzija, Case No. IT-95-17/1-T (Judgment 10 December 1988). Schneider notes that in 176–84 the 'added specificity and addition of body parts narrows the Akayesu definition' (Schneider 2007, 163).

13 Otto Triffterer (ed.), *Commentary on the Rome Statute of the International Criminal Court: Observers' Notes, Article by Article* (Baden-Baden: Nomos Verlagsgesellschaft, 1999), 117–18 (hereafter cited as *Commentary on Rome*). Article 7 reads: '7(1) For purposes of this Statute, "crime against humanity" means any of the following acts when committed as part of a widespread or systematic attack directed against any civilian population, with knowledge of the attack: (a) Murder; (b) Extermination; (c) Enslavement; (d) Deportation or forcible transfer of population; (e) Imprisonment [...] (f) Torture; (g) Rape, sexual slavery, enforced prostitution, forced pregnancy, enforced sterilization, or any other form of sexual violence of comparable gravity; (h) Persecution against any identifiable group or collectivity on political, racial, national, ethnic, cultural, religious, gender as defined in paragraph 3, or other grounds that are universally recognized as impermissible under international law, in connection with any act referred to in this paragraph or any crime within the jurisdiction of the Court; [...] (3) For the purpose of this Statute, it is understood that the term "gender" refers to the two sexes, male and female, within the context of society. The term "gender" does not indicate any meaning different from the above.'

14 *Commentary on Rome*, 173–75: war crimes, grave breaches, and per 8(2)(b) 'Other serious violations of the laws and customs applicable in international armed conflict, within the established framework of international law, namely, any of the following acts', includes at sub-(xxii) 'Committing rape, sexual slavery, enforced prostitution, forced pregnancy, as defined in article 7'.

15 *Commentary on Rome*, 172, 128. Citation in text: Implementation of the outcome of the 4th World Conference on Women, Report of the SG, UN Doc. A/51/322 (3 September 1996), paragraph 9 refers to the Report of the Fourth World Conference on Women (UN Doc. A/CONF. 177/20 (1995)).

16 Richey continues that such an equation, as part of 'how international legal developments regarding war rape during genocide and ethnic cleansing campaigns may be counterproductive for women's human rights'.

17 Richey cited Myra Medible, 'Dominance and Submission in Postmodern War Imagery', *Peace Review* 55, 60 (2005), 17.

18 In the instant case, the prosecution of war crimes, crimes against humanity and genocide for Darfur.

19 Wagner joins with MacKinnon, Richey and others in calling for the assimilation of rape and torture: 'Because of the historical treatment of rape in the course of warfare as a tolerated practice, many may be skeptical of using torture law as a means of prosecuting rape' (237).

20 Citation of Human Rights Watch and Federation Internationale des Ligues des Droits de L'homme, 'Shattered Lives: Sexual Violence During the Rwandan Genocide and its Aftermath' (1996), 16.

21 De Londras cites K. Crenshaw, 'Beyond Racism and Misogyny: Black Feminism and 2 Live Crew', in M. Matsuda, C. Lawrence, R. Delgado and K. Crenshaw (eds), *Words that Wound: Critical Race Theory, Assaultive Speech, and the First Amendment* (Boulder, CO: Westview Press, 1993), 111.

22 De Londres also cites Green for the proposition that, 'The theory of intersectionality is now beginning to gain currency in the context of genocide law as well' as within the context of race discrimination (120).

23 De Londras 2007, 123: 'This decision [i.e. Kunarac at the ICTY] reflects an awareness on the part of the Tribunal of how gender and ethnicity might coincide, and could therefore be said to be an intersectional analysis.'

24 De Londras cites D. Cooper, *Challenging Diversity: Rethinking Equality and the Value of Difference* (Cambridge: Cambridge University Press, 2004), 47–49.

25 Citing television commercial: *How Will History Judge Us?* (CNN television broadcast October 2006), www.savedarfur.org/pages/advertising_campaign; arguing that history will judge us if we allow rape to continue without our intervention.

26 Citation of M. Koskenniemi, '"The Lady Doth Protest Too Much": Kosovo, and the Turn to Ethics in International Law', *Modern Law Review* 65:159, 172–73.

27 'Imagine an international humanitarianism which took a break from preoccupation with the justifications for "intervention". Which no longer imagined the world from high above, on the "international plane", in the "international community"' (Engle 2007, 226, citing David Kennedy, *The Dark Side of Virtue: Reassessing International Humanitarianism* (2004), 351).

Mainstreaming gender into the African Union peace and security agenda

TIM MURITHI

INTRODUCTION

This chapter will assess the extent to which the African Union (AU) has mainstreamed gender equality in its peace and security agenda. It will begin by developing a conceptual framework to articulate the case for incorporating a gendered analysis of the peace and security challenges in Africa. The chapter will then examine the challenge of gender-based violence in conflict situations on the African continent. In particular, it will discuss how Africa has witnessed the confluence between gender, violence, and war in the Democratic Republic of the Congo (DRC) and Darfur. The chapter will assess some of the key international instruments for promoting gender equality, with a specific focus on UN Security Council Resolution 1325 which stipulates the role that women should play in promoting peace and security. This UN Resolution 1325 established the premise upon which the African Union developed its own Declaration on gender equality with a specific focus. The chapter will then briefly assess the efforts that the AU has undertaken to mainstream gender into its peace and security institutions and agenda. In particular, this chapter will argue that the AU gender mainstreaming initiatives have been largely top-down and there is a need for it to take more concrete steps to actualize its normative claims. Specifically, it will assess some illustration of efforts of the limited impact of the AU's norms on its member states, as a result of a lack of an explicit commitment to implementing initiatives to address gender concerns. Ultimately, the paper will propose how further to entrench gender mainstreaming into the work of the AU's conflict prevention and peace-building agenda.

GENDER ANALYSIS AND THE PEACE AND SECURITY CHALLENGES IN AFRICA

The nature of conflicts after the end of the Cold War demonstrate that men and women play a range of roles in war situations (Etchart and Baksh 2005). It is, therefore, necessary to utilize a gendered lens when analysing the implications of these roles (Whitbead 2004). Across the world, men and women in conflict situations can and do simultaneously play the role of perpetrators and victims. However, trends tend to demonstrate the deliberate targeting of women in war situations. Fowzia Musse, for example, has examined the war crimes that have been perpetuated against women and girls in Somalia (Musse 2004). Turshen and Twagiramariya (1998) have also documented the deliberate and premeditated targeting of women in the Rwanda genocide of 1994. Therefore, there is no parity of suffering but rather a discriminate and asymmetrical degree of torment that is endured by women in conflict situations. Indeed, Karin Koen (2006, 3) notes that 'in extreme situations, women have been targets of war, and have been regarded as part of the spoils of war'. Africa is not unique as far as these issues of gender and conflict are concerned. The continent's societies remain afflicted by the continuing impact of gender inequality in the socio-economic and political sphere, which translates into the perpetuation of insecurity for women in conflict and post-conflict situations (Juma 2005). The Mission Statement of the Beijing Platform for Action following the Fourth Conference on Women, in 1995, notes that 'grave violations of the human rights of women occur, particularly in times of armed conflict, and include murder, torture, systematic rape, forced pregnancy and forced abortion, particularly under policies of ethnic cleansing' (United Nations 1995, 11). The case for ensuring a gendered analysis of the peace and security challenges in Africa cannot be understated.

What do feminist perspectives say about the necessity of a gendered analysis of peace and security challenges? Why is a gendered assessment of conflict in Africa necessary? Cynthia Cockburn (1998, 8) suggests that 'feminist work tends to represent war as a continuum of violence from the bedroom to the battlefield, traversing our bodies and our sense of self. We see that the "homeland" is not, never was, an essentially peaceful unitary place'. Heidi Hudson (2006, 2) argues that 'the high incidence of violence against women in Africa can be attributed to an interconnected range of cultural/religious, economic, political, military, and criminal factors'. These lead to a situation in which 'in the name of tradition, different moral standards often apply to men and women' (Ibid.). Indeed, this fact is not unique to Africa but is equally prevalent in other parts of the world. Mary E. King argues that there is 'a new perspective gaining recognition [that] asserts that the empowerment of women is the only way to achieve peace' (King 2005, 31). Ultimately, according to R.W. Connell (2001), it is necessary to adopt 'democratic gender relations' which are premised on the 'move towards equality, non-violence, and mutual respect between people of different genders, sexualities, ethnicities and generations'. The Beijing Platform of Action recognizes that the full participation of women 'in decision-making, conflict prevention and resolution and all other peace initiatives is essential to the realization of lasting peace' (United Nations 1995, 23). Before assessing the efforts that the African Union has taken to fulfil this goal, the next section will assess in more concrete terms some of the challenges relating to gender-based violations in conflict situations in Africa.

THE CHALLENGE OF GENDER-BASED VIOLENCE IN CONFLICT SITUATIONS

The prevalence of gender-based violence and the use of rape in particular as an instrument of war continues to be a major challenge on the African continent (Meintjes et al. 2001). In the ongoing conflict in Northern Uganda, it is estimated that thousands of women have been victims of rape (Ochieng 2003). The rape of women and children in DRC continued despite the efforts of the transitional national government to fight against sexual violence (Action Alert 2004). The World Health Organization's (WHO) Health and Reproduction Programme noted that since the war re-ignited in August 1998, there have been 25,000 cases of sexual violence recorded in South Kivu Province, 11,350 cases in Maniema Province, 1,625 cases in Goma, the capital of North Kivu Province, and some 3,250 cases in Kalemie, a town in south-eastern DRC. A non-governmental organization (NGO) working on the ground, the Nouvelle Dynamique Pour la Jeunesse Feminine, has observed that for the most part, women who have been victims of rape have been rejected by their communities, and many of their husbands no longer want anything to do with them. This leaves the women without an adequate support system and makes them reluctant to denounce the crimes that have been committed against them in order to avoid rejection by their community (Schroeder 2004). On top of that, the possibility of contracting HIV/AIDS from infected combatants is relatively high.

In Darfur, sexual and gender-based violence has been an ongoing violation of women's rights since the war begun in 2003. For example, on 30 November 2004, seven female internally displaced people (IDPs), one of whom was pregnant, were attacked by an armed militia group near the Deraij camp which was 4 km east of the town of Nyala, in the Southern Darfur State. Specifically, 'the seven women and girls were fetching firewood outside the camp where they were reportedly attacked, beaten with guns on their chests and heads, and stripped. The armed militia later took three of them to an abandoned hut where they were raped' (Butegwa 2006). According to Butegwa, 'women and girls are still vulnerable to rape whenever they venture out of the IDP camp in search of water or firewood'. This situation prevails in 2008, in the absence of a comprehensive peace agreement between Darfur and the Government of Sudan. According to Jame Lindrio Alao, a psychologist with the Amel Centre for Treatment and Rehabilitation of Victims of Torture, based in Darfur, 'the majority of the perpetrators are allegedly affiliated directly or indirectly with the Government' (Butegwa 2006). To compound the situation for the victims, rape is not yet recognized as a crime by the national laws of Sudan. Even if a woman claims that she was raped it is more likely that that charge will be changed to one of assault. In addition, the archaic and discriminatory laws in Sudan require four male witnesses to corroborate the charge of rape. This situation is clearly unpalatable for the women and girls who are exposed to the indiscriminate and random violence of the war situation in Darfur. There is self-evidently the need for an overarching framework to compel the Sudanese Government to protect the rights of these women and girls even in a conflict situation.

This is a challenge that the African continent as a whole needs to address, because the situation for women and children in all war-affected areas is precarious. As Lucinda Marshall observes, there is a connection between militarism on the continent and the threat to innocent women and children. According to Marshall (2004, 1), 'the theory of power over an "other" provides the common thread between military campaigns and assaults against women. What this theory says is that it is allowable for a person, ethnic group, government, etc. to get what they want by way of power over an other'. For Marshall (2004, 1), what is observed in most conflict situations is that:

> [...] whether implicitly or explicitly, women are the "other". Consequently, it becomes necessary in the eyes of those who seek power over to control and belittle women, and all aspects of womanhood. In many cultures, women are viewed as the possessions of their men. Therefore, when a woman is raped, it is effectively an attack on the manhood of her man. Using this reasoning, women become the targets of war in order to attack the honor of the men of a particular culture, ethnic group or country. For these reasons, rape and other forms of sexual assault against women are always a part of war and conflict. When women are assumed to be possessions that can be attacked, stolen and dishonored, they become a means of feminizing and degrading the enemy.

Rape is in effect a war-related crime (Sideris 2001). In this regard, the International Criminal Court (ICC) has also defined sexual and gender violence, of all kinds, as war crimes. However, progress is yet to be made by the ICC in investigating these crimes and protecting witnesses and victims. Many women in the DRC, Darfur, Somalia, Northern Uganda, and victims of previous conflict situations in Sierra Leone and Liberia are still waiting to see those who committed these acts punished (Twagiramariya 1998). It is, therefore, vital to indeed consider gender-based violence as a war crime to enforce the necessary international treaties and instruments to ensure the protection of the rights of women and bring justice (Handrahan 2004). The next section will assess some of these instruments, focusing on instruments established by the UN and the AU.

INTERNATIONAL INSTRUMENTS IN THE PROMOTION OF GENDER EQUALITY IN PEACE-BUILDING

The international community has recognized that women are disproportionate victims of the effects of war, yet they have to a large extent been historically excluded from playing a significant role in making, keeping and building peace. On 20 December 1952, the UN General Assembly adopted the Convention on the Political Rights of Women, which outlined their basic rights of political participation including the prerogative 'to exercise all public functions, established by national law, on equal terms with men, without any discrimination' (United Nations 1952). This would by extension include all public functions pertaining to participating in peace processes within the country. On 18 December 1979 the UN General Assembly adopted the Convention on the Elimination of All Forms of Discrimination Against Women which provides 'the basis for realizing equality between women and men through ensuring women's equal access to, and equal access in opportunities in, political and public life' (United Nations 1979). Countries that have ratified or acceded to the Convention are legally bound to put its provisions into practice.

On the specific issue of peace processes, the UN took the issue further with the issuing of UN Security Council Resolution 1265 on the Protection of Civilians in Armed Conflict, on 17 September 1999, which emphasized 'the importance of including in the mandates of peacemaking, peacekeeping and peacebuilding operations special and assistance provisions for groups requiring particular attention, including women' (United Nations 1999, 13). It also requested that the Secretary-General ensure that UN personnel involved in peace operations had the necessary training in 'gender-related provisions' (United Nations 1999, 14). On 31 October 2000 the UN Security Council passed its landmark Resolution 1325 on Women's Inclusion in Peace Processes, which specifically reaffirmed 'the important role of women in the prevention and resolution of conflicts and in peacebuilding and [...] the importance of their equal participation and full involvement in all efforts for the maintenance and promotion of peace and security' (United Nations 2000). This document was significant in the sense that it explicitly codified the universal responsibility of all member states of the UN to ensure that gender issues were mainstreamed in peace and security efforts around the world. Resolution 1325 also explicitly recommends 'the urgent need to mainstream a gender perspective into peacekeeping operations', as well as the necessity to provide 'Member States [with] training guidelines and materials on the protection, rights and the particular needs of women, as well as on the importance of involving women in all peacekeeping and peacebuilding measures' (United Nations 2000, 6). These requirements in effect lay the foundation for subsequent initiatives by sub-regional organizations such as the AU, which adopted its own home-grown declaration on gender equality.

THE AU INSTRUMENTS ON GENDER EQUALITY IN THE PROMOTION OF PEACE AND SECURITY

Article 4(1) of the Constitutive Act of the African Union that formally established the organization, in 2002, adopted as one of its principles 'the promotion of gender equality' (African Union 2000). However, it was only two years later, in 2004, that the AU held its first debate on gender issues at its Annual Assembly of Heads of State and Government, which took place on 6 July, in Addis Ababa, Ethiopia. The specific lobbying and advocacy on the adoption of an AU position on gender equality was undertaken by a number of NGOs co-ordinated by the Senegalese-based women's group Femmes Africa Solidarité (FAS) and the Kenyan-based African Women's Development and Communication Network (FEMNET). This civil society coalition undertook a substantial amount of lobbying work behind the scenes, which laid the foundation for the Assembly of Heads of State and Government to adopt the *AU Solemn Declaration on Gender Equality* (African Union 2004). This turned out to be perhaps a productive and transformative civil society partnership with the AU (Wendoh and Wallace 2005). This Declaration acknowledged the precedent set by the UN Conventions and Resolutions discussed above and noted that 'while women and children bear the brunt of conflicts and internal displacement, including rapes and killings, they are largely excluded from conflict prevention, peace negotiations and peace-building process in spite of African women's experience in peacebuilding' (African Union 2000). The Declaration states that the AU will actively work to accelerate the implementation of gender equality in all its activities. Specifically, the Declaration emphasized that the AU would 'ensure the full and effective participation and representation of women in peace processes including the prevention, resolution, management of conflicts and post-conflict reconstruction in Africa as stipulated in UN Resolution 1325' (African Union 2000, 2). In addition, it committed the member states of the Union to 'initiate, launch and engage within two years (of the signing of the Declaration) sustained public campaigns against gender-based violence' (African Union 2000, 4). The Declaration committed the organization also to implement legislation to enable women to own land and inherit property, improve literacy among women, and generally mainstream gender parity in all spheres of its social, economic and political activities.

The AU has also recognized the importance of upholding the rights of women through its *Protocol to the African Charter of Human and People's Rights Relating to the Rights of Women in Africa*, which was adopted on 11 July 2003, at the Union's Summit in Maputo, Mozambique. Specifically, the Protocol states that 'women have a right to a peaceful existence and the right to participate in the promotion and maintenance of peace' (African Union 2003, Article 10). The Protocol also calls upon the member states of the AU to 'take all appropriate measures to ensure the increased participation of women [...] in programmes of education for peace and a culture of peace' (African Union 2003, Article 10, 2a). The Protocol calls upon 'States Parties [to] undertake to respect and ensure respect for the rules of international humanitarian law applicable in armed conflict situations, which affect the population, particularly women' (African Union 2003, Article 11). It further obligates 'States Parties [to] undertake to protect asylum seeking women, refugees, returnees and internally displaced persons, against all forms of violence, rape and other forms of sexual exploitation, and to ensure that such acts are considered war crimes, genocide and/or crimes against humanity and that their perpetrators are brought to justice before a competent criminal jurisdiction' (African Union 2003, Article 11). The Protocol also legislates for equal pay for equal work and establishes affirmative action to foster the equal participation of women in public office. The Protocol also legislates against female genital mutilation and promotes medical abortions in specific instances.

INSTITUTIONAL STRUCTURES TO PROMOTE GENDER ISSUES WITHIN THE AFRICAN UNION

The AU's gender mainstreaming initiatives have been largely top-down and there is a need for it to take more concrete steps to actualize its normative claims. In terms of the institutional framework to oversee the implementation of these provisions, the Chairperson of the AU Commission relies on advice from an African Union Women's Committee. As far as the governance of the AU in terms of its decision-making structures is concerned, the body has adopted a gender-parity principle with regards to the composition of its Commission, which is the highest executive organ of the Union. Specifically, five of the 10 AU Commissioners are, and will be, women. The organization has also established an AU Directorate for Women, Gender and Development within the Office of the Chairperson of the Commission to oversee the implementation of all its provisions relating to gender mainstreaming on a range of socio-economic and political issues, including peace and security. The AU has established an African Trust Fund for Women, to develop capacity to respond to the Union's initiatives related to promoting gender equality. The AU still has genuinely to focus on the implementation of these provisions in order to manifest its commitment to ending the siege under which women are affected by war and are excluded from participating in the promotion of peace and security.

CASE STUDIES OF GENDER MAINSTREAMING

The AU's preventive diplomacy and mediation institution, the Panel of the Wise, is composed of five distinguished elder statesmen and stateswomen. The Panel of the Wise is a contemporary rendition of the traditional institution of the Council of Elders. However, the Panel is not a direct or an authentic replication of a traditional council format in its authority and remit. Specifically, the AU Panel diverges from the traditional Council of Elders model in one important respect: in the sense that it includes prominent and distinguished women and has, therefore, adopted a gender sensitive re-interpretation of the traditional council of elders. These councils were effectively dominated by men in the majority of traditional settings. There are, of course, an exceptionally small minority of cultures that ascribed a prominent role to women, particularly in matters of war and peace. This innovation of the Panel is important and upholds the AU's stated rhetoric on promoting gender sensitivity. However, it is also vital in order to signify to African societies that there are women of all backgrounds and levels of expertise who are playing a vital role in the promotion of peace on the continent. To date, the Panel members have been involved in preventive diplomacy initiatives relating to Somalia, the Central African Republic (CAR) and South Africa.

In January and February 2008, the then Chairman of the AU Assembly of Heads of State and Government, former President of Ghana, John Kufuor, appointed a former leader of the Mozambican liberation movement, Graça Machel, as a member of the mediation team to respond to the post-electoral violence in Kenya. Machel worked with Kenyan women's networks to set up a process where Kenyan women played a proactive role in post-conflict peacebuilding in the country (Foundation for Community Development 2008, 3).

African civil society has also undertaken the initiative to ensure the implementation of gender mainstreaming in the work of the AU, through the 'Gender is my Agenda' campaign, co-ordinated by a network of NGOs including Femmes Africa Solidarité (Femmes Africa Solidarité 2005). The campaign has called for gender mainstreaming in the peace initiatives being undertaken by the AU. Concrete initiatives have included a campaign to address the plight of the suffering in the Darfur conflict, particularly women and children. In this regard, the African Women Consultation on Darfur was convened between 24 and 25 January 2008, in Addis Ababa, to raise awareness on gender issues pertaining to the conflict, including gender-based violence in the region. The consultation produced the African Women's Declaration on Darfur, which was used as a tool for advocacy initiatives at the 2008 African Union Summit. Indeed, in the margins of the 10th AU Summit in Addis Ababa, in January 2008, approximately 60 women from Darfur lobbied the organization on their plight, notably their living conditions and the violence that they are confronted with as a consequence of the conflict. The AU did not issue a robust statement on Darfur at the Summit; however, it referred to its ongoing commitment to resolving the dispute through the initiatives of the Joint AU and UN Mission in Darfur and the efforts of Djibril Bassolé the Joint UN-AU Special Envoy to Darfur.

Ultimately, the impact of the AU's norms on its member states is not exclusively a result of a lack of an explicit commitment by the organization to address gender concerns at a grassroots level.

MOVING BEYOND THE RHETORIC: THE NON-IMPLEMENTATION OF AU PROVISIONS ON GENDER MAINSTREAMING IN PEACE AND SECURITY

On the basis of the Declaration on Gender Equality and the Protocol on Women's Rights, it is evident that in terms of policy and at a theoretical level the AU has articulated a commitment towards ensuring the well-being and the protection of the equal rights of women. The AU has articulated and placed the rights of women at the centre of the continent's peace and security agenda, in effect recognizing that women are key agents in enabling the continent to transcend its multiple conflict situations. However, this policy is yet to be matched by practice and concrete initiatives to mainstream gender equality, specifically on peace and security issues (Puechguirbal 2005). Specifically, not all aspects of the Declaration on Gender Equality and the Protocol on the Rights of Women are being upheld and implemented in a systematic manner. The reality does not converge with the AU's rhetoric as stipulated in its Declaration and Protocol. Women continue to be excluded from high-level peace negotiations and they are under-represented as Special Envoys of the AU to conflict situations. While there have been increasingly more women serving in AU peace operations, for example in Darfur, they do not occupy a significant number of the leadership positions in these missions. Due to its own internal political dynamics, and administrative and bureaucratic inefficiency, the AU is as yet unable to support capacity development for women across the continent. The AU is, therefore, no different from other intergovernmental organizations in terms of the lackadaisical approach that it takes to implementing its Declarations and Protocols.

Even though the AU has adopted the Declaration on Gender Equality and the Protocol on Women's Rights, and five of the organizations' Commissioners are women, there is still a reluctance genuinely to operationalize the key tenets of gender equality, as well as internalize the principles of UN Security Council Resolution 1325. This is partly due to deeply held cultural beliefs and practices, and the fact that the leadership of the AU is composed of men, and the majority of decision-makers at the level of the Assembly of Heads of State and Government, the Executive Council of Ministers and the Permanent Representatives Committee of Ambassadors are still overwhelmingly male. If there is no genuine political will to implement the principles of the Solemn Declaration and UN Security Council Resolution 1325, faced with such asymmetrical odds in terms of numbers, even the most persistent advocates of gender mainstreaming within these leadership structures, which remain at their core a male fraternity, would find their efforts frustrated (Cockburn 1998). Hudson (2006, 3) suggests that 'most African governments have failed to integrate women into policy formulation, partly as a result of a lack of understanding of gender issues and how to translate these into policy and also because of a reluctance on the part of male power-holders to lose or share deeply entrenched privileges'. The former Chairman of the Assembly of Heads of State and Government for the period 2004–05, the erstwhile President of Nigeria, Olusegun Obasanjo, noted that the majority of African countries, like many other parts of the world, still hold negative attitudes towards women and excluded them from taking part in social, economic and political affairs that affect them. In Africa's current conflict situations, women continue to suffer from the brutality of gender-based violence, and the phenomena of girl soldiers as well as girl-child sex slaves persist. In effect, AU gender mainstreaming at the moment does not sufficiently affect the lives and concerns of ordinary women on the continent. The AU Commission conducted a Gender Audit of its approach to women's empowerment in its policies and programmes and concluded that management 'needed to do more to achieve the AU's commitments and objectives on gender equality' (African Union 2006, 2). Therefore, it is necessary to shift the focus from viewing policies and legal instruments as the solution to social problems on the continent, to analysing the interaction between the policies and the dominant interpretation of social problems.

GENDER MAINSTREAMING AND THE PROMOTION OF STRUCTURAL PEACE IN AFRICA

The United Nations Educational, Scientific and Cultural Organization (UNESCO) Statement on Women's Contribution to a Culture of Peace notes that 'only together, women and men in parity and partnership, can we overcome obstacles and inertia, silence and frustration and ensure the insight, political will, creative thinking and concrete actions needed for a global transition from a culture of violence to a culture of peace' (UNESCO 1995). In addition, the Statement proclaims that 'there can be no lasting peace without development, and no sustainable development without full equality between women and men' (Ibid.). According to Koen (2006, 7), 'if women are to play an equal part in security and maintaining peace, they must be empowered politically and economically, and represented adequately at all levels of decision-making: at the pre-conflict stage, during hostilities and at the point of peacekeeping, peacebuilding, reconciliation and reconstruction'. Specifically, 'in times of conflict and the collapse of communities, the role of women should thus be seen as crucial to preserving the social order. Women also have important roles to perform as peace educators in their families and communities' (Koen 2006, 7). The UNESCO Statement echoes this view when it argues that:

> Women's capacity for leadership must be utilised to the full and to the benefit of all in order to progress towards a culture of peace [...] in such areas such as conflict prevention, the promotion of cross-cultural dialogue and the redressing of socio-economic injustice, women can be the source of innovative and much needed approaches to peacebuilding.

Concretely, the UNESCO Statement is recognizing that without the active participation of women it would be impossible to establish any form of durable peace. Women, therefore, occupy a vital position in terms of achieving the promise of structural peace across the world and Africa is no exception. Structural peace in this instance refers to the dimension in which socio-political structures and institutions are utilized to maintain and sustain peace. As far as Africa is concerned, women's role in the social dimension of society is not in doubt; however, there is a need for a significant paradigm and normative shift to ensure that women play a central role in the socio-political structures and institutions to ensure that structural peace entrenches itself on the continent (Ekiyor 2004). In terms of practical steps to achieve this, the UNESCO Statement calls for the need to take genuine steps to:

- Promote relevant quality education that imparts knowledge of the human rights of men and women, skills of non-violent conflict resolution;
- Encourage new approaches to development that take account of women's priorities and perspectives;
- Seek to reduce the direct impact of the culture of war on women—in the form of physical and sexual violence;
- Promote knowledge and respect for international normative instruments concerning the human rights of women and girls;
- Support national and international efforts to ensure equal access to all forms of learning opportunities, with a view to women's empowerment and access to decision-making;
- Support governmental and intergovernmental structures as well as women's associations and NGOs committed to the development of a culture of peace based on the equality of women and men (UNESCO 1995).

CONCLUSION

This chapter made the case for the adoption of a gender lens in assessing the requirements for sustained peace and security in Africa. The fact that women and girls are increasingly the victims of indiscriminate gender-based violence in conflict situations, as evidenced in the crisis situations in DRC and Darfur, suggests that at the very least they should also be agents in the promotion of peace. The international system has issued a range of UN Conventions and Resolutions which are unambiguous on the central role that women should play in the political, social and economic life of their countries. Specifically, UN Security Council Resolution 1325 was significant in that it stated the need for women to play more significant roles in peace and security across the world. The AU referred to Resolution 1325 when it issued its own Solemn Declaration on Gender Equality, which stipulated a range of recommendations on the role that women should play in promoting peace processes. The AU's Protocol on Women's Rights further outlines vital Pan-African legislation to ensure women's active participation and protection from war crimes. However, it is clear that despite the range of AU political and legal statements on gender equality, their implementation has been cosmetic. While there is a gender parity principle that ensures that half of the AU's Commissioners will always be women, concrete changes in terms of the inclusion of women in peace processes is yet to be genuinely implemented. Specifically, women continue to be marginalized in playing a key role in peace negotiations, whether as part of AU mediation teams, or Special Envoys of the Union, or even as senior representatives of the delegations of member states. Some commentators have attributed this to deeply held negative attitudes towards women that are prevalent equally within the institutional structures of the AU as well as among the governments of its member states. Confronting the impact of gender inequality and continued gender-based insecurity in African societies remains an aspiration rather than a reality. The inclusion of women in efforts of peace-making, peace-keeping and peace-building initiatives would contribute towards redefining gender relations in the promotion of peace and security in Africa.

Sovereignty, subjectivity and human rights: a gender sketch[1]

YOKE-LIAN LEE

INTRODUCTION

This chapter arises from my experience as an academic international lawyer with an interest in human rights law and political feminism. It responds especially to one question that has been of concern to me for a long time: how can legal measures that are often said to be—and in many respects are—progressive in relation to the claims of women to be free and equal subjects seem to work in practice so as to sustain women as unfree and unequal? This problem could, of course, simply be framed in popular terms as the problem of the 'glass ceiling', or as a condition in which rising tides do not in fact make all boats rise higher, but such claims are too simplistic. The literature of international law and human rights law, and particularly women's human rights law does not provide any satisfactory answers to this paradoxical situation. Rather, it tends to treat the law as given, and pays little attention to the inconsistent and often contradictory character of the international legal order in its response to claims about women's oppression—inconsistencies and contradictions that create new sites for the subtle or perhaps 'structural' subjugation of women.

Consequently, one might turn to a range of contemporary theory, namely feminist theory, critical theory, critical/feminist legal studies, international political theory, postmodern and poststructuralist theory, post-colonial theory and so forth. Some of these writings argue that much of the explanation for the subordination of women lies in the international legal doctrine of statehood, derived from the principle of state sovereignty that comes to be deployed as disciplinary power in the production of the modern subject/individual. Michel Foucault's reading of the discursive construction of the modern subject is especially illuminating, and I will use this, alongside Judith Butler's idea of the performative construction of subjectivity, to develop my theoretical framework.

On this basis, the first part of the chapter will focus on the performative conception of identity formation as a discursive practice that produces the normative understanding of sex, gender and women, and especially the individual inscribed as 'woman'.[2] This will lead to an exploration of the gendered nature of international human rights law in which I look at state sovereignty as a performative force in the categorization of the legal subject. This chapter, thus, seeks to 'deconstruct' the legal subject and human rights law. My task is to highlight and pose questions about the repressive sovereign discursive discourses and mechanisms at work in the legalisation of the rights of women.

THE FEMINIST SUBJECT: A BRIEF REVIEW

As an initial point of orientation towards a vast and potentially unmanageable literature, I will now review feminist literature on identity. Through a reading of feminist writing on the conception of 'women' as the subject of theorizing, I will suggest that while there are many potentially useful feminist discourses, three main areas of contemporary debate stand out as most appropriate for this project because they directly address the problem of the subject. First, there are those feminist theorists who call for collective gender identities for women—represented in diverse ways by the work of Catherine MacKinnon and Seyla Benhabib. For MacKinnon, women's identities are not theirs because the female identity is entirely constructed and defined by a patriarchal culture. She argues for the recognition of women as a class or, rather, as an oppressed social class that is different from men. The most important difference between the sexes is the difference in power. Men as a class have the power to dominate women thereby creating and substaining inequality. MacKinnon argues that sexual equality must be constructed on the basis of women's difference from men and that we need to abandon traditional approaches that take maleness as their benchmark. Owing to the unequal power relationship between men and women, MacKinnon calls for legal reforms from the perspectives of women to put an end to the inequality in power and gender blindness (MacKinnon 1983). Benhabib takes a similar line, arguing that women must be united under 'new collective identities' to foster a more inclusive and unified category as 'women'. She further insists that it is absolutely crucial for women to speak and affirm as women under a stable and coherent identity in order 'to become members of a polity' (Benhabib 2002, 357). Only then can women become autonomous agents able to assert and legitimate gender-specific political and legal claims. Benhabib rejects alternative conceptions of female identity, especially those advocated by poststructuralists, and argues that the 'deconstruction' of gender identity would leave women without actual substance and essence to ground 'real' and concrete feminist politics (Benhabib 1989).

Poststructuralist conceptions of gender identity presented by Butler and Foucault form the second set of literature relevant to the current project. Such works are typically committed to the abandonment of a unified and coherent subject and argue that the seemingly free subject already expresses 'the effects of power' in which s/he is also, as Butler argues, 'one who is subjected to a set of rules or laws that precede the subject' (Meijer and Prins 1998, 286). This is what Foucault describes as a 'regulatory ideal'. Central to Foucault's argument is the claim that subjectivity is constructed by power. Power for

Foucault is a crucial precondition for the construction of truth, knowledge and subjectivity. Discussing the construction of the sexual subject, Butler complements Foucault's argument when she writes that the notion of 'sex' not only functions as a norm, but is part of a regulatory practice that produces the bodies it governs' (Butler 1990, 2).

Feminist critical legal theorists such as Drucilla Cornell, Naffine Ngaire and Carol Smart constitute the final group which informs this chapter. They use deconstructive techniques to challenge the power of law and language in the construction of gender identity by drawing attention to the assumptions about feminine identity underlying the conception of womanhood. Cornell calls for a 'deconstructive intervention' in the reaffirmation of the feminine (Cornell 1992b, 280; Cornell 1993). What she proposes is the reinterpretation of the dominant meaning and framing of the feminine, with a particular focus on the 'performative power of language, in and through which gender identity is constituted' (Cornell 1992b, 287). This, she argues, would foster the transformation of gender norms that value masculinity and devalue femininity. Naffine raises the question of what approach feminists should take to the articulation of women as the subject of law. She argues that feminist strategies of pushing woman-specific legal reforms to adopt a more inclusive conception of personhood and to permit and encourage sexual difference is a flawed approach, because women, as the subjects of law, are already tainted by gendered assumptions (Naffine and Owens 1997). Smart also casts doubt on the usefulness of law for emancipatory purposes for women by rejecting a rights-based approach as inappropriate to feminist political activism. Smart cautions against an overemphasis on law, as for her, law functions as a disciplinary discourse or as a system of knowledge. Following Foucault's conception of power and truth, Smart argues that law's power lies in its authority to define what constitutes truth and knowledge, which is then deployed to disqualify other discourses, especially feminist discourse. As Smart further argues, it also follows that we cannot predict the outcome of any legal reform: once the law is enacted, legislation will be governed by state institutional rules, processes and procedures. In this respect, resorting to law and legal reforms, by bringing more law into our everyday life, often ends up empowering the state more than individuals (Smart 1989).

THE CULTURAL AND HISTORICAL CONSTRUCTION OF SEX AND GENDER

Having offered a general survey of the key literature on the subject formation of women, I now turn to address the specific question of gender. Like many other feminist scholars, Joan W. Scott uses 'gender' as a category of historical analysis, defining it as a constitutive element of social relationships based on perceived differences between the sexes, and as a primary way of signifying relationships of power (Scott 1999). She focuses on the ways in which the meaning of gender is constructed. For Scott, gender norms work to suppress and constrain the exercise of freedom by men and women. To explore the contingency of the historical and social construction of gender differences, and to analyse the gender dimensions of society, is to interrogate the claim that gender is the constitutive element in social relations in specific cultural contexts. Scott suggests that earlier gender analyses have revealed that differences between the sexes in terms of social practices focus particularly on the public/private distinction. Historically, at least in liberal theory, women have been associated with the private sphere of the home, while men have been associated with the public sphere of work and politics. However, Scott cautions against universalizing the traditional line of demarcation between the public and private arguing that such a reading tends towards an ahistorical account of gender. She calls instead for the historicization of the public/private division and advances a complex, multifaceted account of gender. For Scott, it is difficult for us to understand gender without employing a historical analysis of the political, social-economic and psychological aspects of the subordination of women. The crucial question that needs to be posed here is this: what are the political and legal implications when women are effectively excluded and naturalized in all significant social relations?

However, before embarking further on this enquiry it is necessary to examine some key terms that are problematic and highly contested. It is not my intention here to give them concrete meanings, but rather to explore their theoretical significance and to interrogate how and in what ways the deployment of meanings within these terms can be utilized and determined for specific political and legal purposes. The crucial point is not to take the meanings for granted, but rather to contest the dominant meanings given by hegemonic forces to generate analyses and critiques as forms of feminist intervention.

'One is not born, but rather becomes, a woman' (Beauvoir 1973, 301): Simone de Beauvoir's renowned statement effectively tried to sever the link between sex and gender. She asserts that what we 'become' is not given by nature. We are not only culturally constructed, we also partake in constructing ourselves—we become a woman or a man. However, while it seems as if we are given the 'choice' to 'become' what we are, this 'choice', according to Butler, 'comes to signify a corporeal process of interpretation within a network of deeply entrenched cultural norms. Gender becomes the corporeal locus of cultural meanings both received and innovated' (Butler 1986, 129). Women are 'Others', as Beauvoir argues: they are defined by a masculine point of view which aims to uphold its own identity through identifying women generally in terms of feminized norms. In this sense, woman becomes the Other because the 'self' is always already dependent on male identity (Butler 2001, 20).

As mentioned earlier, according to Butler the category of 'sex' is from the very beginning a normative term, it functions as a 'regulatory ideal'. Central to Butler's argument is the claim that gender identity is a regulative ideal infused with political and disciplinary objectives. In effect, it is a political operation that constrains and creates gender identity by the exclusion of the other—an exclusion that effectively constitutes and naturalizes sexual difference (Butler and Scott 1992, xiv). Building on these insights, we can therefore see that the foundationalist view of 'sex' or gender identity always results in the constitution of particular meanings for specific ends. For example, Cornell argues that gender hierarchy is a system; sex is given to us by that system and not by a pre-given biological reality. No doubt, biologically women usually do not have male genitals; however, 'it is the meaning given to that fact within the system of the gender hierarchy that continually re-inscribes the de-valuation of women as the castrated other' (Cornell 1992, 78).

THE LEGAL CONSTRUCTION OF SEX AND GENDER

Gender analysis decentres biological explanations that have effectively dominated the way gender relations are understood. The concept of gender helps us better understand and examine meanings that are imposed on bodies, which thus become the presumed location of sexual difference. To deconstruct and problematize the materiality of bodies is to banish the epistemological certainties that have been imposed on it (Butler 1992, 17). The materiality of bodies emerging out of the process of constant reiteration of regulatory norms, of course, has legal as well as political significance. To locate the body or a person within a legal context, Naffine and Owens argue that the legal person is sexed or constructed by dominant cultural norms and assumptions because 'law has always assumed and constituted a subject who is deemed to act in certain ways [....] And law has engaged in this act of creation quite self-consciously, fully aware that it is constituting a subject' (Naffine and Owens 1997, 7).

Feminist engagements with law must first focus on the analytic category of gender in order to reposition the female body as a legal subject. In this way, it helps to reveal the complex and contradictory nature of law which is informed by deeply held dominate (familial) ideology and gendered assumptions. Social and political structures are built on gender hierarchies that produce the gendered legal subjects. Dominant legal conceptions of rationality and truth, in the main, have rendered women's experiences as irrational and unreliable. As Nicola Lacey argues of the effects of law, they 'constitute one of the powerful social practices which in fact produces sexed bodies of particular kinds: not just meaning given to, but also the actual shapes, powers and capacities of human bodies, in their sexual and other sphere of being' (Naffine and Owens 1997, 68). This invokes the enterprise of 'sexing the subject of law' (Ibid.). At the same time, it should be noted that legal discourse is also a site where legal meanings given to women as subjects of law can be reconceptualized, reimagined to foster multiple meanings and to destabilize pregiven meaning of women's identity. In this way, law's sexing of its subjects should be understood as an ongoing, historically contingent process (Naffinc and Owens 1997, 68).

Questioning the subjectivity of women in the field of law is vitally important for feminism. Butler makes this point explicitly: 'The question of "subject" is crucial for politics, and for feminist politics in particular, because juridical subjects are invariably produced through certain exclusionary practices that do not "show" once the juridical structure of politics has been established' (Butler 1990, 2). In order for the law to make representational claims on an individual's behalf, it is essential, as a matter of general principle and also practically, that a legal founding subject must be in place in order for legal procedures to take effect. It is only then that legal rights can be elaborated and legal provisions provided to the subject before the law.

In this sense law has a vital role in constructing its subjects as sexual beings and thereby in asserting the ways in which sexed bodies have particular social meanings (Naffine and Owens 1997, 67). As Lacey argues: 'the repression of the feminine has served the illusion of a unitary and self-identical subject of law'. Moreover, 'the structure of legal subjectivity has depended precisely upon a binary difference of two genres, one of which has been subordinated to the other to the extent of total invisibility: hence the illusion of legal neutrality' (Naffine and Owens 1997, 67). Thus, 'law can envisage woman as the "other" to man, the paradigm subject of legality' (Thornton 1995, xiv). This inevitably leads to questions about the hegemonic force of law; that is, the force of 'law [that] constitutes one of the powerful social practices which in fact produces sexed bodies of particular kinds' (Naffine and Owens 1997, 68).

Until now, I have been addressing the relationship between subjectivity and law in the specific context of struggles over assumptions about gender identities that are expressed in law and in our normative understanding. The question of gender identities provides a way of analysing the subject that has the potential to resist what Foucault describes as 'juridical' or 'sovereign' power. This encourages us to explore the *practice* of sovereignty—how it works and to what extent it has helped to sustain and legitimize a gendered account of international human rights law and, more importantly, what is at stake within a feminist critique of such law.

I now wish to explore what turning to international human rights law might mean for 'women' if we simply take the principle of sovereignty for granted. I am especially concerned with posing questions about the limits of international human rights law in general, and women-specific rights law in particular, and I will do so by examining the ways in which the institutionalized practices of the principle of sovereignty work to enable, sustain and legitimate a gendered conception of international human rights law.

THE QUESTION OF SOVEREIGNTY: SUBJECTIVITY AND INTERNATIONAL HUMAN RIGHTS LAW

Sovereignty has been analysed as norm, as fact, as practice, as belief, as value, as principle, as a neutral concept and as the foundation of modern politics. Indeed, there are numerous contexts in which the concept of sovereignty could be discussed—sociologically, historically, politically, psychologically, anthropologically and juridically. Some discuss sovereignty in a general historical context as an abstract theoretical construct; others, especially international lawyers, 'claim to be talking about sovereignty, while in fact it appears that the subject is statehood or legitimate authority' (Ponzio 1994, 73).

Sovereignty has many contradictory meanings, giving specific expression to the binaries of inside and outside, structure and process, particularity and universality, identity and difference, and space and time. It is both the picture and the frame, and is divided in its loyalties between subject and object.

The dominant realist and legal positivist conception of sovereignty has important consequences in modern political and legal thought and practice, especially the familiar reification of the sovereign nation-state. The state is here given a concrete identity that privileges nationhood as a particular form of human attachment. Global politics is understood as 'international relations'—relations between reified sovereign states—and international law is presented as 'the law of the nations'. However, while the dominant reading of sovereignty went largely unchallenged for much of the Cold War era, there has recently been 'a fundamental explosion of scholarly interest in sovereignty', which casts doubt both on the fixed content of sovereignty and on the assumptions that stem from it: assumptions that have deeply affected the contemporary reading of international law and politics (Biersteker and Weber 1996, 1). Despite countless scholarly attempts to find answers to the question of sovereignty, it is clear that there is no consensus 'as to how the question of sovereignty might or should be answered' (Ashley and Walker 1990, 367). Sovereignty remains contested territory: 'a question, a problem, a contingent political effect whose production, variation, and possible undoing merit the most rigorous analysis' (Ibid., 368).

Yet, the claim to a coherent sovereign voice is crucial for the legitimacy of any political project because:

> [...] a coherent sovereign voice [...] supplies a unified rational meaning and direction to the interpretation of the spatial and temporal diversity of history [that is] in itself regarded as a pure and originary presence—an unproblematic, extrahistorical identity, in need of no critical accounting.
> (Ashley and Walker 1990, 361)

The 'coherent sovereign voice' described by Ashley and Walker gives the principles of sovereignty the 'quality of authoritativeness, the source, or test, or surety of their authoritative nature' (Boyle 1985, 128). The foundations of modern politics and international law are thus conceived and constituted through notions of sovereign power and authority without which claims to legitimacy cannot be grounded. Foundation is always a promise; but it requires one to go beyond legal argument. My aim is to uncover the political and contingent nature of the sovereign state and international legal principles, and specifically the law's function in reifying and justifying social, political, economic and gender inequality. For example, Koskenniemi argues that:

> … the normative issues of international politics through a formal Rule of Law approach will very rapidly show itself as an unsatisfactory argument strategy. Arguments from legal principles are countered with arguments from equally legal counter-principles. Rules are countered with exceptions, sovereignty with sovereignty. Refusing to engage in sociological enquiries about causes and effects and to assess political weights to be given particular arguments, the lawyer can only find himself in an argumentative deadlock
>
> (Koskenniemi 1990, 485)

Hence, 'the very attempt to treat sovereignty as a matter of definition and legal principle encourages a certain amnesia about its historical and cultural specific character' (Walker 1993, 166). The principle of sovereignty takes the nation state as the ultimate authorization of legitimate power and authority within and without a given territory. As Walker puts it: 'The modern principle of state sovereignty has emerged historically as the legal expression of the character and legitimacy of the state' (Ibid., 165). Thus it follows that 'once state sovereignty is defined as a centring of power/authority within a given territory, the way is open for emphasis on other things, like justice and law, freedom and social progress' (Ibid., 169).

The notion of 'politics', at least in its modern form, is organized around the principle of sovereignty within a bounded sovereign territory. In this political space, there exists a sovereign political order and a sovereign individual subject (Edkins et al. 1999, 1–18). Traditionally, conventional international law primarily treats the sovereign state as the primary subject, while universal human rights at the same time assert that individuals are also sovereign subjects. The core problem this gives rise to is the constant 'tensions between power and authority and between sovereign state and sovereign people' (Walker 1993, 170). In the context of international human rights law, these tensions need to be resolved, at least for practical purposes. The liberal solution to this contradiction lies in the principle of democracy, because modern democratic constitutionalism gives rise to 'the precise democratic procedures that might permit some convergence between sovereign state and sovereign people' (Ibid., 170).

Individuals as subjects of rights, in a practical sense, are able to 'exercise their rights' only when states grant human rights to individuals subject to their jurisdiction. Under general international law and, in particular, human rights law, states are under an obligation to take responsibility for the enactment of specific laws and procedures to protect individuals from harm and also to enforce the law for specific remedial actions. However, it should be noted that such procedural law tends to reduce the power and subjectivity of the individual. As a subject before the law, the particularized individual becomes a limited, normative legal subject. Additionally, when procedural law comes into play, state institutions effectively have the power of defining what those rights are. According to Foucault, 'the exercise of power has always been formulated in terms of law' (Foucault 1976, 87), and in this sense, we have not 'escape[d] from the limited field of juridical sovereignty and state institution' (Foucault 1980, 102). Foucault suggests that law and sovereignty in their modern forms not only constitute the sovereign juridical system, but have remained the central component of state power despite the demise of the 'monarchical state'. Hence, in Foucault's words:

> In political thought and analysis we still have not cut off the head of the king. Hence the importance that the theory of power gives to the problem of right and violence, law and illegality, freedom and will, and especially the state and sovereignty [….] To conceive of power on the basis of these problems is to conceive it in terms of a historical form, that is characteristic of our societies: the juridical monarchy.
>
> (Foucault 1976, 88–89)

According to Hunt and Wickham, when Foucault refers to 'the juridical monarchy', he does not mean it in the literal sense of 'monarchs', rather as a form of 'unitary constitutionalism' (Hunt and Wickham 1994).[3] Consequently, in its modern form, the relationship between human rights law and state sovereignty is that, on the one hand, human rights law seems to have the power dramatically to restrict state sovereignty, while on the other hand, its existence seems completely dependent on the power of state institutional structures and procedures to enable, define, limit, sustain and legitimate the realization of those rights in practice. As Kennedy observes, 'human rights strengthens the national governmental structure and equates the structure of state with the structure of freedom. To be free is […] to have an appropriate organised state' (Kennedy 2000, 11).

The duty of states is to ensure that individuals are able to exercise their fundamental human rights entitlements under international law through 'proper' constitutional and institutional mechanisms by means of national legislation and effective legal procedures. However, the legal system, as Foucault and other critical legal scholars argue, always carries with it techniques of marginalization and subordination. Consequently, the law gains the discursive power to normalize, naturalize and depoliticize human subjectivity. In terms of institutional practices through which the principle of sovereignty has been politically inscribed, it has the potential effectively to disable those who are dominated by the mechanisms of law. As a result, individuals as the subject of international human rights law are subordinated under the idea of a single global order based on the principle of sovereignty, and other multiple subjectivities that do not fit within this order become marginalized, distorted or even erased. Ultimately, it seems that law can be read as a kind of sovereign with the power to give or withhold subjectivity. The juridical subject formation of state and individual as international persons in international law is the creation of law, and law thus exercises its power to construct the personality of its subjects. International law's legal framework, based on the principle of state sovereignty, further legitimizes the place of the state as the most important personality in the global order of things. While it seems that human rights law gives the individuals formal equality and inalienable sovereign rights that can readily be interpreted in terms of the universality of human rights, the centrality of the sovereign state in the international domain in fact continues to constrain the growth of other global personalities. The state, thus, continues to be the most powerful subject in international law with the power and authority to continue to define what constitutes human rights violations within the framework of international conventions.

A FEMINIST CRITIQUE OF HUMAN RIGHTS LAW

International law in general and human rights law in particular has been subjected to extensive feminist critique. For example, Charlesworth, Chinkin and Wright have employed feminist legal theory to argue that the structures of international law-making and the content of the rules of international law are gendered—they privilege men while women's interests are largely ignored, and if their interests are acknowledged at all, those interests are marginalized.[4] Central to the feminist critique of the rule of law is that its very conceptualization serves as an instrument of oppression and domination. Slapper and Kelly explain: 'The law is a major vehicle for the maintenance of existing social and power relations [....] The law's perceived legitimacy confers a broader legitimacy on a social system [...] characterised by domination. This perceived legitimacy of the law is primarily based [....] on the distorted notion of government by law, not people' (Slapper and Kelly 2000, 22).

So, to put it bluntly, is law on balance good for women? If not, why not? According to Smart, 'law is powerful in silencing the alternative discourse of women, [but] is far less powerful in transforming society to meet the various needs of all women' (Smart 1989, 81). Often, the law and legal systems have not protected the interests of women effectively. When women have to turn to law to redress their grievances, anticipating that the law and the legal system will deliver equality and justice, they find that the law remains insensitive to many of their concerns (Charlesworth and Chinkin 2000, ix). It thus fails to provide the protection it promises them.[5] For a long time, law has failed to legitimate the claims made by women; this is especially so in areas such as rape and domestic violence (Gibbs 1994, 5). As Susan Gibbs argues: 'the legal system remain[s] permeated by outmoded but nevertheless deeply entrenched attitudes concerning the roles and status of men and women in society. The inherent conservatism of legal institutions, and respect for precedent and established categories, helped to perpetuate the underlying gender bias of the law' (Ibid., ix).

While the position of women within the law has undergone some striking improvements, especially in recent decades, and although the modern liberal state, as Brown argues, may not intentionally set out to oppress women, the gendered structure of liberalism, partially determined by the gendered character of prerogative power, continues to be expressed through the bureaucratic dimension of the state. Its institutions and the organizational structure, its processes and activities serve to legitimate arbitrary state power in policy-making in multiple, diverse, unsystematic and complex ways. Foucault characterizes this form of power as a micro-physics, which like a 'network runs through the whole social body' (Foucault 1980, 119). We can link Foucault's ideas on power to the relationship between multiple dimensions of socially and culturally constructed modes of gender and state power. While the central constitutive principle that modern political power derives from state sovereignty could be understood as a gendered practice,[6] the power of the sovereign state is not always expressed in a coherent way, but in multiple 'disciplinary' forms. As Brown suggests, state power 'is expressed in tangible institutions as well as discourse: bureaucracy's hierarchicalism [and] proceduralism' (Brown 1995, 177).

Feminist critics argue that if law is a gendering practice, then the campaigns for women's legal rights, or indeed legal reform, are at best of limited value and at worst can be positively detrimental to women. The talk of 'women's rights' assumes an essentialist female subject, albeit significant differences among women. This assumption enables contingent social structures to seem unchanging and undermines an appreciation of alternative political possibilities. Given that the formulation of women's rights is based on the principles of equality and non-discrimination, such rights are not only individualistic in nature, and their application of the norms of equality and non-discrimination between men and women focuses on minimum standards of equality in the *public* sphere, usually in terms of political participation, equal pay, institutional access to employment and education. However, women live lives in multiple sites and their concerns and interests may not easily be translated into this narrow framework. Moreover, women's subordination is wide ranging and is not confined to the public sphere. In this sense, rights discourse overly simplifies complex power relations and its promises may be undermined by the structural and institutional inequalities of power (Pateman and Crosz 1986, 192). Adding women's 'voices' to the existing status of international human rights law 'as a part of a broader pattern of processes and structures' does not fundamentally alter the nature, structure and content of international law because international law primarily addresses interstate relations rather than dealing directly with any category of individual persons (man/woman/minority/group). Hence, the usefulness of international law in general and human rights law in particular with regard to the emergence of women's human rights, as Charlesworth and Chinkin suggest, is somewhat limited:

> Although a specialised area of women's human rights law is evolving, and occasionally women are acknowledged in mainstream international law, by and large, whenever women come into focus at all in international law, they are viewed in a very limited way, chiefly as victims, particularly as mothers, or potential mothers, and, accordingly, in need of protection.
>
> (Charlesworth and Chinkin 2000, 48)

International human rights legal instruments in the form of treaties or conventions are the primary sources of 'implicit and explicit principles, norms and rules and decision-making procedures' that are legally binding on all member states that have ratified them (Krasner 1982, 182). The actual practice of integrating women's human rights into the UN human rights system, therefore, remains in the hands of the international community of states. Notwithstanding the series of international human rights conferences of the 1990s[7] and the fact that states 'developed, and agreed to, significant [steps towards] the advancement of women', Riddell-Dixon argues that 'women's rights remain marginalised—outside the central operation of the international human rights regime' (Riddell-Dixon 1999, 150). This is not least because 'women's rights [are] much more complex and [are] generally avoided, except for the purposes of ritual incantation when proved necessary or expedient' (Alston 1994, 379). The fear of 'rocking the boat' often observed in the traditional canon of human rights creates a tendency to adopt a more attractive alternative which, as Alston explains, is 'simply [to] avoid addressing the difficult issue—to cling to the status quo for fear that any attempt at reform or any reopening of the decisions already taken would only threaten achievements which have been so hard won' (Ibid., 380).

The principles of state sovereignty are firmly entrenched in international law, which in turn has served to stabilize and legitimate the traditional foundations of international law where its authority is deeply embedded in treaty law and process doctrines. By and large, therefore, it would seem that mainstream human rights discourse, including women's human rights, has largely internalized the logic of state sovereignty. Although international human rights norms may seem to limit state sovereignty and to prevent derogation, the doctrine of *jus cogens* results in the return of the problem of state sovereignty. As Charlesworth and Chinkin point out:

> The human rights norms that are typically asserted to constitute *jus cogens* are the prohibition of genocide, slavery, murder/disappearances, and the right to life, torture, prolonged arbitrary detention and systematic racial discrimination. All these violations of human rights are of undoubted seriousness but the silences of the list indicate that women are peripheral to the understanding of fundamental community values. For example, prohibition of sex-based discrimination is not generally understood as a basic norm.
>
> (Charlesworth and Chinkin 2000, 120)

Moreover, although it appears that the international community has accepted human rights norms, the gap between theory and practice remains wide. As Knop points out, even though 'states consent to be bound by [human rights norms,] the acceptance of women's rights of non-discrimination or the right to be free from gender-based violence as *jus cogens* would not ultimately alter state sovereignty' (Knop 1994, 154).

International norms can only be established through state practice, but under treaty law all multilateral treaties come with a 'claw back' clause where states have the right to make a reservation, objecting to certain provisions of the treaty when they become parties to that treaty.[8] Thus, article 29 of the Women's Convention states that: 'a State Party has the right to make a reservation in accordance with paragraph 2 of this Article'. It is worth mentioning in this context that the Women's Convention is the treaty that has the most reservations made by states parties. According to Amnesty International:

> [...] many of the reservations entered by countries are contrary to the spirit and purpose of the Convention. Some of the reservations are so wide ranging that they are difficult to review or challenge. Others are based on conflicts between Shari'a law and CEDAW, although other countries in the region have not made similar reservations, suggesting that different interpretations of Shari'a exist. National legislation has also been used as a basis for reservations, although such a use of national legislation is clearly proscribed by international law.
>
> (Amnesty International 2004)

Hence, even in its own terms the universalization and internationalization of women's human rights law has been only partially effective, and has often been frustrated by the powers concealed by the principle of state sovereignty.

Smart argues that certain essential characteristics of law make it too limited in its potential for radical change. Liberalism demands a continuing practical engagement with law reform, but a critical examination of law reveals its 'central role in reproducing aspects of women's oppression' (Smart 1986, 110). The source of women's oppression derives from a number of factors, and social and institutional structures based on patriarchal practice obviously remain one of the main obstacles to change. Any particular legislation tends to solidify at a particular point in time. Law itself, therefore, often presents a barrier to further reform because existing enforcement structures tend to assimilate women's rights within a male-dominated structure. Cornell clearly understands this liberal institutionalist approach, arguing that liberal feminist legal struggles simply reproduce the worst of patriarchy's notions of identity, gender, authority and power (Cornell 1992a).

If the sovereign state is an historically constituted entity that is infused with inherent contradiction at all levels, then it cannot be a neutral mediator in social relations. The state 'may legitimately be seen as the initiator of important dynamics and as a place where interests are constituted as well as balanced', and 'the various components of the state encounter the same pushes and pulls, blurring out boundaries, and the possibility of domination by social organisation' (Peterson 1992, 3). Derrida takes the institutional organization of law as the 'force of law', claiming that the role of 'law is always an authorised force, a force that justifies itself or is justified in applying itself, even if this justification may be judged from elsewhere to be unjust or unjustifiable' (Derrida 1992, 5). Exposing the internal contradiction of the rule of law is accordingly an essential part of a feminist strategy aimed at undermining vested power relations.

SOVEREIGNTY, SUBJECTIVITY AND LAW: THE FEMINIZED VICTIM-SUBJECT

While acknowledging that women have traditionally been excluded from much of international law, to maintain that women should be included does not fundamentally alter the international legal order. Moreover, the gender bias embedded in particular social structures often assumes the form of norms which seemingly remain beyond the reach of international law. When supposedly gender-neutral international human rights law is applied, it is often infused with contradictions, discriminatory to women, and riddled with practical and legal problems. In this sense, the potential usefulness of women's international human rights law is severely circumscribed. The notion of placing one's trust in women's international human rights law to deliver justice to women is at best utopian and at worst a form of implicit co-option in the perpetuation of women's oppression. These international women's human rights represent women as a specific category of victim-subjects, often subjugate other subjectivities and almost always reinforce gender and cultural essentialism. The international women's human rights movement is characterized by its primary focus on the issue of violence against women and their victimization. This category of 'women' as victim-subjects presupposes that there is a collective identity derived from a single group with common experiences, attributes and oppression as victims. Postcolonial feminists have criticized Western forms of feminism, notably radical/liberal feminism for its false universalization of female experience (Spivak 1988; Mohanty 1991; Kapur 2002, 2005). The attribution of the label of victim mostly functions to legitimize claimants' political, legal and social status, and in practice the label of victim is the straightforward translation of any feelings of social difficulties into a ready-made category that casts women as paradigmatic victim-subjects.

To address women's oppression (as well as men's), we need to address many current social issues (in particular race, gender, sex, class and sexual orientation) and acknowledge that they are inextricably interconnected. Historically, legal discourse has been unable to recognize the alternative claims of women—claims that are not framed in human rights language—let alone to remedy their problems. The failure to acknowledge the multiple ways in which women are constructed limits the ability of legal discourse to offer a solution. Legal discourse will only respond to claims that are authorized, acknowledged and empowered. Changing legal discourse means transforming the ways that claims are made, authorized, acknowledged and communicated in the field of law.

As subjects of law, women are unrecognized as anything other than a figure within legal discourse. As Butler argues, the legal discursive construction of the subject fails to take account of the differences between women, and gender identity is a regulative ideal that is both political and disciplinary in nature and is, in effect, a political operation that constrains and constitutes gender identity by the exclusion of the other (Butler and Scott 1992, xiv). As Kapur argues, 'feminist politics in the international human rights arena has promoted this [helpless] image of the authentic victim subject while advocating for women's human rights [...] has reinforced the image of the women as a victim subject, primarily through its focus on violence against women (VAW). The focus on the victim subject in the violence against women project reinforces gender and cultural essentialism in the international women's human rights law' (Kapur 2002). Feminist legal politics, therefore, is forced to construct an essentialized subject—and the notion of this subject precedes the actual human persons who come before the Court.

Can an individual be a subject of law? On this question Boyle argues that 'the subject is loaded up, consciously or unconsciously, with a particular set of qualities or attributes' (Boyle 1991). Women, especially from the post-colonial world, are cast as either oppressed victims, incapable of decision-making, or helpless objects. According to Boyle, the casting of women as the paradigm victim-subject has meant that 'de-personalised subjects rely on their supposed universality for their epistemological and rhetorical utility. But a truly universal subject is, by definition, contentless. Self-interest is an empty term, until you have defined what a self is and the kind of things it is interested in. [It follows that] legal interpretation will not, cannot, be "objective", it must come from some unanswerably authoritative subject' (Boyle 1991).

At this juncture, it becomes clear that global feminism has unwittingly allowed itself to be co-opted by the structure and substance of human rights law. Moreover, the assumption that a 'transcendental' subject is required to give authority to the law has further helped to foreclose other forms of contestation. Such a categorization of women becomes a hegemonic move that serves to normalize and universalize the subjectivities of 'women' in formulating international women's human rights discourse without sufficiently taking on board the critique of gender essentialism. Consequently, international women's human rights law in the end creates new sites for the subtle subjugation of the very subject such law aims to help. Part of the problem is that the very category of international women's human rights law ultimately rests on deeply entrenched forms of inclusion and exclusion and, as such, legitimates the subjugation of other subjectivities in the name of an essentialized female subject of law (Engle 2005). Unfortunately, then, it appears that the women's human rights project does not offer a sufficiently complex understanding of the ways in which women's lives and experiences are mediated by ethnicity, religion, class/caste, gender and sexuality. Consequently, as Brown argues, the inscription of gendered identity as having distinguishing attributes in legal discourse depoliticizes those identities and can reaffirm historical injustices (Brown 1995, xi). Hence, the eradication of all forms of violence against women in and through human rights law has, as Kapur argues, 'deprived human rights discourse of some of its political character, for law is a discourse where power relations are obscure and political claims are decontextualised and deradicalised'(Kapur 2006, 110).

CONCLUSION

While acknowledging the harmful effects of the exclusion and marginalization of women, feminists' efforts in casting 'women's rights as human rights' can, at times, offer the marginalized and oppressed female subjects a powerful language to challenge and contest their exclusion and marginalization. Violence against women is obviously harmful to women, and, according to legal logic, 'if there is an injury there should be a remedy' (MacKinnon 1985, 170). The Committee on the Elimination of Discrimination Against Women, makes it clear in its General Recommendation 19 (11th session, 1992), paragraph 9 that, '[u]nder general international law and specific human rights covenants, States may also be responsible for private acts if they fail to act with due diligence to prevent violations of rights or to investigate and punish acts of violence' (UN Doc. HRI/GEN/1/Rev. 6 at 243, 2003). Hence, states are under an obligation to take responsibility for the enactment of specific laws and procedures to protect victims of domestic violence and to enforce laws to punish the offender effectively. As indicated earlier, the sovereign state is the most powerful of all persons recognized in international law having the authority to make, interpret and enforce international law, and as international law structures women's international human rights, the system of sovereign states hence underpins women's human rights law. International human rights are built on the positivist international legal order.[9] As a branch of international law, human rights law shares the statist principles of sovereignty, territorial jurisdictional divisions, non-intervention and consensual norms expressed in traditional international law. Under positivist law, the source of human rights is to be found in the enactments of a system of law sanctioned by the sovereign state. Hence any global discussion of remedial action or measures to deliver universal values such as justice, freedom and democracy would certainly be ineffectual without states partaking in their implementation at the domestic level (Kennedy 2000). However, as we have seen, this has its considerable limits.

The question of gender identities provides a valuable methodological framework for analysing a way to think about the subject that at least has the potential to resist the temptation to neutralize the legal subject. By examining the role of law in the constitution of the subjectivity of the sovereign state and gender identities enables us to pose a range of questions about the limits of international law in general and human rights law in particular, thereby enabling us to understand the ways in which the practice of the doctrine of sovereignty works and to what extent it has helped to sustain and legitimize a gendered account of international human rights law and what is at stake with a feminist critique of such law.

NOTES

1 My thanks to Patrick Thornberry, R.B.J. Walker, T.J. Lustig and John Horton for their helpful comments.
2 'Woman' cannot be pinned down to a given 'essence'.
3 Broadly includes terms such as: the Crown; Parliamentary sovereignty; the President or the Constitution.
4 See, for example, Charlesworth et al. 1991; and Charlesworth and Chinkin 2000.
5 See, for example, McColgan 2000.
6 For further details see Walker 1992, 179–202.
7 The 1990 Summit for Children, the 1992 Conference on the Environment and Development, the 1993 Conference of Human Rights, the 1994 Conference on Population and Development, the 1995 Summit for Social Development, the 1995 Conference on Women, and the 1996 Conference on Human Settlements.
8 See *Vienna Convention on the Law of Treaties* (1969), Section 2 Reservation.
9 According to Akehurst, during the 17th century, especially after the death of Grotius, the conception of law was largely based on positivism, 'that is, man-made; consequently, law and justice were not the same thing, and law might vary from time to time and from place to place, according to the whim of the legislator. Applied to international law, positivism (as this new theory was called) regarded the actual behaviour of states as the basis of international law'. See, Peter Malanczuk, *Akehurst's Modern Introduction to International Law*, seventh revised edn (London: Routledge, 1997), 16.

A–Z Glossary
Gender and politics

Compiled and edited by Yoke-Lian Lee

Abortion

Abortion demonstrates perhaps better than any other moral issue the impossibility of segregating 'personal' from 'social' ethics, morality from law, or religious from secular moral argument.

In ethical discourse, abortion is understood as the deliberate choice to terminate a pregnancy through an action that either directly destroys the foetus or causes its expulsion from the uterus before viability. Such a choice is obviously a highly personal moral action. In abortion a person or persons cause the death of another individual member of the species (although whether the foetus is likewise a 'person' is a hotly contested and crucial point in the abortion debate). At the same time, moral views of abortion are closely entwined with social concerns about sexuality, **gender** and family, and about the social institutionalization of various forms of homicide. The relative influence of these social factors in shaping evaluations of abortion has varied historically as well as culturally.

Historical variation no less characterizes Christian approaches to abortion. While until recent decades the Christian churches had always condemned abortion, they did so for different reasons and in different degrees, and theological opinion has been offered in support of different exceptions to the general prohibition. In contemporary Christianity, the social and political emancipation of women and their struggle for sexual equality have created strong challenges to the traditional presumption against using abortion as a means of avoiding motherhood. The conflict over gender, sexuality and abortion in the public sphere has also led to debates about the proper relation between law and morality, and about the legitimacy of the involvement of religious bodies in policy formation.

Lisa Sowle Cahill in Paul Barry Clarke and Andrew Linzey (eds), *Dictionary of Ethics, Theology and Society* (Routledge, 1996). Religion Online. Taylor & Francis. www.routledgereligiononline.com:80/Book.aspx?id=w004 (accessed 15 October 2009).

Absence/presence

A first definition: cinema makes absence presence; what is absent is made present. Thus, cinema is about illusion. It is also about temporal illusion in that the film's narrative unfolds in the present even though the entire filmic text is prefabricated (the past is made present). Cinema constructs a 'reality' out of selected images and sounds.

This notion of absence/presence applies to character and **gender** representation within the filmic text and confers a reading on the narrative. For example, an ongoing discourse in film on a central character who is actually off-screen implies either a reification (making her or him into an object) or heroization of that character. Thus, discourses around absent characters played by the young Marlon Brando, in his 1950s films, position him as object of desire, those around John Wayne as the all-time American hero. On the question of gender = presence, certain genres appear to be gender-identified. In the western genre, women are, to all intents and purposes, absent. We 'naturally' accept this narrative convention of an exclusively male point of view. However, what happens when a western is centred on a woman, for example Mae West in *Klondike Annie* (Raoul Walsh 1936) and Joan Crawford in *Johnny Guitar* (Nicholas Ray 1954)? Masquerade, mimicry, cross-dressing and gender-bending maybe, but also a transgressive (because it is a female) point of view—absence made presence.

Another definition: woman as absence (as object of male desire), man as presence (as perceiving **subject**). The woman is eternally fixed as feminine, but not as subject of her own desire. She is eternally fixed and, therefore, mute.

Susan Hayward, *Cinema Studies: The Key Concepts* (London and New York: Routledge, 2000), 1–3.

Accountability

Accountability in the modern state has two major meanings, which overlap. Firstly there is the standard meaning, common in democracies, that those who exercise power, whether as governments, as elected representatives or as appointed officials, are in a sense stewards and must be able to show that they have exercised their powers and discharged their duties properly. Secondly, accountability may refer to the arrangements made for securing conformity between the values of a delegating body and the person or persons to whom powers and responsibilities are delegated. Thus in the United Kingdom the government is said to be accountable to Parliament in the sense that it must answer questions about its policies and may ultimately be repudiated by Parliament. In 1979, for example, the Labour Government headed by James Callaghan was defeated by a majority of one in a vote of no confidence, precipitating a general election. In the United Kingdom the Parliamentary Commissioner for Administration (popularly known as the Ombudsman) is thought to have improved the accountability of the administration by the scrutiny of administrative methods and inquiries into complaints against government departments. Ultimately, of course, governments in democracies are accountable to the people through the mechanism of elections.

Accountability is not confined to democratic forms of government, although it is in democracies that demands for greater accountability are generally heard. Any delegation of power will usually carry with it a requirement to report on how that power is exercised, and any institution seen as having power may be required to justify its operations to a superior authority. Thus it would be possible to speak of a dictatorship or of a totalitarian regime making the press, the universities or the trade union movement accountable to the government. With an increased interest in **human rights** and democracy throughout the world, and especially in the new Eastern European democracies, electorates desire accountability more than ever. It is often linked with the idea of 'transparency' in government, the ability to know exactly what elected officials are doing.

David Robertson, *A Dictionary of Modern Politics*, third edn (Europa Publications, 2002). Politics Online. Taylor & Francis. www.routledgepoliticsonline.com:80/Book.aspx?id=w007 (accessed 15 October 2009).

Activism

Activism is any intentional action or activity by an individual or a group of actors in order to achieve social or political change. Whether an activity is perceived as activism or not, primarily depends on its purpose and/or the intention of the actor rather than of the type of activity itself. Types of activism are differentiated according to the strategies used (e.g. internet activism, non-violent confrontation, community-building), their objective or substantial aim (e.g. feminist activism or veganism) or the types of activists involved (e.g. student activism or youth activism). Activism can address social entities at all levels of social organization. In contrast to concepts like social or political movements, activism does not require actors to be organized. Thus activities by social movements can be understood as activism, while not all forms of activism require the establishment of a movement.

Lena J. Kruckenberg

Aesthetics, feminist

The making of art and the experience of art are often held to be gendered in a significant way. This must be taken into account, it is argued, if we are to understand art fully. Feminist aesthetics is not a way of evaluating art or our experience of art; rather, it examines and questions aesthetic theory and attitudes concerning **gender**.

Although feminist work in literature criticism, **film theory**, and art history is well established, feminist aesthetics is a relatively young discipline, dating from the early 1990s. Because of its relatively recent beginning, feminist aesthetics is still a discipline without a canon. In fact, several writers resist the idea that feminist aesthetics should have a canon at all, since they believe that work in this field necessarily needs to develop as women artists and theorists do themselves. Moreover, since it draws upon several brands of **feminism** and feminist work in other disciplines, feminist aesthetics is rarely concerned to respect disciplinary boundaries. Further, its primary task is to broaden our concept of what counts as art—and enable the discipline to include more varied perspectives of artists, art appreciators, and the wider contexts in which art develops.

Although feminist aesthetics begins with the recognition that gender matters in art, the study of feminist aesthetics is of art history or feminist art criticism, each of which also begins with the same assumption. The fact that women are oppressed as the **subject** of art does play a part in the acknowledgement that gender is influential, but it is not necessarily all that matters. There is a bias in painting (and print media) toward female subject (often nude women perceived as passive and wanting to be looked at) and male artists (always in control, always doing the looking). What this feminist view of art history produces is our recognition of the 'male **gaze**', which is a significant part of feminist aesthetics. In this case, feminist aesthetics has contributed something that traditional aesthetics has not so much got wrong, as overlooked entirely.

Both feminist aesthetics and feminist art criticism have focused on the unbalanced relationship between the subject and object of aesthetics contemplation—and both want to initiate an important blurring of distinctions between them. Further, there is an emphasis on the aesthetics dimensions of everyday life and the importance of seeing art as a process or activity rather than a product. Feminist analyses attempt to link aesthetic judgement and the resultant implied meaning and value of works of art to beliefs and desires in everyday life. That is, we need to consider art and the aesthetic within their own context. It is only here, in this complicated nexus of circumstances, that we can fully understand the significance of art.

Cheris Kramarae and Dale Spender (eds), *Routledge International Encyclopedia of Women: Global Women's Issues and Knowledge*, Vol. 1 (New York & London: Routledge, 2000), 24–25.

Affirmative action

Once the need to provide legal recourse against **discrimination** became widely accepted in Western societies, from about the mid-1950s onwards, a new problem occurred. How far, and in what ways, could a state take positive action to remedy the consequence of past discrimination and inequalities? Policies intended to make up for a history of discrimination, for example, the setting aside of places in educational institutions for people of particular backgrounds, came to be known as positive discrimination, or affirmative action. There are, inevitably, philosophical, and therefore legal, problems associated with affirmative action. For example, a state might want to remedy past discrimination against racial minorities in access to higher education by having a minimum quota of places that must be filled by members of such a minority. The result might be that some members of this minority are accepted instead of more qualified members of the dominant racial group, which could itself constitute racial discrimination; in a classic case on these lines in 1978 (*Bakke v. Regents of the University of California*) the US Supreme Court ruled that such direct quotas were discriminatory. However, so complicated is the issue, that *Bakke* has long been seen as one of the least satisfactory and least clear of all pronouncements in civil rights law. In general, affirmative action has come to take the form of making special efforts to recruit the disadvantaged, or to train them to increase their chances of succeeding in direct competition with others, rather than directly giving them easier access to jobs or educational places. In 2003 the US Supreme Court revisited this issue at a time when increasing public dissatisfaction with the effects of affirmative action, and a conservative government in Washington, led many to expect the ruling to curtail even measures which were thought to be unaffected by the *Bakke* decision. *Bakke* had allowed some engineering of admissions in pursuit of socio-economic diversity, a goal which has been accepted largely without argument as valid and not constitutionally forbidden. This much was confirmed by the new decision, but any further move towards quotas intended to remedy past injustice was again stopped.

David Robertson, *A Dictionary of Human Rights*, second edn (Europa Publications, 2004). Politics Online. Taylor & Francis. www.routledgepoliticsonline.com:80/Book.aspx?id=w006 (accessed 15 October 2009).

African Charter on Human and Peoples' Rights

The Charter was issued by the Organization of African Unity (OAU, now the **African Union**) in 1981, and entered into force in 1986. It took its emphases from the OAU's own Charter, the United Nations (UN) Charter and the UN Universal Declaration of Human Rights. The African charter is one of a series of regional **human rights** documents encouraged by the UN as part of a general strategy for enforcing human rights world-wide, the most effective of which is the **European Convention on Human Rights**. Although the very universality of the original UN Charter implies that human rights are generally valid, there is an acceptance that regional world cultures may evaluate, and even partially define, such rights in different ways. The specific thrust of the Charter of the OAU was to bring its commitment to 'eradicate all forms of colonialism from Africa' to bear on the definition and support for human rights. Thus the enumeration of rights, though not very different in detail from what one would find in any classic listing, was set against a background that recognized two points missing in, for example, the European Convention. First, some tension seemed to be recognized, though it was posited to be a fruitful tension, between peoples' rights and individual human rights. The Preamble recognized that: 'fundamental human rights stem from the attributes of human beings which justifies their national and international protection and on the other hand that the reality and respect of peoples rights should necessarily guarantee human rights', and: 'that it is henceforth essential to pay a particular attention to the right to development and that civil and political rights cannot be dissociated from economic, social and cultural rights in their conception as well as universality and that the satisfaction of economic, social and cultural rights is a guarantee for the enjoyment of civil and political rights'.

David Robertson, *A Dictionary of Human Rights*, second edn (Europa Publications, 2004). Politics Online. Taylor & Francis. www.routledgepoliticsonline.com:80/Book.aspx?id=w006 (accessed 15 October 2009).

African Union (AU)

In May 2001 the Constitutive Act of the African Union entered into force. In July 2002 the African Union (AU) became fully operational, replacing the Organization of African Unity (OAU), which had been founded in 1963. The AU aims to support unity, solidarity and peace among African states; to promote and defend African common positions on issues of shared interest; to encourage **human rights**, democratic principles and good governance; to advance the development of member states by encouraging research and by working to eradicate preventable diseases; and to promote sustainable development and political and socio-economic integration, including co-ordinating and harmonizing policy between the continent's various 'regional economic communities'.

Activities (African Union—AU), in Europa World online (London: Routledge), www.europaworld.com/entry/wb02062.io.txt.111 (accessed 15 October 2009).

Agency

The notion of agency is both a facet of and a deliberate corrective to the more traditional philosophy topic of action. In Anglo-American philosophy, acting is usually discussed in terms of the possibility of free will, the nature of rationality and motivation, and the capacities of persons; the social and political contexts in which action takes place are often overlooked. By contrast, agency is concerned with the social conditions for and requirements of action, as well as with the internal and external barriers to action. As such, agency is a **subject** of central importance to feminist theorists and activists seeking to identify the causes of women's subordination and oppression, and possibilities for their self-realization and freedom.

Contemporary discussion of agency in philosophy and political thought can be traced to developments in social theory during the 1980s and 1990s. In particular, Michel **Foucault**'s writings on the nature and effect of social power and relations of power; material, culture and discursive practices, including 'disciplinary' practices of the modern state; and the 'technologies' of the 'self' in the constitution of modern subject have all been important to feminist discussions of subjectivity, oppression and agency. Feminist thinkers such as

Jana Sawicki and Sandra Barky have employed Foucault's thought to explore the ways in which modern states shape women's identities both as specific socio-historical subjects and as self-disciplining, feminine subjects through social practices and discourses.

The issue of agency has come to the fore in recent debates between feminists sympathetic to **postmodernism** and psychoanalytic theory, and feminists critical of postmodernism. Judith **Butler** argues in her highly influential *Gender Trouble* that identities are discursively produced through **gender** 'performances'; her controversial suggestion that the very notion of 'woman' is overly essentialist has generated the criticism that without this category we cannot name and work to transform sex-based oppressions. Seyla Benhabib and Nancy Fraser question the usefulness of **deconstructionist** thought for **feminism** (especially the work of Jacques Derrida), and disagree with postmodernists' suggestions that feminists should dispense with Enlightenment conceptions of autonomy, rationality, truth and the idea of a unified subject. Postmodern feminists counter that deconstructive approaches do not efface the possibility of agency or social change, but merely offer a more accurate and radical picture of the contingency and instability of social identities and relations.

Monique Deveaux in Lorraine Code (ed.), *Encyclopedia of Feminist Theories* (London & New York: Routledge, 2000), 15.

Akayesu

This is the name of the former mayor of a town in the Taba region of Rwanda, an area that saw the massive and systemic perpetration of **rape** as an instrument in the **genocide** in 1994. The name has become famous as an **International Criminal Tribunal for Rwanda** (ICTR) case in which for the first time under international criminal law an accused was charged with rape as a crime of genocide. Jean-Paul Akayesu allegedly did not himself commit sexual assault but oversaw the soldiers who did. The trial began in 1997 and ended with Akayesu's conviction in 1998 on all counts of genocide, including the rapes.

Tawia Ansah

American dream – *see* Gangster/criminal/detective thriller/private-eye film

Androcentrism

This is the view that male behaviour and characteristics are central, are the norms. This view so permeates society that female behaviour is understood and seen as deviate, that is, deviating from the male norms.

Androcentric society values characteristics that are associated with men and maleness. Thus, competitiveness and aggressiveness are highly valued and rewarded, whereas characteristics associated with women-caring and co-operation are devalued.

Androcentric societies are organized on the assumption that the male is the norm. This is clearly illustrated by the ways in which paid employment is organized in advanced industrial societies: full-time work with a lifelong and uninterrupted commitment to the labour market is assumed to be the norm. Women who have domestic responsibilities and cannot conform to this expectation are disadvantaged both in the jobs for which they are seen as eligible in the labour market and in their promotion prospects in employment.

Androcentric texts and scholarship present women as absent or silent or threaten them in stereotyped ways. Hartmann, for example, points out the inherent **sexism** in Marxist theory. Hoagland points to the way in which androcentric rhetoric is used in sociobiology. Forenza demonstrates that theological texts present women as absent or silent or in their traditional roles. In this way women have been 'hidden from history'.

Cheris Kramarae and Dale Spender (eds), *Routledge International Encyclopedia of Women: Global Women's Issues and Knowledge*, Vol. 1 (New York & London: Routledge, 2000), 59.

Anthropology

Anthropology is distinguished by its combination of fine-grained, qualitative, descriptive field methodology (ethnography) and cross-cultural analysis (ethnology). While not a necessary feature of late 20th-century anthropology, historically another distinguishing characteristic has been a focus on non-Western peoples as objects of study. In Europe, ethnography and ethnology constitute the discipline. In the Americas, anthropology also encompasses archaeology and biology or physical anthropology, including primatology.

Anthropology's contributions to feminist analysis have been grounded in ethnographies. Some ethnographic studies have undermined the assumption of universal women's experience. Anthropology has generated sustained theoretical debates about the nature and significance of such universals, particularly the **division of labour** by sex; distinct public/private spheres; male dominance; and the categorical symbolic association of nature/culture with dichotomous sex/**gender** identities. Feminist anthropology has participated in feminist critiques of the methodology of science and in the development of experimental forms of writing to convey women's experiences, including life histories, biographies and autobiographies, and fictions. The most significant contribution that anthropology has made to feminist thought is undoubtedly the development of a social concept of sex/gender system.

Kamala Visweswaran (1997) uses changing concepts of gender within anthropology to divide the history of feminist anthropology into four periods. Until 1920, gender was equated with biological sex, and biological essentialism justified fixed divisions of labour along lines of sex. Following the development of first-wave **feminism**, beginning in the 1920s with the work of scholars such as Margaret Mead (Visweswaran 1997, 601–2), sex and gender were analytically distinguished, and considerable attention was paid to the social construction of gender. None the less, the focus of feminist ethnography during this period was on women, conceived of as a unitary subject. In the 1960s and 1970s, as second-wave feminist **activism** emerged, a new generation of feminist anthropologists devoted considerable effort to the formal definition of sex/gender systems in which the facts of biology were the basis of culturally-specific gender systems. Critiques of the universality of 'woman' emerge within anthropology in the 1980s and 1990s as part of broader discussions of neglected aspects of **difference**, inspired by cultural studies, **postcolonialism**, and gay and lesbian studies. This latest phase of feminist ethnography has confronted many issues raised by third-wave feminist activism.

Henrietta Moore argues that universal male dominance and a division of society into domestic and public spheres, associated, respectively, with women and men, are embedded assumptions of anthropological kinship theory, also critiqued by Collier and Yanagisako (1987). Similarly, Micaila di Leonardo (1991) argues that from 1920 to 1970 anthropological analyses of social organization and kinship assumed the existence of universal male dominance and a normative male viewpoint. Di Leonardo characterizes the anthropology of this period, which includes the ethnographic contributions to sex and gender studies of Margaret Mead, as 'pre-feminist'.

Explicitly, feminist anthropology emerged in the 1970s with the publication of two significant collections of essays. Edited by Rayna Rapp Reiter, *Toward an Anthropology of Women* provided the site for Gayle Rubin's explication of the sex/gender system. This construction carried forward Margaret Mead's definition of gender as only loosely constrained by biological sex, and thus subject to many different social forms. It also incorporated the persistent anthropological assumption, present from the late 19th century on, that sex was a necessary and biologically given ground for gender. Thus, while other societies might construct maleness and femaleness differently, they would always construct them as linked, dichotomous categories determined by sex.

In *Women, Culture and Society,* Rosaldo and Lamphere (1974) provide a venue from a set of arguments that present universal male domination of women as founded on specific culture classifications which follow universal structural principles, and women's position in society as universally tainted by close association with a domestic sphere. Sherry Ortner in her famous essay, 'Is Female to Male as Nature is to Culture?' (67–87), argues that cross-culturally women's biological role in reproduction has led to their structural association with nature, balanced by an equation of men with culture. Women's close association with the domestic sphere reinforces their categorical symbolic position, since the domestic sphere is the origin of challenges to social order. Michelle Rosaldo, in the same volume (17–42), further explicates the presumed universal nature and origins of the domestic/public dichotomy.

With these two central propositions established in the feminist anthropology of the 1970s, the basis for the universal domination of women by men is explicated. The public sphere, which is exclusively the domain of men, is where society was constituted, institutionally through politics, and symbolically as the opposite of nature. This notion that men everywhere were dominant over women was not new in the anthropology of the 1970s. Visweswaran (1997, 569, 600–1) identified Elsie Clews Persons as establishing the idea of women's experience of universal **patriarchy** as the coherent ground for cross-cultural analysis in the early 20th century. However, the feminist anthropology of the 1970s takes asymmetric relations of power as the justification for regarding all women, cross-culturally, as inhabiting identical **subject** oppositions, even though anthropology there might otherwise be demonstrated stark differences between women in different cultures.

The propositions that women inhabit a domestic sphere and are excluded from a public sphere, that this dichotomy reflects a symbolic association of women with nature because of the biological facts of reproduction, and that men exercise universal dominance justified by their symbolic association with culture and order, and institutionally enabled by their control of the public sphere, while framed by feminist anthropologists, are in fact congruent with the dominant symbolic structuralism and functionalist analysis of social organization that had been developed in anthropology between 1920 and 1970, as Moore (1988) notes. Once clearly articulated, these ideas were subject to intense debate by feminist anthropologists in an influential series of edited volumes. One strand of debate questions the identification of male dominance by pointing to the presence within societies of multiple arenas of distinction, and documenting shifting hierarchies among men and women. Other critiques present analyses of cultures in which men and women were not symbolically associated with culture and nature. Some of these studies begin to problematize the assumption that women and, although seldom explicit, men, are unitary categories. The domestic/public dichotomy is repeatedly challenged. These varied critiques are perhaps most cogently illustrated in the work of Marilyn Strathern (1987). By the end of the 1980s, few feminist anthropologists could argue for the real universality of the dichotomies men/culture/public/dominant versus women/nature/domestic/subordinate.

Challenges to the unitary nature of 'woman' as a subject emerge from these feminist anthropological debates, but they also are seen in anthropology during the 1980s and 1990s as a result of critiques of ethnography from postcolonial studies and gay and lesbian studies. Third-wave feminist critiques, particularly indigenous feminist theory and First Nation feminism, directly challenges feminist anthropology's assertion of a unified female subject. Di Leonardo (1991) argues that under these pressures, anthropology, and feminist anthropology in particular, have begun to reconceptualize their subject of study as gender difference, rather than gender **identity**. Feminist anthropologists question the biological essentialism of the sex/gender dichotomy.

Late 20th-century anthropology witnessed the growth of a body of explicitly feminist ethnography, often biographical and experimental in form. Such ethnographical experiments reflect, even if indirectly, feminist standpoint theory. They place in sharp relief contradictions in the position feminist ethnographers assume in speaking as women and for (other) women, Visweswaran (1997) argues that experiments with the form of ethnographies are not exclusively products of **postmodernism** in anthropology. She notes the use of autobiography, fiction and other narrative forms in the ethnography of the period between 1920 and 1960. She draws particular attention to the fact that feminist ethnographies written between 1960 and 1980, under the influence of second-wave feminism, are notable for their narrative and reflexive questioning of identification between female ethnographer and female ethnographic subjects.

Di Leonardo (1991) characterizes feminist anthropology in the 1990s as sharing five main assumptions. It rejects social evolutionism, particularly the idea that different contemporary peoples can be used as proxies for past stages of human development. It emphasizes historical context, often a context of colonialism. It problematizes apparent stable categories, exemplified by the realization (following **Foucault**) that diverse forms of sexuality have specific histories. Late 20th-century feminist anthropology builds on this insight to question the terms of debate about human universals, and is committed to resist essentializing. Feminist anthropology assumes that gender is inextricably linked to, or embedded in, other institutions and discourses, and does not assume that gender has a primary or overarching status. As a result, feminist anthropologists study the simultaneous construction of **race**, gender, **sexuality**, age, **ethnicity** and **class**. Feminist anthropology in the 1990s transformed the question of male domination into one of the existence and nature of social stratification, without assuming the form it might take, and without assuming that social stratification is stable. Finally, di Leonardo suggests that late 20th-century feminist anthropology, in common with other anthropologies, is concerned with a reflexive and critical awareness of the 'social location' of the ethnographer and the ethnographic subject. (*See also* **Categories and dichotomies**.)

Rosemary A. Joyce in Lorraine Code (ed.), *Encyclopedia of Feminist Theories* (London & New York: Routledge, 2000), 21–24.

Aristotle

Aristotle (384–322 BC) was a thinker of the classical Greek period whose political theories, like those of Plato, set the bounds of political discourse throughout the Middle Ages; his work still exercises a profound influence on modern political and social thought. Aristotle's political ideas are more immediately acceptable to the modern Western mind than Plato's because he comes closer to approving of democracy. However, even Aristotle saw direct democracy as the least undesirable of existing types of government, rather than as the best obtainable form. Like most Greeks of his period he would have preferred a mixed government with important elements of aristocracy intermixed with popular rule. (In this context it should be remembered that the original meaning of 'aristocracy' is 'the rule of the best', not 'the rule of the well born'.)

An important aspect of Aristotle's thought, which derives from his interest in marine biology, was his use of biological analogies in discussing social life. Following Plato, he took an essentially functionalist approach to social and political institutions, believing that political life, being natural, takes certain natural forms, and that individuals, therefore, have natural and fitting places in society from which it would be both immoral and 'disfunctional' for them to depart. Aristotle's direct impact on European social thought began with his reinterpretation by the late medieval Catholic church and Thomas Aquinas's development and interpretation of his ideas into the Catholic doctrine of natural law, from which our modern inheritance of natural rights derives. Aristotelian views appear in contemporary moral philosophy, with special emphasis on his concern for education and the training of moral instincts. (*See also* **Hobbes, Thomas**.)

David Robertson, 'Aristotle', in *A Dictionary of Modern Politics*, third edn (Europa Publications, 2002). Politics Online. Taylor & Francis. www.routledgepoliticsonline.com:80/Book.aspx?id=w007 (accessed 15 October 2009).

B

Beauvoir, Simone de

Simone de Beauvoir (1908–86) was born in Paris into a comfortable bourgeois family. She was educated at the Sorbonne in Paris, where the philosopher Jean-Paul Sartre, her lifelong intellectual companion, was a fellow student. Beauvoir taught philosophy during the 1930s at schools in Marseilles, Rouen and Paris, and from the 1940s began to publish—novels, essays, philosophy, articles, autobiography—becoming an internationally famous writer, celebrated as a thinker and feminist. She died in 1986, and was buried next to Jean-Paul Sartre in Paris's Montparnasse Cemetery.

Beauvoir's major feminist work is *The Second Sex*, a wide-ranging, existential analysis of women's situation that is possibly the most influential feminist text of the 20th century. It was first published in France in 1949 in two volumes, under the title *Le Deuxième Sexe*, and was an immediate publishing success, not least because of its open treatment of women's sexuality. A shorter version in English appeared subsequently, in time to influence the 1960s women's movements in America and Great Britain.

When Simone de Beauvoir wrote *The Second Sex* in the late 1940s, much had been achieved politically by European and American feminists, and many of the legal disabilities of the past had been overcome. Women in many countries, for example, could vote and own property, and had access to higher education and to the professions. After the two World Wars, women also had more freedom socially and sexually. However, Beauvoir argued that despite all these gains women were not emancipated from men and remained in a subservient relationship. In *The Second Sex* she focused on women's situation, using a mix of history, **anthropology**, myth, ethnography, biology, literature and sociology to examine why women were effectively the inferior, the second sex: submissive, uncreative and unfree.

Philosophically, Beauvoir adopted an existentialist perspective in *The Second Sex*. Basic to her argument was the existential ethical concept that freedom is the most desirable of human conditions, particularly the freedom to choose. She adapted the existential categories developed by Sartre in *Being and Nothingness* (1943), in particular the notions of **Subject** and Object or Self and **Other**. These categories ultimately derive from the philosophy of Hegel, which sees the purpose of existence, for the individual and for humanity, as the achievement of self-understanding. Central to this process is defining and understanding oneself in terms of the 'other', or that which is not the self, that which is secondary, inessential and inferior. Modern and most historical societies, Beauvoir argued, objectified woman as Other and man as Self. Humanity had been defined as male and the human condition as masculine, with woman defined always in relation to man. The source of woman's subservience and enslavement, she suggested, lay in woman's 'otherness' in relation to man. Only man had the freedom to choose, to set himself up as essential and Subject, while woman as a consequence became inessential and Object.

The Second Sex is a massive text, dealing with a multitude of topics. It explores biological, psychoanalytic and Marxist explanations of women's destiny; reviews the history of relations between the sexes from primitive forms of society to modern times; and looks at the sexuality of women and its relationship with women's 'otherness'. Beauvoir also looks at myths relating to woman, particularly those relating to motherhood; considers the representation of women in male authors' novels; and reviews the evolution of contemporary women's situation from formative years to woman as wife and mother. The special situation of lesbians, independent women and career women are considered, and Beauvoir analyses at length the role of prostitutes, a role she saw as one where women might, in certain circumstances, use 'otherness' to exploit men.

In analysing why women were Other, Beauvoir rejected explanations of women's subordination offered by theories of **biological determinism**. Biologically, because women have a reproductive and rearing role, Beauvoir conceded it is difficult for women to be free. However, she argued that a woman need not be defined by her womb; it was possible for a woman to have a life beyond her reproductive function. Freud and psychoanalytical theories also did not provide satisfactory explanations of women's 'otherness'. Women's physiology and lack of a penis did not mean that women were inferior or envious of men, or suffered from a castration complex. The 'prestige of the penis', Beauvoir argued, was an aspect of power relations and the 'sovereignty of the father'. Beauvoir was also sceptical of the Marxist view that in a capitalist society everything, including women's oppression, derived from economic relations and the **hegemony** of the ruling class, and that with the advent of socialism the subordinate position of women would be transformed. Despite being a firm socialist, she saw women's situation not as a consequence of private property and **capitalism**, but due to male dominance over the female Other.

Central to *The Second Sex* was an examination of women's situation as wives and mothers. Women did not dispute male sovereignty in marriage because of their economic dependency and their reproductive function. Though Beauvoir recognized that the traditional form of marriage was in a period of transition, she maintained that within the institution of marriage women remained subordinate, secondary and parasitic, and that equality in marriage would remain an illusion as long as men retained economic responsibility. Beauvoir felt that women of talent were lost to humanity because they were engulfed in the repetitive routines of housework. She had a particular horror of cleaning, and said that 'few tasks are more like the torture of Sisyphus' than housework.

Motherhood, Beauvoir appreciated, might be for many women a supreme and happy stage in their life history, and in maternity women might be said to fulfil their destiny, but she maintained all women did not enjoy maternity and that pregnancy and motherhood were variously experienced. Some women enjoyed pregnancy, but for others the experience was one of nausea, discomfort or painful trauma. Beauvoir was sceptical of the sacred character of motherhood, and pointed out that it was only married mothers who were glorified, while unwed mothers were usually considered disreputable. She recognized that some women found their whole existence justified in fecundity, but Beauvoir considered the notion that having babies made women into full, free human beings to be illusory. She thought that good mother love was a conscious attitude, a moral free choice and not an instinct, and that there was such a thing as a bad mother. A mother's attitude depended on her total situation, and though circumstances had to be unfavourable not to be enriched by a child, Beauvoir suggested perils in motherhood, such as the mother as slave to the child and as left behind as the child transcends mother love. Beauvoir thought that to be a good mother a woman had to be well balanced, with interests and a life beyond child-rearing. She thought women who undertook paid work outside the home might be the best mothers.

Beauvoir saw the difficulties for women of reconciling work and maternity, and the 'slave labour' nature of women's work outside the home. She saw clearly that child-care outside the home was needed, and she had a very robust attitude for a French woman of her time towards **contraception** and **abortion**, and their role in permitting women to have the freedom to choose maternity or not. She advocated contraception, legal abortion, easy divorce and, indeed, artificial insemination, so that women might maximize their freedoms and choices. Paid employment outside the home, provided it was not exploitative, was a vital means to women's independence. However, such things in themselves would not be enough to change women's situation. Attitudes and understanding must also change.

Beauvoir saw women's 'femininity' as supporting male sovereignty and insisted that there was no ready-made essence of femininity—it was a myth. Civilization, not biology, had constructed the feminine. *The Second Sex* is not an assault on **masculinity**, but it can be said to be about femininity as a social construct, and a major theme is women's submission in their formative years to the feminine **gender** role, and the limitations and burdens of that role within the male-dominated power structures of the family. For Beauvoir, femininity was artificially shaped by custom and fashion, and imposed from without. She described how women learned to assume the female gender, and summarized her view in what is probably her most famous and widely quoted sentence: 'One is not born but rather becomes a woman' (Beauvoir 1952, 295).

Beauvoir concluded that there was no eternal hostility, no battle of the sexes between man and woman. Sexuality was not destiny, a woman's ovaries did not 'condemn her to live her life for ever on her knees' (736). There was no eternal feminine, there was no eternal masculine. 'New' woman needed an accompanying infrastructure of moral, social, cultural and attitude changes, as well as economic opportunity. Men and women should recognize each other as equals. 'New' woman needed equilibrium, a free exchange between sexes. They should be in perfect equality. Though she thought a range of relations was possible between men and women, Beauvoir's ideal was the balanced couple; a couple not living as a closed cell, but each integrated individually into society. Such a couple would display 'equality in difference, and difference in equality' (740) by mutually recognizing each other as Subject. Thus, the slavery of half the human race would be abolished and the human couple would find its true form.

A new women's movement—sometimes called 'second-wave **feminism**' or 'women's liberation'—began to develop during the 1960s, then grew explosively from the end of the decade. The major texts of the period, such as Betty Friedan's *The Feminine Mystique* (1963), Kate Millet's *Sexual Politics* (1970), Germaine Greer's *The Female Eunuch* (1970) and Shulamith Firestone's *The Dialectic of Sex* (1970), all owe a good deal to *The Second Sex*, in which Beauvoir anticipated many of their themes.

Since this initial phase, feminism has fragmented into a multitude of forms with many varieties—liberal, liberal socialist and **radical feminism**; psychoanalytical feminism; female supremacism; New Right feminism; eco- and anarcho-feminism; poststructuralist feminism and post-feminism, among others—and inevitably Beauvoir has been criticized by some. She has been accused of failing to celebrate women's nurturing and caring role, and of having little sympathy with women's reproductive function. It is said that she wanted women to be more like men. However, despite such criticisms, *The Second Sex* is widely recognized as the seminal text of the women's movement of the late 20th and early 21st centuries.

Ian Adams and R.W. Dyson, 'Simone de Beauvoir (1908–86) and Second Wave Feminism', in *Fifty Major Political Thinkers*, first edn (Routledge, 2003). Politics Online. Taylor & Francis. www.routledgepoliticsonline.com:80/Book.aspx?id=w063 (accessed 15 October 2009).

Becoming woman

We owe the concept 'becoming woman' to Simone de **Beauvoir**. She begins the second volume of *The Second Sex* (1948) with the now much-cited phrase, 'one is not born a woman: one becomes one'. Second-wave Anglophone feminists have most commonly construed this statement to mean that being 'a woman', or being 'feminine' is not a natural attribute. Having a female-sexed biological **body** does not automatically make one a women; rather one becomes a woman through a process of initiation into a socially constituted **identity**. In short, sex (anatomical differences between women and men) and **gender** are not synonymous.

Although Beauvoir did not use the term gender, the distinction between sex and gender is implied in her observation that 'every female human being is not necessarily a woman' (Beauvoir 1952). Beauvoir sets out to describe what she call 'the lived experience of becoming a woman'. She begins with those early childhood experiences through which girls learn to feel that they are inferior to boys, and continues through accounts of sexual initiation, marriage, motherhood, to reveal how women acquire a sense of inferiority to men and frequently come to accept their situation of oppression rather than assert freedom.

However, the phrase 'becoming woman' is open to further interpretations, which later feminists have explored. The verb 'to become' is ambiguous, being both passive and active in its implications. Passively construed, it may imply that one is *made* a woman by being forced to comply with norms of femininity that are not of one's choosing. It may also be read to imply that social construction, social pressures, or else what feminists influenced by **Foucault** have called 'disciplinary practices' (e.g. Ramazanoglu 1993), have the effect of constituting some persons as 'women' without their being consciously aware of that process.

However, construed in a more active sense, 'becoming woman' can imply a volitional engagement in the creation of one's gender identity. Beauvoir herself suggests that women are often complicit in their oppression, choosing to assume 'feminine' characteristics because there are benefits—such as male protection—associated with them or because the costs of resisting are high. More recently **Butler** (1990) has suggested that 'becoming woman' can also be construed as an ongoing 'performance' in which—albeit under duress—female gender is produced and sustained over time by the repetition of certain styles of action that signify femininity. (*See also* **Beauvoir, Simone de**.)

Sonia Kruks in Lorraine Code (ed.), *Encyclopedia of Feminist Theories* (London & New York: Routledge, 2000), 40–41.

Binary opposites – *see* **Categories and dichotomies**

Biological determinism

Biological determinism is the belief that human behaviours may be attributed to a person's underlying essential genetic make-up. Popularly discussed in debates about nature versus nurture or heredity versus environment, deterministic thinking has its roots in evolutionary theory and dates back to the middle-to-late 19th century. Deterministic theory rose to prominence after Charles Darwin's theory of evolution by natural selection was popularized, and they became widespread in the USA and Europe in the mid-19th and early 20th centuries. Applications of biological determinism to social and cultural problems have usually coincided with periods of social upheaval (Bem 1993). Biological determinism has been used to discredit social movements such as anti-slavery women's rights, and women's suffrage. Theories rising from biological determinism include eugenics, social Darwinism and socio-biology.

Francis Galton applied—actually, misapplied—evolutionary thinking to heredity when he invented eugenics. Although Galton appreciated the interaction of heredity and environment, he still thought that selective mating would result in a superior populace—one that did not include racial mixing (Pearson 1996). The Nazis applied such biological ideas to social and cultural institutions in the mid-20th century. Biological differences assumed enormous significance and resulted in **gender** segregation in schools as well as a continuum of racial superiority, which placed some races below animals. This was the worst manifestation of biology as destiny.

Throughout history, women have been subjugated in various cultures because of their assumed inherent nature. Seen as passive, nurturing and dependent, women have been denied education, the right to vote, and other means for social and cultural advancement. Education was thought to be damaging to women's reproductive system. Even many 19th-century suffragists used deterministic beliefs about innate racial differences to promote their cause.

The sexual **division of labour** has long been assumed to be natural and universal, although, for example, Indian women contribute widely to agriculture and dairy farming. Recent studies reveal that in rural India agriculture is the major occupation of working women—a fact that often does not appear in statistics because of a sexist definition of 'work'.

In the late 19th century, Herbert Spencer used evolution to justify a conservative political and social agenda know as social Darwinism. Spencer used biology to justify restrictive Victorian constraints on women's roles. Although social Darwinism is no longer popular in its original version, a new form became popular in the 20th century—sociobiology. In the mid-1970s E.O. Wilson first published on sociobiology, which casts the world in biological universals largely free from the influence of environment, learning, or culture.

This persistent manifestation of biological determinism reinforces cultural and racial stereotypes and continues to case women as passive, manipulative and dependent. In the wake of the modern feminist movement, feminist critiques of science have included both critics and adherents of biologically deterministic thinking. Though no one perspective of biological determinism appears to be totally correct, women and society continue to benefit from the ongoing critique.

Barbara I. Bond in Cheris Kramarae and Dale Spender (eds), *Routledge International Encyclopedia of Women: Global Women's Issues and Knowledge* (New York & London: Routledge, 2000), 108–9.

Body, the

In the dominant Western tradition, the body has long been the unspoken of abstract theory, dismissed from consideration as the devalue term of the mind/body split that marks the post-Cartesian modernist period. It has been seen simply as a material and unchanging given, a fixed biological entity that must be transcended in order to free the mind for the intellectual pursuits of fully rational subjectivity. However, not surprisingly the ability to effect such transcendence has been **gender** marked as an attribute of men alone, such that women remain rooted within their bodies, held back by their supposedly natural biological processes and unable to exercise full rationality. In consequence, **feminism** has been deeply concerned with the body—either as something to be rejected in the pursuit of intellectual equality according to a masculinist standard, or as something to be reclaimed as the very essence of the female. A third, more recent alternative, largely associated with feminist **postmodernism**, seeks to emphasize the importance and inescapability of embodiment as a differential and fluid construction, rather than as a fixed given.

Where the body is viewed through conventional biological taxonomies, it is taken for granted that sexual **difference** is an inherent quality of the corporeal, and that male and female bodies may each be known universally. In terms of the historical disempowerment of women, the justificatory linking of the female to the body has been centred largely on the reproductive processes. The very fact that women are able in general to menstruate, to become pregnant and give birth, to lactate, is enough to suggest a potentially dangerous volatility, in which the body is out of control, beyond, and set against, the force of reason. In contrast to the apparent self-containment of the male body, which may then be safely forgotten, the female body demands attention and invites regulation. The age-old relation between hysteria and the womb (*hystera* in Greek) is just one example of how femininity itself becomes marked by the notion of an inevitable irrationality. Women just are their bodies in a way that men are not, biologically destined to inferiority. However, at the same time that women are seen as more wholly embodied, and hence inferior, the boundaries of that embodiment are never secured. As the processes of reproduction make clear, the body has a propensity to leak, to overflow the proper distinctions between self and **other**, to contaminate and engulf. Thus women themselves are, in the masculinist imagination, not simply lesser beings, but objects of fear and repulsion. The devalued body is capable of generating deep anxiety; and indeed, as Susan Wendell (1996) points out, feminists, too, have difficulty in accepting lack of control in conditions such as disability.

So powerful are such ideas that many feminists have been reluctant to theorize the female body, preferring instead to deny the links that have worked so efficiently in the interests of **patriarchy**, and even those contesting the determinism of biology have seen it in a decidedly negative light. At the beginning of the second wave of feminism, Simone de **Beauvoir** famously likened it to 'a carnivorous swamp' (Beauvoir 1952), and Shulamith Firestone (1979) looked forward with optimism to a time when the recipient advanced reproductive technologies might free women from the 'oppressive "natural" conditions' of procreation. Against such somatophobia, which to an extent mimics the masculinist fear and rejection of the body, other feminists have responded by celebrating their bodyliness, particularly with regard to reproduction. The uniquely female capacity to give birth 'naturally' has been taken up as the centre of women's power, to be jealously guarded against the incursions of biotechnology, while more generally women are urged to take control of their own bodies in the face of the medical establishment. In addition, for many feminists, the maternal body has come to figure the claim that women have a unique ethical sense that lays stress on caring, relationality and responsibility in contradistinction to the masculine goods of autonomous rights and duties which equally figure the separation of body and mind, and body and body. This stress on the embodied nature of sexual difference runs two related risks: on the one hand it may uncritically universalize the male and female body, while on the other it appears to reiterate the biological essentialism that historically has grounded women's subordination.

The intention to take up positively the issue of female embodiment has, then, been highly controversial, and nowhere more so than in the response to the work of Luce Irigaray. Her project of rewriting sexual difference beyond the binary of male:female—a binary which positions the sexed body as static, ahistorical and determinate—nevertheless makes constant reference to the anatomical differences between the sexes. Her concern is to revalue the way in which femininity is inscribed onto the female form in a culture in which **masculinity** is in retreat from the body and where disembodiment is privileged. Irigaray (1985) places great emphasis on the multiple forms of female embodiment, the self-touching 'two lips' that characterize female morphology, and on the fluidity that marks the inherent excess of the feminine that is uncontainable within binary sexual difference. Hers, though, is not a 'real' biology, as much as the imaginative redeployment of a contested terrain that takes on board the force of psychic investments and insists on the sexed specificity of corporeality. In being strongly influenced by psychoanalysis, Irigaray engages with a body that, though material, is never given, but always filtered through and constructed by a set of discursive strategies. As with many other contemporary feminists, and notably Liz Grosz, she is concerned with the irreducible interplay of text and physicality which posits a body always in process, never fixed or solid, and never one.

In large part, the enormous proliferation of feminist theorizations of the body had been mobilized by the response to the insights of **poststructuralism**/postmodernism, which ironically in their masculinist forms often have been accused of an indifference to materiality. With the demise of the belief in a given reality, what feminists have seized on is not simply that the body is a discursive construction, but that the notion of 'the' body is untenable. There are only multiple bodies, marked not simply by sex, but by an infinite array of differences—**race**, **class**, sexuality, age, mobility status are commonly invoked—none of which is solely determinate. Inspired by the work of theorists such as **Foucault**—who sees the body as a text variously inscribed by history, and most recently in the interests of **capitalism**—feminists have undertaken to extend that analysis to take account of patriarchy. The discursive operations that construct the useful, manipulable body—what Foucault calls the 'docile body'—have been a fertile ground for feminist understandings that make clear the links between the everyday body as it is lived, and the regime of disciplinary and regulatory practices that shape its form and behaviour. Theorists such as Susan Bordo (1993) and Sandra Bartky (1988) have been in the forefront in analysing how the processes of surveillance and self-surveillance are deeply implicated in constituting a set of normativities towards which bodies intend. The practice of diet, keep-fit, fertility control, fashion, health care procedures and so on, are all examples of disciplinary controls that literally produce the bodies that are their concern. Given the pre-eminent position of the discourse of bio-medicine in such a schema, it is incumbent on feminists, and particularly those associated with the longstanding women's health movement, to rethink the traditional claims of medical practice to cure and care, not simply in terms of control, but also in terms of the constitution of the body.

It is, then, the forms of materialization of the body, rather than the material itself, which is the concern of feminism. As Judith **Butler** puts it: 'there is no reference to a pure body which is not at the same time a further formation of that body' (Butler 1993, 10). Butler's notion of performativity is highly relevant here in theorizing the ways in which the deployment of the body, especially in terms of gendered sexuality, is both open and constrained. The normative binaries of male/female, health/ill-health, heterosexual/homosexual and so on, that are used to characterize embodiment may be exposed in their instability—but also paradoxically confirmed—by the performativity of abject bodies. Butler's argument is at times highly abstract, but she never loses sight of the body as a lived entity, or that 'language and materiality are fully embedded in each other' (Ibid., 69). That move away from a purely textual analysis is even more evident in the feminist take-up of the phenomenology of Merleau-Ponty, in which the being-in-the-world of the **subject** is intricately and irreducibly bound up with the constitution and extension of the body. In the phenomenological tradition, the structure of the self is indivisible from its corporeal capacities, but what feminist theorists importantly add is an emphasis on the differential forms of embodiment that confound normative boundaries. The work of Iris Marion Young (1990b) and, increasingly, Liz Grosz (1994b), is deeply concerned with the processes of embodied subjectivity as it evolves within temporal and spatial parameters. The contrast with the Cartesian mind/body could not be clearer.

The body, then, has become the site of intense inquiry, not in the hope of recovering an authentic female body unburdened of parochial assumptions, but in the full acknowledgement of the multiple and fluid possibilities of differential embodiment. As Grosz puts it: 'the stability of the unified body image, even in the so-called normal subject, is always precarious. It cannot be simply taken for granted as an accomplished fact, for it must be continually renewed' (Ibid., 43–44). (*See also* **Butler, Judith**.)

Margrit Shildrick in Lorraine Code (ed.), *Encyclopedia of Feminist Theories* (London and NY: Routledge, 2000), 63–65.

Borders

Borders are usually defined as territorial lines that demarcate one sovereign nation-state from another, or, within the nation-state, one legal jurisdiction from another. Given men's virtual monopoly of government decision-making historically, one could say that borders are lines men have drawn. Rather than view borderlines as fixed and immutable, scholars increasingly treat borders as politically constructed, in order to explore when, why and how they are drawn, and the consequences of peculiar borderlines. In *Imagine Communities* (1983), Anderson transformed thinking about nations, seeing them as 'imaginations' that were made coherent by schools, languages and compulsion.

Increasingly, borders are used in metaphorical ways to differentiate cultural and linguistic identities. Sometimes these identities are imposed on people (under apartheid, South Africa corralled its 'Bantu-speakers' into 'Bantustans'), but at other times people claim **identity** for themselves (such as Chicanos and Chicanas or Latinos and Latinas in the USA, versus Hispanics or North Americans). Identities are likewise imposed on or claimed by diverse women, either within nation-states or across territorial and cultural borderlines.

Until relatively recently, few scholars theorized about or pursued research on women's containment within or across borders. Research on women, **gender** and **feminism** blossomed only a quarter of a century ago, and when it did, writers often operated within disciplines that necessarily relied on territorial borders as units of analysis. Just as political science focuses on nation-states, their territorial borders, and relations between, among and within them, the sub-fields of comparative politics and international relations (IR) also confined their analysis to nation-states, and IR paid little attention to women.

In comparative women's studies, writers often take the nation-state as the unit of analysis. Many fine collections exist, including the 43-country study, *Women and Politics Worldwide* (Nelson and Chowdhury 1994). These works are important for contextualizing global movements and ideologies; *The Challenge of Local Feminisms* (Basu 1995) is an example. Some collections have focused on women's participation in transitions to democracy, for instance, in *Women and Democracy: Latin America and Central and Eastern Europe* (Valenzuela et al. 1998).

Owing to feminist theorizing in international relations, gender has become visible in constructions of nationalism. For example, in wordplay involving domestic imagery and the stark language of apartheid, Christine Sylvester situates women's homelessness and the homeland men have made (Sylvester 1994a). Women assume **agency** in the nuanced constructions of *Feminist Nationalisms* (West 1997).

Where does this leave women's transnational political agency across borders? At one level, those who analyse global movements and international organizations provide the broadest panorama (see, among many, Baden and Goetz 1997). However, the scope of such analysis does not extend to everyday transnationalism, such as migration and regional 'free trade' schemes, as some issue-oriented (Pettman 1996) and regional political economy analyses (Staudt 1998) have done.

Conceptually, analysts need to attend more to cross-border networking and organizing with regard to specific issues and areas.

The European Union (EU) and North American Free Trade Agreement (NAFTA) provide manageable regional units of organizing within transnational political units, ranging from narrow (NAFTA) to broad (EU) in policy and legal leverage. These regional communities show how new borders are drawn from finance, commerce and occasionally migration, and old borders lose some of their traditional meaning. Still, national capital-to-capital organizing offers challenges regarding cultural, linguistic and national identities that continue to affect women. Electronic communication, especially bilingual communications, eases cross-border exchange, but the machinery is selectively available—that is, available mainly to the privileged. Ultimately, it is the person-to-person, cross-border organizing among women regarding health, immigration, **human rights** and social justice that demonstrates the challenges and opportunities of contesting the national political machinery that men have made. Borders are policed, even militarized, with some ferocity, and immigration policies continue to draw real and metaphoric lines between 'natives' and 'foreigners'.

Analysts cannot assume that women's solidarity will transcend borders. Rather, women are implicated in nationalism (McClintock 1996), along with actions that seemingly protect jobs, encourage cheap consumer goods in the global economy, and mute or affirm national and cultural differences. Meanwhile, globalization in the new millennium will probably continue to be associated with simultaneous bordering, debordering and rebordering.

Kathleen Staudt in Cheris Kramarae and Dale Spender (eds), *Routledge International Encyclopedia of Women: Global Women's Issues and Knowledge* (New York & London: Routledge, 2000), 119–20.

Bosnia

Part of the former Yugoslavia, Bosnia-Herzegovina had three ethno-religious groups residing there: Bosnian Muslims, Croats, and Serbs. After Yugoslavia began to break up, a process that started in the 1980s, civil war broke out in the early 1990s. The war was partly fuelled along ethnic lines. Bosnia is the site of what is believed to be the first instance of **genocide** in Europe since the 1940s. The civil war lasted from 1992 to 1995. The charge of genocide was levelled in the first months of the war with the discovery of detention camps, most famously in the town of Omarska, where pictures of emaciated prisoners released in the media recalled similar images from Auschwitz and other earlier death camps. The massacres in the town of Srebrenica, toward the end of the war, are also cited as evidence of an ongoing intent of the Serb forces to destroy the Muslim (and Croat) populations.

Tawia Ansah

Butler, Judith

Judith Butler's (1956–) significant contributions to feminist theory are her engagement with the Foucauldian account of the **body** as the inscribed surface of regulatory discourses, an elaboration of **gender** as the performative stabilization of sexual **difference**, and considerations of the exclusion inevitable to any **identity** category. The popularity of Butler's work indexes an increasingly philosophically-minded **feminism** in the Anglo-American academy.

Gender Trouble: The Subversion of Identity (1990) is one of the most widely read and controversial critiques of 'woman' as a universal and stable subject grounding feminism. Following Michel **Foucault**'s claim that juridical systems of power produce the **subject** they ostensibly represent, Butler argues that the subject 'woman', understood to ground the feminist enterprise, is instead a representational product of feminism. This argument and its ethical implication—that feminism must reckon with the exclusions foundational to any prescriptive description of identity—corroborates long-standing social difference critiques of the implicit attributes of the woman *supposed* to be the *subject* of feminism ('able-bodied', white, heterosexual, middle class).

Butler is best known for her formulation of 'gender performative', which articulates models of social life as a series of pre-scripted or ritual performances, with psychoanalytic descriptions of femininity as 'masquerade', and philosophical account of linguistic performativity. Employing Monique Wittig and Adrienne Rich's critiques of compulsory heterosexuality, Butler describes a 'heterosexual matrix'—a hegemonic, discursive/epistemic model of gender intelligibility—through which stable relations between sex ('male', 'female'), heterosexual object choice, and gender ('masculine', 'feminine') are produced and maintained. Compulsory 'gender performativity' produces the naturalness of 'sex', while gay and lesbian parodies of gender, Butler argues, 'implicitly reveal the imitative structure of gender itself—as well as its contingency' (Butler 1990, 137).

Though Butler's target audience is feminist, *Gender Trouble* has been hailed as a major contribution to the nascent field of queer theory. Popular applications of Butler's work largely emphasize the theoretical valences of gender 'performance', valorizing (queer) texts that denaturalize 'sex' (camp, drag). Her *Bodies That Matter* (1993) responds to criticisms that *Gender Trouble* evacuates the materiality and historicity of the body, leaving gender simply a matter of choice. *Bodies That Matter* underscores the constraints upon performative utterance (symbolic intelligibility, for instance), and situates gender performativity in relation to speech act theory.

Julia Creet in Lorraine Code (ed.), *Encyclopedia of Feminist Theories* (London and NY: Routledge, 2000), 69–70.

Buzkashi

Often deployed by anthropologists, and then journalists, as a metaphor for Afghanistan's political chaos, *buzkashi* is the Persian word for a horseback game involving a headless goat or, in the modern version, a headless calf.

Corinne Fowler

C

Capitalism

This usually refers to a form of economic or social organization characterized by the pervasive commodification of property, labour and knowledge. It has been a focus for feminists in many different ways. In the 1970s, when Marxist vocabulary was a kind of lingua franca among left-leaning intellectuals, socialist feminists tried to construct a theoretical understanding of **patriarchy** by using the concepts associated with the Marxist analysis of capitalism as a theoretical template. Alongside this fairly abstract analysis, those concerned with shifts in **gender** relations, including women's place in economic organizations, have necessarily had to understand changes in the relationship between local economics and the world market. These and other aspects of feminist analyses of capitalism and its relation to gender are usefully reviewed in Barrett (1988) and Andermahr et al. (1997) among others. However, for a number of reasons analyses that focus on capitalism fell out of favour in the 1990s.

Marxist analysis of capitalism: In classical Marxism, capitalism has a two-fold character. Although it is defined by a unique type of exploitation—the construction of human labour as a commodity for which the worker is paid, but at less than the value of that labour—it also has a progressive role in the evolution of human historical development. Because production under capitalism is organized around the continuing accumulation of capital, as against other social or spiritual goals, it tends to revolutionize the forces of production through scientific and technological development. As it expands, it subordinates other, earlier modes of production, either appropriating their surplus or replacing them with capitalist forms of production. However, according to orthodox Marxists the means of production created by capitalism will eventually outgrow their foundation in the exploitative capitalist relations that gave birth to them. Thus, although capitalism alienates human labour and blocks the free development of human potential, class conflict engendered by the oppression of the proletariat world led the working classes to join forces to overthrow bourgeois rule and created a socialist society in which the fruits of capitalist development could be enjoyed by all.

Ambivalence about capitalism in Marxism was also present in the ways it envisioned the relationship between capitalism and what second-wave **feminism** came to call gender relations. On the one hand, in the *Communist Manifesto* Marx drew a parallel between the power relation between wives and husbands as social categories and the oppression of the proletariat by the bourgeoisie, seeing the monopolization of property ownership as crucial in both cases. However, Marxists have also tended to see capitalism as sweeping away all pre-existing hierarchies and divisions, including women's oppression, and in 19th-century Europe the new economic independence of women employed in the factories spawned by the industrial revolution seemed evidence that this was occurring. Since then, many theorists, feminists and others, have tried to posit a relationship between capitalism and sexuality. Marcuse (1964), for instance, initially assumed that capitalism required the repression of sexual pleasure, but later argued that sexuality was deployed by capital in the interest of expanded consumption. In contrast, others have argued that because modern urban capitalist development has made it possible for individuals to lead lives outside marriage, it is linked to the development of modern homosexual lifestyles and communities. In the 1990s analysis of the commodification of sexuality also made reference to the expansion of capitalist investment into new spheres (Hawkes 1996), but here (as was increasingly the case elsewhere) capitalism has tended to be defined in terms of commodification and market exchange rather than in terms of the social relations of labour.

Feminist analysis of capitalism: To begin with, second-wave feminist theory in the West turned to highly abstract analyses of capitalism as a mode of production with its own laws of development. The most influential text, initially, was Engel's late 19th-century work *The Origin of the Family, Private Property, and the States* (Engels 1972), which argued that modes of production incorporated systems of *re*production as well as systems of production, thereby creating a space for feminists to bring the analysis of sexual partnerships between men and women within Marxist theory. Feminist theorists extended the concept of reproduction to include not only biological, generational production but also women's domestic labour, which reproduced the labour force on a day-to-day-basis. They also drew heavily on the French Marxist philosopher Louis Althusser's analysis (Althusser and Balibar 1970) of capitalism's dependence on its political and ideological levels, not merely its economic laws, in reproducing itself as a mode of production. Thus even though women may not have been as involved as men in capitalist relations in the workplace, their position in society was none the less determined by capitalism. The state, which was seen to reflect the interests of capital, played a particularly important role in cementing women's dependence on, and subordination to, men. However, the question whether the patriarchal basis of women's oppression was an aspect of capitalism or a separate but interacting system (dual system theory) was much debated. Though it would be hard to say that the exact nature of the relationship between different aspects of gender relations and capitalism was ever resolved, these debates, which now seem rather arcane, put the situation of women at the centre of theories of social formation rather than at their margins.

Since the collapse of the socialist regimes in Eastern Europe and the marginalization of Marxist social theory more generally, capitalism as a focus for analysis has faded into the background. Even feminist researchers concerned with workplace issues have argued that the concepts Marxists developed to understand women's entry into industrial production under capitalism, such as the concept of the reserve army of labour, are not well suited to understanding women's concentration in reproductive labour or the service sector (Benhabib and Cornell 1987). Studies of large-scale change have deployed concepts like globalization or post-Fordism (Fordism being a technological system that increases efficiency by breaking down and interlocking production operations, as on an assembly line, to mass-produce goods) or post-industrialism rather than focusing on capitalism as such. As examples, one could compare writing from the 1970s and early 1980s on what was called 'gender development', in which the expansion of the capitalist world system was seen as a crucial determinant of transformations in gender relations, with later work.

The marginalization of capitalism as an analytic focus has been particularly pronounced in Western feminist thought, where the 'cultural turn' privileging of the local, the personal and the textual as against the structural and economic, play and choices as against constraint and oppression, has been particularly pronounced (Barrett 1988). Foucauldian definitions of power, which refuse to see it based in any one sphere (such as economy), have also turned attention away from relations of production. This analytic refocusing obviously has complicated roots and is linked to economic and social changes, not just shifting intellectual currents. However, although calls for a feminist revival of materialist as against cultural studies may be justified, real changes in economic life and women's role in it make a simple return to the analysis of gender and capitalism where it left off very unlikely. For instance, the proliferation of cultural and intellectual activities and their centrality to economic life make it difficult to contrast the cultural and the economic in any simple way. Indeed, the centrality of commodification to the expansion of feminist culture—through the purchase and sale of commodities such as books, journals, theatre and cinema, and the development of electronic communication—suggest that feminist theory will have to take note of how feminism is itself linked to, if not dependent on, developments in late capitalism.

Carol Wolkowitz in Cheris Kramarae and Dale Spender (eds), *Routledge International Encyclopedia of Women: Global Women's Issues and Knowledge* (New York & London: Routledge, 2000), 139–40.

Cartesianism, feminist critiques of

The Cartesianism criticized by contemporary feminists is the potent legacy from Descartes that survives in present-day philosophy. Feminists looking at this Cartesian legacy have found a transformation of the philosophical conception of the self and its relation to the world. They have focused in particular on the consequences of Cartesian dualism, the isolating of a rational self, independent of the sensual, emotional **body**, and on an epistemological outlook that grounds knowledge of acts of a self whose first act is to know itself in isolation from other selves and from the world. Feminists have discussed the contribution of these views to the support of an ideology that privileges an autonomous rational **masculinity** over a relational, emotional, corporeal femininity. Cartesianism in Descartes's own day seems to have struck women somewhat differently, encouraging them to see themselves, although deficient in education, as possessing rational faculties.

Margaret Atherton in Lorraine Code (ed.), *Encyclopedia of Feminist Theories* (London and NY: Routledge, 2000), 72.

Categories and dichotomies

All perception and thought involves and depends upon categories. There are categories of concrete things, such as Living and Non-Living, Furniture and Fruit, and Chair/Table/Bed, Orange/Apple/Pear. Animal and Plants are categories: Furry and Scaly across the category Animals. There are abstract categories, such as Cause, Democracy, ideas with criss-crossing sub-categories such as Western or Confucian, Ancient or Modern). Theories (scientific, mathematical, social) involve distinctive central categories. Physics has the categories of Time, Matter, Sub-atomic particle, Energy. Biology has the categories of Species, Endocrine. In the case of feminist theory, the **gender** categories Woman and Man are central for analysing political, social and historical situations.

The categories Woman and Man are troublesome, partly because they are also the central categories in the 'theory' of male supremacism or **patriarchy** (they embody ideas and beliefs that made patriarchy seem 'right' or 'natural'). As used in patriarchal thought, these categories are often dichotomous or binary; feminists have seen this as oppressive to women and are concerned not to replicate such usage themselves.

Categories are *dichotomous* when they divide a certain domain into two groups that are mutually exclusive and exhaustive, and the two groups are each other's opposites; for example, Reality being divided into Nature and Culture, where every phenomenon is classed as one or the other, and the qualities of Nature and Culture are seen as opposite.

Mutually exclusive means that no items can be in both categories; *exhaustive* means every item has to be in one category or the other. If a pair of categories is mutually exclusive and exhaustive, they must be *absolute opposites*, meaning there is no spectrum of intermediate cases between them. *Polar opposites* such as Hot and Cold are at the ends of a spectrum of intermediate cases. Strictly speaking, categories should only be called 'dichotomous' if they involve absolute opposites, but people often refer to polar opposite categories as 'dichotomous'.

A *binary* category divides the universe, or a domain, into those things which are its members, and *everything else*, presented simply as 'not this category'; for example, the category Numbers simply contains all numbers, and leaves everything else in the universe undifferentiated as Not Numbers.

In some contexts, the category Man (or the category of things male or masculine) functions as a binary category, leaving Women (or things female or feminine) with no specific characterization, merely as *not-Men/not-male/not-masculine.* Man is functioning as a binary category, and women disappear into *not-Man* in all the situations where men's experience or subjectivity is taken to be what experience or subjectivity *is*, and a person has to have that experience or subjectivity, or be just off the map. If membership in the category A is thought of as the member's **identity**, then using the category A gives no corresponding or parallel identities to any of the things outside A. If the A-identity qualifies A-members for respect, dignity, privileges and so on, then such categorization leaves all others off the map, without identity or value. This not only deprives those who are not A-members of something valuable, it masses them all together, relative to A, as having no distinctions among themselves and having no identities or name of their own; this, in turn, works against their being able to form political resistance to their exclusion and deprivation.

In other contexts, the categories Woman and Man seem to function as dichotomous, when, for instance, people think of women and men as having distinctive kinds of sexuality, work styles, or personality (e.g. relational/performance, passive/aggressive), and see these as in some sense opposite.

Some people think that mutually exclusive and exhaustive opposite genders can be equal in status (and the above definition of *dichotomous* leaves this possibility open). They believe that the opposites can be complementary and the two categories interdependent, with neither being dominant or subordinate. However, many feminists think either that this is impossible in contemporary societies, or that *dichotomy* necessarily implies hierarchy, i.e. one group being in some sense dominant or primary, the other subordinate or secondary.

If you think dichotomous categories can only be hierarchical and that genders are dichotomous categories, then you will not be able to put the idea of gender and the idea of equality together. You would think that as long as there are genders, there will be inequality—domination, oppression. So, instead of thinking the goal of **feminism** is gender equality (which many people do think), you might begin to think about creating a world in which there is no gender. Another option might be to think about the possibility that genders which are dichotomous (and hence, hierarchical) in patriarchal social orders might be differently constructed in a non-oppressive social order, so that they are two categories, but are not dichotomous. If practices and usage were changed so gender categories were not dichotomous, then one could imagine an individual being both genders, or being neither, and one might imagine there being three or more gender categories.

Words have meanings by virtue of contrasts with other words: for example, *red,* by contrast with *blue, yellow, round, tall…, or car,* by contrast with *bus, house, boat, lawnmower*.... As these examples suggest, though, it does not have to be a dichotomous contrast or a binary contrast. Even a word that apparently has a dichotomous contrasting term, such as *husband* (versus *wife*), may in fact have its meaning through many non-dichotomous contrasts such as the whole array of other kin terms: *brother, uncle, mother, father, cousin.* We can recognize that categories operate by way of contrast with and differentiation from other categories, without assuming that all categorizing is dichotomous or binary, hence, likely to be oppressive. Categorizing is unavoidable. Categorizing in ways that are oppressive is avoidable.

Marilyn Frye in Lorraine Code (ed.), *Encyclopedia of Feminist Theories* (London & New York: Routledge, 2000), 73–74.

Circumcision – *see* **Clitoridectomy**

Class and feminism

It cannot be overemphasized how central the issue of class has traditionally been to **feminism**, and it may be argued that attitude toward class is the point from which the various formations of feminism have taken their departure. Feminists disagree about the importance of the role that class plays in the oppression of women. The political task for feminists has been to determine where to draw the line between the oppression of women's experience as a result of class disempowerment and the oppression they experience as a result of **patriarchy**. Roughly speaking, Marxist feminists deny there is a system of oppression called patriarchy; they argue that women's oppression is caused by the stratification of society into classes and that the oppression of women commences with the advert of the privatization of the ownership of the means of production, the accumulation of capital by the capitalist class at the expense of the working class, and the resultant lack of democratic control of the means of life. Oppression specific to women exists only insofar as women's practices become different from men's under the capitalist mode of production. What is primary to this view is that women are women only insofar as they are human beings and, because there is no immutable or essential 'human nature', there is no essential 'women's nature'. Rather, human nature is formed by the predominant type of labour in which human beings engage in any historical epoch. Women are not an exception in this, and thus 'women's nature' is manifold, reflecting the fact that women from different classes have engaged in varying types of labour over the course of history. Hence, women from the same class have more in common with one another as 'workers' than they do as 'women'.

Oppression: class versus patriarchy. In direct opposition to the Marxist view, liberal feminists deny that class plays any oppressive function in the social subjugation of women. In the liberal feminist analysis, patriarchy is the only structure that causes the oppression of women. Between these two poles there exists a range of opinions. For example, some feminists maintain that class in the Marxist sense exists and provides the correct analysis of capitalist society, but that it has not played the principal role in the oppression of women. Others reject the primacy of the category of 'class' in favour of a feminist analysis of oppression that posits a **difference** between women's and men's way of experiencing the world. Feminists who subscribe to the latter belief are likely to be interested in what they consider the specificity of female experience.

How do Marxist feminists argue for the oppression of women in terms of class? The fact that the (lower and middle) working classes have possessed only their physical and mental abilities to labour to sell in the marketplaces has meant that workers have been forced into commodity production in order to live. This is because much of a worker's day is taken up with selling her or his activities, which include all of those things that, together, maintain the worker's ability to take part in productive activity in return for wages in the marketplace: the cleaning of the home, the washing of clothes, the preparation of food, entertainment, sleeping and the procreation of children. The family unit, in this view, is simply a socially constructed way of forming a cohesive economic unit within the constraints of **capitalism**, but its construction is a result of the part orchestrated for women to play in capitalist society rather than a 'natural' family arrangement. The term *family* for Marxist feminists, refers to nothing more than a system of association and kinship dependent on the social and economic function that it serves. The Marxist feminist sees nothing 'natural' about heterosexual coupling in which children are produced to be raised by their biological parents on the basis of the blood tie.

This analysis poses a problem for Marxist feminists, who face the question of why it is women *in particular* who stay in the home while men take part in commodity production in the public sphere. If capitalism is **gender**-blind, this argument would suggest, then why have men not stayed at home? Why has the distribution of women and men over reproductive and productive activities not been equal? If women have almost exclusively performed reproductive activities rather than been involved in productive activities because their proximity to their children constitutes an economically convenient arrangement, then why do women stay at home longer than this formative period of child development in which they must be at home, for example, to breast-feed their babies?

The Marxist feminist argues that though it is true that women have stayed in the home, this is more likely to be a function of their class position. Working-class women simply do not stay at home after having their children, and most do, in fact, labour in the marketplace just as men do. If working-class women do stay at home until their children are of school age, then this is only because child care is unaffordable to them. Only bourgeois women face this type of exclusion from productive activity in that they are productively redundant in the eyes of the marketplace because they do not contribute to commodity production. Bourgeois women, according to Marxist feminists, become objects that belong to capitalist men because, unlike working-class women, they are supported by and dependent on capitalist men in living directly off the profits of the working class generated by their husbands. This economic enslavement of bourgeois women to bourgeois men gives bourgeois men a position of dominance over bourgeois women that working-class men have never had over working-class women, and this in turn leads to a particular sexist formulation of what Marx calls 'false consciousness' in the distorted view of women's nature that it creates. Noteworthy here is the fact that Marxist feminist political **activism**, through 'wages for housework' campaigns, has sought to address the 'invisibility' of women's domestic labour and to expose the implicit dependency of capitalism on the reproductive labour that women perform.

It might seem that the Marxist feminist position would be simple enough to disprove by appealing to the fact that women's oppression predated industrial capitalist society as we know it, or that disadvantage of class is at least partially neutralized for the women's struggle because of the integration of global feminist discourse. However, a refutation is not quite so easily achieved as this, because capitalist society has deep historical roots from which we may chart a continuous development from the ancient slave societies. Both historically and politically, it is perilous to assert that the forms of oppression that characterize capitalism's predecessor modes of production (such as feudalism) can be clearly distinguished from the forms of oppression that women experience today.

Melissa White in Cheris Kramarae and Dale Spender (eds), *Routledge International Encyclopedia of Women: Global Women's Issues and Knowledge* (New York & London: Routledge, 2000), 184–86.

Climate change

Due to natural causes and human activity, there have been increased levels of greenhouse gases (carbon dioxide and methane in particular), which have led to an increase in average temperature at Earth's surface; an increase of 0.6 degrees in the past 35 years, estimated to rise by about 0.17 every decade (Homer-Dixon 2006, 161). With temperatures projected gradually to rise to levels that will significantly alter ecosystems, attention is given to a range of problems including sea level rise, water shortages and droughts, crop failures, increased flooding and extreme weather events. In the past eight years or so, the debate has shifted from the question of whether or not anthropogenic climate change is real, to debates about the scope of likely impacts and what is to be done by whom, when and how.

Sherilyn MacGregor

Clitoridectomy

Also known as female genital mutilation (FGM), or genital cutting, clitoridectomy includes any procedure carried out with the express intention of injury to and partial or total removal of the clitoris for cultural, religious or other non-medical purposes. FGM also refers to such practices carried out on the entire external female genitalia including the clitoris. FGM is occasionally described as female circumcision and is seen as a rite of passage into womanhood for girls, but in practice can be carried out at any time from infancy to the age of 15. It is usually performed by traditional midwives but the World Health Organization (WHO) reports that it is increasingly performed by trained medical professionals. FGM has been classified into four types by WHO. These include: *Clitoridectomy*: partial or total removal of the clitoris and occasionally the folds of skin surrounding the clitoris; *Excision*: partial or total removal of the clitoris and labia minora, with or without the excision of the labia majora; *Infibulation*: narrowing of the vaginal opening by creating a seal through, cutting and repositioning the inner and occasionally the outer labia. This can be done with or without the removal of the clitoris; *Other*: various other harmful practices for non-medical reasons under this heading include pricking, piercing, incising, scraping and cauterizing the genital area.

Corinne Fowler

Commission on the Status of Women (CSW)

Functional Commission of the Economic and Social Council (ECOSOC), established by ECOSOC in 1946 by Res. 11(II). The mandate of the Commission was to report to ECOSOC on the promotion of women's rights—political, economic, social and educational—and make recommendations on problems requiring immediate attention. The Commission was to act as a preparatory committee for the UN's World Conferences on Women.

The Commission originally had 15 members, but it was enlarged on several occasions and in the late 1990s had 45 (13 member states from Africa, 11 from Asia, four from Eastern Europe, nine from Latin America and the Caribbean, and eight from the group of Western European and other states).

In 1983 ECOSOC established within the Commission a five-member Working Group on Communications on the Status of Women to consider the communications received, including the replies of governments, and bring to the attention of the Commission those that appeared to reveal a consistent pattern of injustice and **discrimination**.

Jan Osmańczyk and Anthony Mango (ed.), *Encyclopedia of the United Nations and International Agreements* (Routledge, 2003). Politics Online. Taylor & Francis. www.routledgepoliticsonline.com:80/Book.aspx?id=w032 (accessed 15 October 2009).

Committee on the Elimination of Discrimination against Women (CEDAW)

CEDAW is the body of independent experts who monitor the implementation of the Convention on the Elimination of All Forms of Discrimination against Women.

CEDAW consists of 23 independent experts on women's issues from around the world. Countries that have become party to the treaty (states parties) are obliged to submit regular reports to the Committee on how **discrimination** against women is implemented. During its sessions the Committee considers each state party report and addresses its concerns and recommendations to the state party in the form of concluding observations.

In accordance with the Optional Protocol to the Convention, the Committee's mandate is very specific: a) receive communications from individuals or groups of individuals submitting claims of violations of rights protected under the Convention to the Committee; and b) initiate inquiries into situations of grave or systematic violations of women's rights. These procedures are optional and are only available where the state concerned has accepted them.

The Committee also makes general recommendations on issues affecting women and urges states parties to devote more attention to the issues. General recommendations are directed to states and in accordance to the articles or themes in the Conventions.

www2.ohchr.org/english/bodies/cedaw/index.htm

Concubinage

Cohabitation between sexual partners outside the *civil law.* Compare with *marriage.*

Glen Newey

Consciousness-raising

This is a central activity of the women's liberation movement, enabling women as a group to share problems, experiences and feelings. It allows women to recognize that what they perceive as personal problems—'problems that have no name'—are shared with others. It also enables women to realize that what they think of as resulting from their own personal inadequacy, or their own inability, may be a result of living in a patriarchal society—that the personal is often political.

Consciousness-raising can be seen as enabling women to overcome false consciousness—to throw off the man-made model of women, come to a realization of their own potential, and move from self-deluded dependence to autonomy and self-reliance. This view is reflected in the 'consciousness-raising novels' of the 1970s—for example, Marilyn French's *The Women's Room* (1977). Liz Stanley and Sue Wise (1983) point out that consciousness-raising is not just a series of stages—going from a prefeminist consciousness to one of true understanding, a feminist consciousness—but an ongoing process. It is a process that brings about personal and collaborative change as opposed to structural change. The need for ongoing discussion—in small groups or informally—is, they suggest, central to being a feminist.

Pamela Abbot in Cheris Kramarae and Dale Spender (eds), *Routledge International Encyclopedia of Women: Global Women's Issues and Knowledge* (New York & London: Routledge, 2000), 221–22.

Contraception

Contraception aims by various means to prevent conception or pregnancy after sexual intercourse. Some have distinguished contraception, as affecting the act of sexual intercourse, from sterilization, which affects the sexual faculty or power to procreate, but the more common understanding today views sterilization as a form of contraception when it is used to prevent pregnancy. **Abortion** disrupts an already existing pregnancy or conception and thus differs from contraception, although abortion, too, is used for fertility control.

The ancient world of both East and West knew the reality of contraception either by avoiding insemination in the female vas or by employing potions or magic. However, the term *contraception* is of 20th-century origin. Despite the long recognition of contraception and some use of it, the practice of contraception became widespread throughout the world only in the 20th century and especially in recent times. The widespread use and acceptance of contraception today truly constitutes a revolution. Many different factors help to explain this. New contraceptive techniques have been developed especially in more recent times—the condom (originated in the 17th century but manufactured on a wide scale only in the 20th century); male and female sterilization (started at the turn of the 20th century but much easier and more readily available today); and the pill and the IUD (very recently). Contemporary science continues to look for new and better methods of contraception.

The increased life expectancy of all human beings, massive improvements in infant and child health care and development, the requirements of an increasingly industrialized society, the growing acceptance of sexual relations apart from procreation, and especially the changing role of women in society, have all contributed to the growing acceptance of contraception. Contemporary discussion has sensitized the whole world to the need for population control—a position first proposed by Thomas Malthus at the very end of the 18th century. Malthus's solution at that time advocated moral restraints such as delayed marriage and not contraception. In our day the most common forms of contraception are sterilization, both male and female (comparatively simple procedures have made sterilization much more popular today), the pill and the IUD (medical considerations have slowed the use of these methods), diaphragm, condom, periodic abstinence or natural family planning, foams, and coitus interruptus.

Charles E. Curran in Paul Barry Clarke and Andrew Linzey (eds), *Dictionary of Ethics, Theology and Society* (Routledge, 1996). Religion Online. Taylor & Francis. www.routledgereligiononline.com:80/Book.aspx?id=w004 (accessed 15 October 2009).

Contractarianism

This is a family of views that seek to justify morality or political institutions by reference to rational agreement. We are, according to this tradition, to think of morality or of legitimate states as objects of some sort of social contract. The general idea is that a morality or a form of political organization is to be justified by being shown to be the outcome of the rational agreement of the individuals over whom it has authority. This general idea may take many different forms, and we need to distinguish between different sorts of contractarian theory, as well as between different purposes to which it may be put.

Christopher W. Morris in Lawrence C. Becker and Charlotte B. Becker (eds), *Encyclopedia of Ethics*, second edn (Routledge, 2001). Religion Online. Taylor & Francis. www.routledgereligiononline.com:80/Book.aspx?id=w020 (accessed 15 October 2009).

Convention Against Torture and Other Cruel, Inhuman or Degrading Treatment or Punishment

The Convention Against Torture and Other Cruel, Inhuman or Degrading Treatment or Punishment was adopted by resolution 39/46 of 1984 at the 39th session of the United Nations General Assembly and came into force on 26 June 1987. Its mandate is to encourage states to operationalize measures necessary to prevent **torture** from occurring within their **borders**. The Convention also requires that states should not return persons to their country of origin if there is a reasonable expectation that those persons may face the threat of torture. The convention defines torture as 'any act by which severe pain or suffering, whether physical or mental, is intentionally inflicted on a person for such purposes as obtaining from him or a third person, or for any reason based on **discrimination** of any kind, when such pain and suffering is inflicted by or at the instigation of or with the consent or acquiescence of a public official or other person acting in an official capacity. It does not include pain or suffering arising only from, inherent in or incidental to lawful sanctions.' The Committee Against Torture (CAT) monitors the implementation of the Convention by states parties, through a 10-member committee of independent experts. To this end all states parties are required to submit regular reports upon which the committee will comment and make recommendations in its concluding observations.

Tawia Ansah

Crime against humanity

In international criminal law, a crime against humanity is an international crime and is distinguished from a domestic crime on the basis that its breach is of concern to the whole of humanity. The victims are not only those persons directly affected by the commission of the offence, but all of humanity. Such crimes include murder, extermination, enslavement, deportation, imprisonment, **torture**, **rape**, and persecution on political, racial or religious grounds.

Crimes against humanity are distinct from war crimes because they are not restricted to periods of armed conflict. It was not until the end of the Second World War that vague references were formally categorized as a new species of international crime. The Charter of the International Military Tribunal for Nuremberg (1950) expressly created the category of Crimes Against Humanity, and provided sanctions for their breach. The crimes were to attract individual criminal liability for the perpetrator.

The drafters of the Nuremberg Charter had to grapple with the non-intervention principle, which provided that international law had no application to events that occurred within the **borders** of a country. In other words, international law only applied to events that had occurred during the course of an international event such as an armed conflict. This meant that the crimes committed by the Nazis against their own people prior to the outbreak of the war could not be punished by an international tribunal. The Nuremberg Charter broke new ground in creating a category of international crime that was punishable under international law.

For a crime to be classified as a crime against humanity it must be directed at a civilian population. There is also a need for the crime to exhibit the characteristics of system or organization and be of a certain scale and gravity. The crime cannot be the work of an isolated individual acting alone. It must be shown to be part of a wider plan or policy, but there is no requirement that the crime be carried out pursuant to the policy of a state. In other words, the perpetrator of a crime against humanity must know that his/her act is part of a widespread or systematic attack against civilians, even if a perpetrator is motivated by personal reasons for committing the crime.

The widespread or systematic nature of such crimes must be distinguished from random acts of violence unconnected to any system or organization. The term ‘widespread’ refers to frequent, large-scale action carried out collectively with considerable seriousness and directed against a multiplicity of victims. The term ‘systematic’ refers to activities that are thoroughly organized and that follow a regular pattern of abuse.

Grant Niemann in Martin Griffiths (ed.), *Encyclopedia of International Relations and Global Politics* (Routledge, 2005). Politics Online. Taylor & Francis. www.routledgepoliticsonline.com:80/Book.aspx?id=w163 (accessed 15 October 2009).

Critical theory

The establishment of the Institute for Social Research in Frankfurt, Germany on 22 June 1922 marks the beginning of Frankfurt School critical theory. The principal members of the School have been the founders of the Institute, Max Horkheimer (1895–1971) and Theodor Adorno (1903–69); Herbert Marcuse (1898–1979), the major 'New Left' theorist of the 1960s; and Jurgen Habermas (1929–), the foremost critical theorist of recent times.

Their writings develop an approach to society that is faithful to the spirit but not to the letter of Marxism. Classical Marxism or historical materialism used the 'paradigm of production' to analyse particular social systems and to comprehend human history. This paradigm maintains that the forces of production (technology) and the relations of production (class relations) provide the key to understanding political systems and historical change. In particular, class conflict has been the greatest influence on how societies have developed.

Karl Marx (1818–83) and Friedrich Engels (1820–95) argued that the struggle between the bourgeoisie (the class that own the means of production) and the proletariat (the class that has to sell its labour-power in order to survive) is the central dynamic in capitalist societies. They believed that class conflict would destroy and lead to a socialist system in which the forces of production would be used to benefit the whole of society rather than to maximize bourgeois profit. They also had a vision of global political progress in which the whole of humanity is eventually linked in a socialist world order. Crucially, Marx and Engels thought that the purpose of social inquiry was to promote the emancipation of exploited members of the proletariat. Marx maintained that 'philosophers have only interpreted the world: the point is to change it'. This, in a nutshell, is the commitment to emancipatory social science that is defended by the Frankfurt School.

Its members sought to preserve this conception of social inquiry while breaking with what they saw as the fatal limitations of the paradigm of production. It was plain to Horkheimer and Adorno in the 1930s that the stress on the centrality of production and class conflict could not explain violent nationalism in the fascist societies, the rise of totalitarian state power and the outbreak of total war. Their writings displayed increasing pessimism about the possibility of emancipation. Nowhere is this more striking than in Adorno's claim that human history has led from the slingshot to the A-bomb. To them, the promise of emancipation, which had united the members of the Enlightenment (such as Kant) with their successors (such as Marx and Engels), seemed impossible to fulfil in the modern era. This was to agree with Max Weber's bleak vision in which society is increasingly dominated by pressures to administer society more efficiently and economically.

Later members of the Frankfurt School sought to recover the emancipatory project without relapsing into classical Marxism and without neglecting the dangerous side of modernity. Herbert Marcuse (1898–1979) analysed how

capitalism created 'one dimensional man' caught up in the satisfaction of manufactured material needs; but he believed that the student movement of the 1960s and the struggles for national liberation and socialism in the Third World represented a major political effort to create the free society. Habermas has focused on how efforts to administer capitalist societies have led to what he calls the 'colonisation of the life world'—that is to the encroachment of administrative rationality on everyday life—but he sees in the social movements that promote human security, equality for women and environmental restoration the promise of a new kind of society that replaces the quest to control nature and administer society with the struggle to enlarge human freedom. (*See also* **Capitalism**.)

Andrew Linklater in Martin Griffiths (ed.), *Encyclopedia of International Relations and Global Politics* (Routledge, 2005). Politics Online. Taylor & Francis. www.routledgepoliticsonline.com:80/Book.aspx?id=w163 (accessed 15 October 2009).

D

Darfur

This is a region of the Sudan that, since 2003, has seen much violence: militia, with government support, have razed whole villages and pillaged livestock, causing deaths estimated in the hundreds of thousands and the displacement of at least 2.5m. people. Darfur has been described by many, including the International Criminal Court's Office of the Prosecutor, as the most recent instance of **genocide**. However, others, including a UN Commission of Experts Report, have determined that it is not.

Tawia Ansah

Decolonization

The process whereby a colonial society achieves constitutional independence from imperial rule. It is the reverse of colonization—a process whereby one state occupies the territory of another state and directly rules over its population. Although it has a very long history (the Greeks, for example, set up colonies around the Mediterranean several hundred years before Christ), it is the period of European expansion into Africa, Asia, the Americas and the Pacific between the 15th and the early 20th centuries that is generally associated with colonialism as a system of rule.

Martin Griffiths and Terry O'Callaghan, 'Decolonisation', in *International Relations: The Key Concepts*, second edn (Routledge, 2002). Politics Online. Taylor & Francis. www.routledgepoliticsonline.com:80/Book.aspx?id=w043 (accessed 15 October 2009).

Deconstruction

Deconstruction is usually associated with poststructuralist continental philosophy; it draws upon Nietzsche, Saussure and Roland Barthes among others. Its best known representative is Jacques Derrida, but while formally recent, its roots are deep, as deep as the moment of the expulsion of the poets from the Republic. Deconstruction is a mid-to-late 20th-century phenomenon, but it is no more than the return of the poets who have been biding their time for two and a half millennia and, having returned, have sufficient vigour to expel the philosophers and the theologians and reclaim the Republic. Such a dramatic reversal of tradition could not be placed with one or even a few people; it is rather a consequence of a reversal of several related features of Western intellectual life. Prime among these are the death of the subject, the death of God and the linguistic turn. (*See also* **Subject, the death of the**.)

Paul Barry Clarke and Andrew Linzey (eds), *Dictionary of Ethics, Theology and Society* (Routledge, 1996). Religion Online. Taylor & Francis. www.routledgereligion online.com:80/Book.aspx?id=w004 (accessed 15 October 2009).

Determinism – *see* **Biological determinism**

Difference

Difference is an important issue in contemporary feminist debate. It refers to sexual difference and to discussions in linguistics, psychoanalysis, politics and science.

The term *difference* occurs in the work of Swiss linguist Ferdinand de Saussure (1857–1913). In his *Course in General Linguistics* (1988), Saussure described language as a convenient system that we utilize to structure and transmit our experience of the world. He argued that words do not contain meaning, but that meaning arises from the compositional differences between words. The French critic Jacques Derrida (1930–2004) developed Saussure's theory to suggest that meaning is also a process of deferral. Linguistic meaning not only results from the differences between words but also is a ceaseless and unstable interaction between both present and 'absent' differences. Derrida coined the term *difference* to denote this. He argues that language continually evokes different meanings that exceed and disrupt any intended meaning.

The French philosopher and historian Michel **Foucault** (1926–84) extended this notion that meaning is a result of difference to explore its formative role in the construction of **identity**. Foucault suggested that perceptions of sameness and difference organize the way we think, speak and define ourselves in relation to others. In the 19th century, the German philosopher Georg Hegel (1770–1831) provided a metaphor for the way difference is ordered through opposition and hierarchy in his model of a master and slave. The model refers to the procedure whereby a master defines himself in relation to his slave, good is designated with reference to evil, white in contradistinction to black. The French feminist philosopher and writer Simone de **Beauvoir** (1908–86), in her pioneering study *The Second Sex* (1949), developed the paradigm to argue that man has appropriate, negate, and made us of woman's difference in order to guarantee his position as master.

Psychoanalysis has contributed to the debate on difference. The French psychoanalyst Jacques Lacan (1901–81), drawing on the pioneering work of Austrian Sigmund Freud (1856–1939), argued that language structures identity and that our perception of difference, including sexual and **gender** difference, are culturally determined.

The issue of difference has been discussed by feminist commentators. The feminist critic Hélène Cixous (1938–), Julie Kristeva (1941–), and Luce Irigaray (1930–) insist that men have repressed or employed women's differences to establish patriarchal rule. In her study *Speculum of the Other Woman* (1974), the French philosopher and psychoanalyst Luce Irigaray argues that our conception of difference derives from a single, male view. She examines the premises of Western **metaphysics** and concludes that our entire system of thinking has been determined with the result that women exist within it only as the inverted **other** of men. She argues that this bias is encoded in language, reducing women to silence. The Bulgarian-born linguist and psychoanalyst Julia Kristeva believes that the monotheism of Western culture is sustained by differentiating between the sexes, since only by designating another sex can **patriarchy** institute the 'one law' on which it depends. The French critic and writer Hélène Cixous also takes this view, and in her essay 'Sorties' (1975) she argues that women must inscribe their differences in order to shatter the patriarchal state.

This insistence that the exploration of difference will provide an impetus for change has in turn been challenged. Anglo-American **feminism**, with its background in the grassroots women's movements that emerged in the aftermath of the US civil rights campaigns in the 1960s and 1970s, has intended to promote sexual equality rather than sexual difference, and especially the view that linguistic revolution will instigate change, and detracts from the struggle to end women's legal and economic oppression.

There have been various attacks based on biological reduction or essentialism. Essentialism refers to the belief that there are innate physiological differences between the sexes and that these give rise to deferent perspectives and patterns of behaviour.

Other critiques maintain that sexual difference can be understood only in the context of the differences that operate between women, such as the differences of **race**, class, wealth, education, political persuasion and sexual preference. The Indian critic Gayatri Chakravorty Spivak (1941–), for example, has criticized Western feminism for its exclusion of black women, its insularity, and its assumptions concerning those women it perceives as inhabiting a 'third world.' Spivak points out that these differences, too, have been appropriated or ignored. With regard to sexual orientation, critics such as the Italian-born Teresa de Lauretis (1938–) have argued that lesbianism must be considered in terms of difference. Her view—that lesbian sexuality constitutes an important source of identity—is in contrast to broader definitions, such as that expressed by the North American writer Adrienne Rich (1929–), who argues for a 'lesbian continuum' of relationship between women. Other feminists have identified differences in wealth, class and education as crucial factors in distinguishing between women's experiences and opportunities.

In science, difference is similarly an important issue. Recent research has suggested that the traditional allocation of X and Y chromosomes (whereby a woman has two X chromosomes and a man one X and one Y) is overstated and that sexual difference derives from a single gene (SRY). This gene produces a protein responsible for transforming otherwise female-destined embryo into a male. Breakthroughs in genetic, surgical and hormone engineering have further reduced the boundaries of differences. The realities of foetal implants, drug therapy and the plastic construction of sexual organs seem set to challenge our ideas of sexual difference.

Susan Sellers in Cheris Kramarae and Dale Spender (eds), *Routledge International Encyclopedia of Women: Global Women's Issues and Knowledge* (New York & London: Routledge, 2000), 378–80.

Difference, politics of

The politics of **difference** has several meanings attached to it in ongoing debates on multiculturalism, **feminism** and social justice. If we define 'politics' as those processes through which social power is organized, 'politics of difference' emphasizes the fact that these processes are shaped by a multiplicity of relational differences between individuals and groups of individuals. Accordingly, 'politics of difference' is often juxtaposed to 'politics of universalism', which indicates the liberal principles of equality of recognition and treatment despite (and in face of) existing differences, e.g. **gender**, **race**, ethnicity or sexual orientation. In contrast, 'politics of difference' proceeds from an understanding that differences between groups and individuals need to be recognized and their particular needs accommodated in order to realize substantial equality and justice—even if this requires differential treatment for different groups on the basis of their distinct features. Such an understanding has its roots in the assumption that supposedly neutral, difference-blind conceptions of citizenship and equality tend to obscure ruling norms and dominant institutions that privilege some groups and render others as deviant. 'Politics of difference' is thus related to the concepts of 'recognition' and 'redistribution'. Depending on the differences considered—cultural differences, ethnic differences, socioeconomic differences, gender differences, or differences in capabilities to name a few—'politics of difference' can merge with 'identity politics' or 'interest politics' in some conceptualizations, while being opposed to them in others. Its theoretical localization also depends on how the respective social groups are defined, on differences between them, as well as within them. Notwithstanding that it is frequently used by many authors, 'politics of difference' remains a term often undefined as such. Iris M. Young's work is an exception. Conceptualizations of politics of difference are central to the development of her theory. She differentiates distinct forms of 'politics of difference' according to the understanding of what constitutes the social groups, and the issues of justice emphasized by them. (*See also* **Biological determinism**.)

Lena J. Kruckenberg

Disciplinary power – *see* **Foucault, Michel**

Discourse analysis

Discourse analysis is a qualitative method that aims at revealing the ways in which communication legitimizes or maintains ideology. This work offers a means for understanding how language reflects and reproduces biased pattern of thought, decision-making and distribution of power. Discourse analysis examines the language used within a given text or interaction: it assumes that relations of power are embedded in language and attempts to deconstruct the way in which this occurs. Individual analyses of discourse might take a variety of approaches to identifying pattern of articulation and absence, including the observation of dominant metaphors and analogies, the recognition of cultural myths and stereotypes, the examination of available frames of meaning, or simply an analysis of central terms or clusters of key terms within a text. Such work is designed to reveal values, perspectives and interests that are not explicitly stated in the text but that convey subtle ideological messages such as who or what is dangerous or threatening. Discourse analysis is used by feminists to reveal the subtle way in which communication transmits familiar ideas about **gender**, such as the idea that women are emotional and men are rational.

Discourse analysis is based on the assumption that a sense of reality is constructed through the use of language and is often interested in revealing how racism, **sexism**, class differences and other inequalities are subtly furthered. Thus, discourse analysis can be used to reveal gender or racial bias within a language or within a particular usage pattern or text, such as the observation that women and girls are often in a grammatically 'passive' position while males are grammatically 'active', or the finding that mainstream news makes use of traditional ideas about **rape** to structure its coverage of sexual assault. In its emphasis on the deconstruction of ideological power within texts, this work can sometimes make use of highly abstract scholarly jargon that is difficult even for readers with graduate-level education. As such, it has been accused of being detached from or unrelated to practical issues of social change. Discourse analysts claim that only limited social change can take place without accompanying change at the level of language and language use, and thus assert that their work is sort of praxis aimed at the transformation of elements of social life such as that which a group accepts as reality, truth, normality and knowledge.

Lisa Cuklanz in Cheris Kramarae and Dale Spender (eds), *Routledge International Encyclopedia of Women: Global Women's Issues and Knowledge* (New York & London: Routledge, 2000), 201.

Discrimination

Discrimination connotates, in a morally neutral sense, differential treatment resulting from the perception of a difference. However, the term usually describes a situation in which individuals (or groups of individuals) are treated differently in an unfavourable way notwithstanding the fact that there is no difference that is of moral relevance, or a legitimate purpose for doing so. Discrimination can be directed against individuals as well as groups. When individuals are discriminated against, they are often perceived as merely belonging to a group defined by characteristics such as **gender**, sexual orientation or ethnicity that are defined as deviant and inferior as such. Discrimination may be open or more subtle, it can be intended or appear as an unintended consequence. It can be institutionalized across different sectors of society and through the rules governing interactions within them. Thus, discrimination can be situational and restricted to a limited range of social relationships, and can also encompass instances where members of certain groups are denied equal standing in society generally, and are assigned an inferior social status, which impacts all aspects of their daily existence. Discriminatory attitudes and practices are often deeply entrenched and produce social strata, the very existence of which reinforces and 'legitimates' further discrimination. Discrimination on the basis of sex, for example, was justified as reasonable for centuries as it contributed to the stabilisation of traditional roles of men and women.

Lena J. Kruckenberg

Discursive formations – *see* Foucault, Michel

Division of labour, sexual

The sexual division of labour relegates women to a dual form of exploitation. One facet is the fact that the overwhelming majority of unpaid work needed to maintain a household and care for children is performed by women. The other facet is apparent where women have gained access to waged work: their remuneration is generally less than that of men for the same work, while they tend to be concentrated in less prestigious occupations. In households where women do not perform waged labour, there is in effect a hidden tax—the wage paid to a male head of household is expected to support an unwaged worker. Where women work out of the household, social and cultural norms often result in a 'second shift' being performed by women: namely, the upkeep of the household.

Globally, what has been referred to as the New International Division of Labour (NIDL) has seen many women in Less-Developed Countries (LDCs) become enmeshed in a system of global commodity production. Women are perceived as ideal for light manufacturing and assembly production for their putative 'nimble fingers' and docility in the face of exploitation.

Christopher May in R.J. Barry Jones (ed.), *Routledge Encyclopedia of International Political Economy*, first edn (Routledge, 2001). Politics Online. Taylor & Francis. www.routledgepoliticsonline.com:80/Book.aspx?id=w053 (accessed 15 October 2009).

Doing difference

This is a term coined by Candace West and Sarah Fenstermaker which refers to an understanding of '**difference**' as an 'ongoing interactional accomplishment' (West and Fenstermaker 1995, 8). West and Fenstermaker thus transcend conceptualizations of 'differences' as individual properties or traits, and embed them in the social contexts in which they are 'accomplished' by social actors. This broader conceptualization of difference (which includes but not exclusively encompasses **gender**, **race** and class) is derived from an ethnomethodological understanding that properties of social life need to be researched in the ways in which they are actually constructed in interactions. This perspective leads on to questions as to how an 'objective' or 'factual' character of difference is achieved, how 'appropriate' activities relating to perceived differences become established, and how differences are used to hold individuals accountable. Notwithstanding that 'differences' such as gender, race or class vary in their particular characteristics and consequences, West and Fenstermaker understand them as being similar in the ways in which they function as 'mechanisms' in the production of inequality and the exertion of power (Ibid., 9).

Lena J. Kruckenberg

Domestic sphere – *see* Patriarchy

Durable inequality

A term coined by Charles Tilly (1999, 2000, 2005), which denotes systematic and persistent unequal relations between individuals and groups of individuals, where interactions create advantages for one group over another. These asymmetrical relations rest upon (perceived) categorical differences such as man/woman, black/white, citizen/foreigner, rather than continuous individual differences in, for example, performance or other attributes that are often thought to account for them. Durable inequalities emerge when individuals who control access to resources and opportunities use readily available categorical distinctions—such as man/women or majority/minority—to solve immediate organizational problems, such as when deciding whom to employ for a particular job. Durable inequalities hence rely on the institutionalization of categorical pairs and related '**social boundaries**'.

Lena J. Kruckenberg

E

Ecofeminism/feminist environmentalism

This is a branch of feminist theory that has as its central concern the connections and intersections between the historic domination/exploitation of the environment by humans and the oppression/exploitation of women by men. Key theorists include historian Carolyn Merchant, philosophers Val Plumwood and Karen J. Warren, and scientists Donna Haraway and Vandana Shiva. Ecofeminist theorists have developed important, critical analyses of the **gender** politics of the environment, but have unfortunately been associated with some essentialist ideas about women's unique relationship with nature that have led feminists to avoid it (Davion 1994).

Sherilyn MacGregor

Ecological modernization

Ecological modernization is an environmental discourse that advocates the use of 'technological advancement to bring about better environmental performance' and economic efficiency in a win-win situation (Schlosberg and Rinfret 2008, 256). It has a supply side focus and sees great hope in the idea that co-operation between government and business will solve environmental problems (Hajer 1995). It is a discourse that has become dominant in the past decade in Europe and is catching hold in the USA (Schlosberg and Rinfret 2008).

Sherilyn MacGregor

Ecomaternalism

This is a term used to describe the rhetorical-political connections that women often make between their caring and mothering roles/practices/values and their concern for the environment. Women's maternal role is often used as a justification for their involvement in environmental **activism** on 'quality of life' issues like toxic contamination of the ecosystem (MacGregor 2006).

Sherilyn MacGregor

Empiricism, feminist – *see* **Philosophy of science, feminist; Feminism and international relations**

Environmental privatization

The process of placing greater responsibility for environmentally responsible behaviour on the private sphere of the household, than on the regulation of institutions and corporations in the public domain.

Sherilyn MacGregor

Environmental security

Environmental security is a discourse that is underpinned by pessimistic Hobbesian predictions that **climate change** will inevitably lead to conflict over scarce resources between and within states (Homer-Dixon 1999). There has been growing interest in recent years in presenting climate change as a serious threat to national and global security (see Elliott 2004). For example, in 2004 the United Kingdom's Chief Scientist, Sir David King, made the connection clear, saying that climate change is a worse threat than terrorism (Connor 2004).

Sherilyn MacGregor

Environmentality

Invoking a Foucauldian analysis of governmentality, some environment theorists have referred to the internalization of the sense of duty to live in a more sustainable way as a form of 'environmentality' (Luke 1995). It is a disciplinary process through which people are made into good green subjects by adopting the environment values of government so that the state can govern 'at a distance' thereby meeting its policy goals with minimal (costly) intervention or coercion.

Sherilyn MacGregor

Epistemology – *see* **Philosophy of science, feminist**

Essentialism – *see* **Biological determinism**

Essentialist debate, environment

Feminists historically have steered clear of environmentalism because of its focus on 'nature', which treads too close into essentialist, biological determinist territory for feminist comfort (for an excellent discussion see Alaimo 2000). Second-wave **feminism** was characterized by an '**identity** crisis' that (to simplify) pitted 'cultural' feminists who celebrate women's unique qualities against poststructuralist critics of the very concept of woman as essentialist. Third-wave feminism has all but killed off the essentialist and 'cultural' factions, and now focuses on identity and **difference** issues with varying levels of commitment to **poststructuralism** and postcolonialism. Ecofeminists are often accused (rightly in some cases and wrongly in others) of advancing some claims that could be called essentialist. The celebration of women's unique knowledge of and connection to the environment—and the concomitant **positioning** of women as saviours of the imperilled planet—by some ecofeminist scholars (e.g. Mies and Bennholdt-Thomsen 1999; Salleh 1997) is admittedly problematic.

Sherilyn MacGregor

Ethnic cleansing

When ethnic populations are minorities in territories controlled by rival ethnic groups, they may be driven from the land or (in rare cases) systematically exterminated. By driving out the minority ethnic group, a majority group can assemble a more unified, more contiguous and larger territory for its nation-state. The term 'ethnic cleansing' was coined in the context of the dissolution of Yugoslavia in the 1990s. It is a literal translation of the expression *etnicko ciscenje* in Serbo-Croatian/Croato-Serbian. The precise origin of this term is difficult to establish. Mass media reports discussed the establishment of 'ethnically clean territories' in Kosovo after 1981. At the time, the concept related to administrative and non-violent matters and referred mostly to the behaviour of Kosovo Albanians (Kosovars) towards the Serbian minority in the province.

The term derived its current meaning during the war in **Bosnia** and Herzegovina (1992–95). As military officers of the former Yugoslav People's Army had a preponderant role in all these events, the conclusion could be drawn that the concept has its origin in military vocabulary. The expression 'to clean the territory' is directed against enemies, and it is used mostly in the final phase of combat in order to take total control of the conquered territory.

Analysis of ethnic cleansing should not be limited to the specific case of former Yugoslavia. This policy can occur and have terrible consequences in all territories with mixed populations, especially in attempts to redefine frontiers and rights over given territories. There is a new logic of conflict that relies on violent actions against the enemy's civilian population on a large scale, rather than on war in the traditional sense, i.e. between armed forces.

It is important to underline that the policy of ethnic cleansing fundamentally represents a violation of **human rights** and international humanitarian law. Only when the means and methods of ethnic cleansing policies can be identified with genocidal acts, and when a combination of different elements implies the existence of an intent to destroy a group as such, can such actions represent **genocide**. Ethnic cleansing lacks the precise legal definition that genocide has, although it has been widely used in UN General Assembly and **UN Security Council** Resolutions, documents of special *rapporteurs*, and the pamphlets of non-governmental organizations.

Some suggest that ethnic cleansing is merely a euphemism for genocide. However, there would seem, to be a significant difference between them. The former seeks to 'cleanse' or 'purify' a territory of one ethnic group by use of terror, **rape** and murder in order to convince the inhabitants to leave. The latter seeks to destroy the group, closing the **borders** to ensure that none escapes. This observation should not be taken to imply that ethnic cleansing is not a barbaric international crime. It is most certainly punishable as a **crime against humanity**. (*See also* **Genocide**.)

Martin Griffiths (ed.), *Encyclopedia of International Relations and Global Politics* (Routledge, 2005). Politics Online. Taylor & Francis. www.routledgepoliticsonline.com:80/Book.aspx?id=w163 (accessed 15 October 2009).

Ethnography – *see* **Anthropology**

European Convention on Human Rights

The European Convention on Human Rights, the full title of which is the European Convention for the Protection of Human Rights and Fundamental Freedoms, is a document sponsored in 1950 by the Council of Europe. It represents an unprecedented system of international protection for **human rights** and enables individuals to apply to the courts for the enforcement of their rights. It came into operation in 1953. All the member states of the European Union (EU) are signatories of the Convention and, in accepting it, the EU has also accepted the role and predominance in this area of both the associated European Commission of Human Rights and the **European Court of Human Rights**. The EU is not a signatory to the Convention although concerns about its democratic credentials and the meaning of citizens' rights have led to some pressure to accede. However, this would require an amendment to the Treaty on European Union. Instead, a Charter of Fundamental Rights was drawn up and proclaimed in 2000.

David Phinnemore and Lee McGowan, 'European Convention on Human Rights', in *A Dictionary of the European Union*, second edn (Europa Publications, 2004). Politics Online. Taylor & Francis. www.routledgepoliticsonline.com:80/Book.aspx?id=w009 (accessed 15 October 2009).

European Court of Human Rights

The European Court of Human Rights, which is based in Strasbourg, operates under the aegis of the Council of Europe. It hears cases concerning individuals and practices in those states that are party to the **European Convention on Human Rights**. Cases may be brought by individuals or by the European Commission of Human Rights. All the member states of the European Union (EU) have ratified the Convention, and the EU has accepted the jurisdiction of this court in the sphere of **human rights**.

David Phinnemore and Lee McGowan, 'European Court of Human Rights', in *A Dictionary of the European Union*, second edn (Europa Publications, 2004). Politics Online. Taylor & Francis. www.routledgepoliticsonline.com:80/Book.aspx?id=w009 (accessed 15 October 2009).

F

Family division of labour – *see* **Division of labour, sexual**

Female genital mutilation – *see* **Clitoridectomy**

Femininity – *see* **Masculinity; Beauvoir, Simone de**

Feminism

Feminism is an interdisciplinary body of knowledge based primarily on the experiences and lives of women. It encompasses a variety of theoretical approaches. These include liberal, radical, standpoint, socialist, postcolonial and postmodern feminisms. All these approaches are seeking to understand the sources of women's oppression and prescribe strategies for ending it. Liberal feminists see the source of female subordination in legal constraints that block women's advancement in the public sphere. All other approaches identify deeper structures of inequality—variously associated with **patriarchy**, relations of production and reproduction, class and **race**—which contribute in different ways to women's oppression. The term 'feminism' is also used to refer to political and social movements working to end women's subordination.

Ann J. Ticker in R.J. Barry Jones (ed.), *Routledge Encyclopedia of International Political Economy*, first edn (Routledge, 2001). Politics Online. Taylor & Francis. www.routledgepoliticsonline.com:80/Book.aspx?id=w053 (accessed 15 October 2009).

Feminism and international relations

A simple definition of **feminism** means the study of and movement for women not as objects but as subjects of knowledge. Until the 1980s, and despite the inroads of feminism in other social sciences, the role of **gender** (i.e. the relationship between sex and power) in the theory and practice of international relations (IR) was generally ignored. Today, this is no longer the case as a number of feminist thinkers have turned their critical sights on a field that has traditionally been gender blind. Over the last decade, feminism has emerged as a key critical perspective within the study of IR. The initial impetus of this critique was to challenge the fundamental biases of the discipline and to highlight the ways in which women were excluded from analyses of the state, international political economy, and international security. One can now distinguish between at least two main types of feminism in the study of IR.

The first wave of feminist scholarship in the 1980s is now called *feminist empiricism*, in which IR scholars have sought to reclaim women's hidden voices and to expose the multiplicity of roles that women play in sustaining global economic forces and state interactions. For example, women's participation and involvement facilitate tourism, colonialism and economically powerful states' domination of weak states. The maintenance of the international political economy depends upon stable political and military relations among states. In turn, the creation of stable diplomatic and military communities has often been the responsibility of women (as wives, girlfriends and prostitutes). Feminist empiricism exposes the role of women and demonstrates their importance in a wide variety of arenas. In case one might think that the role of women is marginal to the real business of the international economy, it should be noted that Philippine women working abroad as domestic servants annually contribute more to the Philippine economy than do the national sugar and mining industries.

A second focus of feminist research has been directed at deconstructing major discipline-defining texts and uncovering gender biases in the paradigmatic debates that have dominated the field since its inception in 1919. Sometimes referred to as *standpoint feminism*, this type of feminist scholarship argues for the construction of knowledge based on the material conditions of women's experiences, which give us a more complete picture of the world since those who are oppressed and discriminated against often have a better understanding of the sources of their oppression than their oppressors. Whilst feminist empiricism exposes the role of women in IR, standpoint feminism alerts us to the ways in which the conventional study of IR is itself gendered.

Despite the rise of feminism in the field, there remains a major imbalance between male and female academics in IR, and many feminists attack the ways in which men's experiences are projected as if they represent some universal standpoint. According to standpoint feminists, the major Western intellectual traditions of realist and liberal thought have drawn from culturally defined notions of **masculinity**, emphasizing the value of autonomy, independence and power. Those traditions have formulated assumptions about interstate behaviour, security, progress and economic growth in ways that allegedly perpetuate the marginalization and invisibility of women.

Feminism is a rich, complicated and often contradictory body of research in the study of IR at the end of the 20th century. In a broad sense, feminism is an umbrella term. It embraces a wide range of **critical theory** aimed at examining the role of gender in IR. However, there is **liberal feminism**, **radical feminism**, Marxist feminism, post-Marxist or socialist feminism, postmodernist feminism, and the list continues. Given the commitment by all feminists to some kind of ethic based on equality between men and women, their work is sometimes equated with idealism, and they have themselves been criticized for ignoring men in their zeal to promote the emancipation of women. It remains to be seen how feminist scholarship evolves to include a broader agenda of questions about gender in IR theory and practice. (*See also* **Gender and international relations.**)

Martin Griffiths and Terry O'Callaghan, *International Relations: The Key Concepts*, second edn (Routledge, 2002). Politics Online. Taylor & Francis. www.routledgepoliticsonline.com:80/Book.aspx?id=w043 (accessed 15 October 2009).

Feminism and Psychoanalysis – *see* **Film theory, feminist**

Feminist anthropology – *see* **Anthropology**

Feminist film theory – *see* **Film theory, feminist**

Feminist geographies – *see* **Geographies, feminist**

Feminist identity – *see* **Butler, Judith**

Feminist philosophy of science – *see* **Philosophy of science, feminist**

Feminist standpoint epistemology – *see* **Philosophy of science, feminist; Feminism and international relations**

Film theory, feminist

Although there were women film-makers back in the 1920s and even earlier making statements about the suitability of the camera to a woman's expression of her own subjectivity, feminist film theory did not come about fully until the late 1960s. This second wave took the world of academia and journalism in the Western world and Australia by storm and very quickly began to generate texts related to women's issues and, just as importantly, to disciplines taught within academia—including film studies. Indeed, no discussion of film (or television for that matter) can ignore feminist film theory which, since the early 1970s, has had such a strong impact on film studies—starting with issues of **gender** representation.

Feminist film theory 1968–74: Although we use the term 'feminist film theory', Annette Kuhn (1982, 72) rightly points out that there does not exist one single theory, rather a series of 'perspectives'. The logic for this can be found in two occurrences directly linked to the second wave of **feminism**. The first is a contemporary occurrence, an effect of the 1960s. The second is based in a longer look at history. Women (mostly students), radicalized during the 1960s by political debate and so-called sexual liberation, reacted against being placed second after men in the intellectual—and at times violent (as in the USA)—pursuit of political change. The failure of radicalism, which culminated and then crashed in 1968, to produce any substantive change for women led them to form **consciousness-raising** groups that effectively galvanized women into forming a women's movement. This rejection of male radicalism in intellectual terms can be interpreted also as a rejection of the pursuit of 'total theory' as exemplified by the totalizing effect of structuralism (the debate of the 1960s)—although, as we shall see, feminists did not reject some of the fundamental principles of structuralism. Indeed they were party to moving that debate on to the more pluralistic one of **poststructuralism**.

The second occurrence was in some ways an outcome of the first. Looking at what had happened to them in recent history, women saw the need to look longer and further at women's history and from a number of perspectives. Annette Kuhn's definition of feminism, in this instance, most aptly sums up how feminism and feminist film theory after it are composed of a number of tendencies, so resisting the fixity of one 'single theory': '[feminism is] a set of political practices founded in analyses of the social/historical position of women as subordinated, oppressed or exploited whether within dominant modes of production (such as **capitalism**) and/or by the social relations of **patriarchy** or male domination' (Kuhn 1982, 4). By extension, feminist film theory, then, is political, and as early as its first period it set about analysing, from different perspectives, dominant cinema's construction of women. As we shall see, at this juncture, the differences in perspectives were particularly marked between feminists on either side of the Atlantic.

The effect of this first period of feminist film theory was to shift the debate in film theory from class to gender. Feminist film critics examined the question of feminine **identity** and the representation of women in film images as the site/sight or object of exchange between men. At this point the focus of these analyses was exclusively Hollywood cinema and had the sobering effect of dismissing films, which previously had been elevated to auteur status, for their male-centred point of view and objectification of women. Exemplary of this approach was Molly Haskell's *From Reverence to Rape: The Treatment of Women in the Movies* (1974). It is noteworthy that the person to coin the term auteur theory was Andrew Sarris, Haskell's husband. In her book Haskell, an American journalist, made two very important points. First, she suggested that film reflects society, and vice versa and in so doing reflects the ideological and social construction of women who are either to be revered (as the virgin) or reviled (as the whore). Second, in her analyses of the 1930s and 1940s Hollywood cinema she made the important distinction between melodrama and women's films—the latter, as she pointed out, being made specifically to address women. This second point, that of the central role of the female protagonist *and* female spectator, led to a renewed and different focus on genre and the possible aesthetic and political consequences of gender **difference**. This very crucial point was one that was developed in the second period of feminist film theory. However, for the moment let's focus on the first point, since this was the one that revealed differences in perspectives held by feminists on either side of the Atlantic.

In the USA in the early 1970s, Molly Haskell, Marjorie Rosen and Joan Mellen, were the three leading feminists writing on the representation of women in cinema. The approach they adopted was sociological and empirical—which was consonant with the state of the art in film criticism in the USA at that time. Their critical approach was intended to expose the misrepresentation of woman in film, which it did, but in a specific way. Consistently with a sociological-empirical approach which aims to ascribe—to fix meanings based on fact—they also assumed a presumed feminine essence repressed by patriarchy. That is, their analyses presumed a predetermined sexual identity, difference. In simple terms, what was being said was 'the facts show that women get represented in images as virgin or whore because that's how patriarchal society represents women to itself'.

With hindsight, this conclusion now seems quite reductionist, but the fact remains that these findings constituted a first important stage. Meantime in Europe (particularly in the United Kingdom) this **essentialist debate** represented three major problems for feminist film theorists, who included Claire Johnston, Laura Mulvey, Pam Cook and Annette Kuhn. These feminists were influenced by the more scientific and definitely anti-empirical approach offered by semiotics and structuralism. They argued that the essentialist debate assumed, first, that all women possessed an innate ability to judge the authenticity of the representation of women in film and, second, that all women film-makers were feminists. Most critically of all, they pointed out that a belief in a fixed feminine essence meant legitimating patriarchy through the back door. By accepting the fixed essence of a woman as predetermined, 'given order of things', implicitly what was also being accepted was the 'naturalness' of the patriarchal order (Lapsley and Westlake 1989). The British feminists, arguing along Althusserian lines, also pointed out that film was as much productive as it was a product of ideology and, furthermore, that—in the same way that Althusser theorized that the **subject** was a construct of material structures—so, too, was the spectator a constituted subject when watching the film. The time was ripe for moving on from a causal and reductionist debate. Historical materialism (an analysis of the material conditions within historical contexts that placed women where they were), semiotics and psychoanalysis were the tools invoked to investigate beyond the current superficial findings and reflectionist statements produced by the American authors.

These criticisms notwithstanding, Haskell, Mellen and Rosen's works represented a significant first benchmark that had important outcomes. Their 'saying that it was there' made it clear that the next step was to find out 'how it got there', the better to change 'it'. Their work led to the definition of three basic approaches to achieve this uncovering and the subsequent changing of the way women are constructed in film. There was a need for, first, a theoretical analysis of the way in which mainstream cinema constructs women and the place of women; second, a critical analysis of the work of women film-makers; and third—as a conjuncture of these two points—the establishing and implementation of feminist film practices.

The final point that needs to be made about this period is that women were becoming a presence in all areas of the cinematic institution. The year 1972 witnessed the first feminist film festivals in the USA and the United Kingdom. Women film-makers formed collectives to encourage women into film-making (for example the London Women's Film Group, formed in 1972). *Women and Film*, the first feminist film journal, was also published in the USA between 1972 and 1975. The next period then had a powerful heritage already as it set its own trail-blazing practices.

Feminist film theory 1975–83: In 1972 Claire Johnston, Laura Mulvey and Linda Myles organized the Women's Event at the Edinburgh Film Festival. To accompany the event Johnston edited a pamphlet, *Notes on Women's Cinema*, published in 1973, which contained her own ground-breaking essay, 'Women's Cinema as Counter-Cinema'. This text was one of the first to make clear that cinematic textual operations could not be ignored. To change cinema, to make women's cinema, Johnston argued, you had to understand the ideological operations present in actual mainstream film practices. The task was to determine both the 'how' of female representation and the 'effect' of female **positioning** in the process of meaning construction. Under the 'how', of first importance was a reading of the iconography of the image. How was the female framed, lit, dressed and so on? Under the 'effect', of primary significance was the female's positioning within the structure of the narrative. In the first instance what was required was a reading of the image of sign, a need to understand the denotation and connotations of the image. In the second instance the psychology of the narrative had to come under scrutiny: why, repeatedly, do so many narratives in mainstream cinema depict the woman as the object of desire of the male character embarked upon his Oedipal trajectory? Camera-work and lighting make it clear that she is a figure upon whom he can fix his fantasies. These, then, are the textual operations constructing ideology which, argues Johnston, we must come to understand through a deconstruction of the modes of production, which in turn leads to an exposing of the cinematic and narrative codes at work. Only then can an oppositional counter-cinema become a possibility. It is worth quoting from her conclusion to make the point that the cinema she envisaged was not one that relied on self-reflectivity or foregrounding the conditions of film production alone (what she refers to here as political film) but one that would draw on female fantasy and desire:

> A strategy should be developed which embraces both the notion of films as a political fool and film as entertainment. For too long these have been regarded as two opposing poles with little common ground. In order to counter our objectification in cinema, our collective fantasies must be released: women's cinema must embody the working through of desire: such as objective demands the use of entertainment film. Ideas derived from the entertainment film, then, should inform the political film, and political ideas should inform the entertainment cinema: a two way process.
>
> (Johnston 1976, 217)

The formidable recipe of foregrounding film practices and female subjectivity would, Johnston believed, effect a break between ideology and text—a dislocation that would create a space for women's cinema to emerge.

The other ground-breaking essay of this period was Laura Mulvey's *Visual Pleasure and Narrative Cinema* (1975). Mulvey's focus was less on textual operations within the film and more on the textual relations between screen and spectator. The important text is perceived as *the key* founding document of psychoanalytic feminist film theory. In this essay Mulvey seeks to address the issue of female spectatorship within the cinematic apparatus. She examines the way in which cinema functions, through its codes and conventions, to construct the way in which woman is to be looked at, starting with the male point of view within the film and, subsequently, the spectator who identifies with the male protagonist. She describes this process of viewing as acopophilia—pleasure in viewing. However, she also asks what happens to the female spectator, given that the classic narrative is predominantly that of the Oedipal trajectory and since that trajectory is tightly bound up with male perceptions and fantasies about women. Mulvey can conclude only that she must either identify with the passive position of the female character on screen, or, if she is to derive pleasure, she must assume a male positioning. This deliberately polemical essay met with a strong response by feminist critics. The next section details the development of that debate. For now, let's examine the third aspect of the theoretical work during this period: textual analysis. First, though, a brief synopsis of the relevance of psychoanalysis to feminist film theory.

The introduction into feminist film theory of psychoanalysis represented a major departure from the first, sociologically based period. It made it possible to address feminist issue such as identity and memory; it also led to the discussion of femininity and **masculinity** as socially constructed entities as opposed to the more simplified binary divide along biological lines into female and male. This important distinction meant that it was now possible to analyse sexual difference rather than just assume it as a predetermined reality. If it was socially constructed, then that construct could be deconstructed and changed. Difference could be analysed in terms of language rather than in terms of biologism. Until Jacques Lacan's analyses insisted on the importance of the

Symbolic (that is, language) in the construction of subjectivity, feminine sexuality had languished under the Freudian principle of penis envy—that is, of lack. Given that Sigmund Freud privileged the penis, it is hardly surprising that he never managed to theorize the female Oedipal trajectory satisfactorily. Why should he, since it would mean a loss of power? According to a Freudian analysis, then, power relations are based in sexual difference. This ignores the social material conditions that construct human sexuality. This is why the shift from sex to language, which Lacan's approach did much to assist, meant that the ideological operations of language (patriarchal language) could come under scrutiny. By bringing structuralism and psychoanalysis together, feminist film theorists had at their disposition an incisive analytical tool. The text was ready for deconstruction!

In the USA, particularly, feminists turned their attention to analyses of the textual operations in film and their role in constructing ideology. Essays in the influential journal *Camera Obscura*—launched in 1976 by a breakaway group from the earlier *Women and Film* collective—were instrumental in revealing the ideological operations of patriarchy; and through application of structural-psychoanalytic theories they demonstrated how narrative codes and conventions sustain patriarchal ideology in its conditioning and control of women. Note that, for the moment, historical materialism has disappeared from the theoretical triumvirate. That was to go on the back-burner until the third period. Currently, textual analysis focused on narrative strategies and, in particular, picked up the question of genre and its role in structuring subjectivity—a point which, as we have seen, Haskell had raised in drawing the distinction between melodrama and women's films.

The intervention of psychoanalysis and feminism now brought melodrama and women's films into the critical limelight, focusing on the discourses that construct the symbolic place of 'women and the maternal'. The western, too, was investigated. The dominance of male-defined problematics within the narrative, the role of women as mere trigger [sic] to male action and choice and, finally, the counter-Oedipal trajectory of so many of its heroes—all these narrative strategies are exposed. Why, indeed, do so many gunmen ride off into the sunset leaving 'the women' behind? Film noir was another genre which under scrutiny could not hide the dominance of male subjectivity whose **gaze**, motivated by the fear of castration, either fetishizes the 'threatening', dangerous woman into the phallic and therefore unthreatening **other**, or seeks to control and punish the perceived source of this fear.

Another part of the feminists' strategy in their textual analyses was to read against the grain and, in particular, to foreground sexual difference. Horror and film noir, because of their voyeuristic 'essence', were prime targets of investigation, but so too were melodrama and women's films because of their focus on the family and the so-called 'women's space'. Thus, for example, in a film noir or horror film a women's annihilation, which on the surface might appear as a result of her own actions, may in fact be read as the male repressing the feminine side of his self by projecting it on to the woman and then killing her. Hitchcock's *Psycho* (1960), for example, plays on this sexual identity in crisis in a number of very complex ways. Norman Bates constantly reminds himself that women are filth—who, of course, he must eradicate; he is also the voyeuristic gaze of mainstream cinema looking through numerous holes (peeping Tom) at the woman's body and, finally, the male gaze that must annihilate that which he must not become—woman and filth.

Melodrama is historically perceived as a theatre that reflects 19th-century bourgeois values whereby the family at all costs will prevail, remain united and in order. When the genre was taken up by early cinema this reflection continued. Superficially, by the 1930s and 1940s, melodrama and women's films appeared not to challenge the patriarchal order. However, in their narrative construction these films gave space for a woman's point of view. It was in this respect that a reading against the grain was possible; harmony in the family had to be restored, but what caused the initial disorder? Most often conflict between the two sexes. As Laura Mulvey (1977, 39) points out, melodrama is the one genre where ideological contradiction is allowed centre screen: 'Ideological contradiction is actually the overt mainspring and specific content of melodrama [....] No ideology can ever pretend to totality: it searches for safety-valves for its own inconsistencies'. Melodrama is just one of those safety-valves. Interesting in this context are the implications of the setting specific to melodrama: the domestic sphere, the woman's space or place and the family. The male character at the centre of the conflict has to resolve it within that sphere. 'If the family is to survive, a compromise has to be reached, sexual difference softened, and the male brought to see the value of domestic life [....] The phallocentric misogynist fantasies of patriarchal culture are shown here to be in contradiction with the ideology of the family' (Mulvey 1977, 40). Because it is the female character's subjectivity to which we are privy, it is her emotions with which we first identify. Unlike the western or gangster movie and the dominance of male-defined problematic that only the hero's action can resolve, here we are confronted with 'the way in which sexual difference under patriarchy is fraught, explosive and erupts dramatically into violence within its own private stamping ground, the family' (Ibid., 39). There is something tantalizing in the implications of this reading against the grain that goes something like this: 'If melodrama and women's films are the only site where narrative strategies expose the contradictions in patriarchal ideology and if the domestic sphere is the private stamping-ground of patriarchy, then how much indeed the rest of mainstream cinema must work to assert in public spaces that patriarchal ideology is without contradiction. *And* how "dangerous" it would be if female subjectivity crept out of the private domestic sphere and exposed those contradictions'. If melodrama did not exist, Hollywood would have had to invent it. (See also **Gangster/criminal/detective thriller/private-eye film**.)

Susan Hayward, *Cinema Studies: The Key Concepts* (London and New York: Routledge, 2000), 112–16.

Fortress Europe

This is a term often used in political economy to describe the difficulty of penetrating European markets created by the European economic zone. It was first applied to Europe by American politicians and businessmen attempting to enter European markets. Within the area of migration studies it has come to refer to the increasing geospecific exclusionary practices fostered under legislation for free movement in Europe that restricts migrants from certain geographic regions outside Europe.

Foucault, Michel

Paul-Michel Foucault (1926–84) was born in Poitiers, the son of a wealthy surgeon. He had a somewhat troubled youth and attempted suicide several times, but went on to study philosophy, psychology and psychopathology. After working for a period with the mentally ill, he taught at a number of universities in France and abroad, culminating in his appointment to the illustrious Collège de France in 1970, choosing the title of Professor of the History of Systems of Thought.

Like most post-war French intellectuals, Foucault began as a Marxist, being briefly a member of the Communist Party. Disillusioned, he was strongly influenced in the 1950s by the philosophy of Nietzsche and by French structuralism. Structuralism was an intellectual movement that developed within linguistics and then spread to **anthropology**, literary studies and a variety of other disciplines. It saw human activity, social organization and, above all, language as governed by deep internal structures, complex sets of rules, which unconsciously work to severely constrain what human beings can think and do. On this view, our natural assumptions about our freedom to think and act and confer meaning are illusory. This anti-humanist stance was retained when structuralism evolved into the more fluid and ambiguous **post-structuralism**. Anti-humanism was endorsed by Foucault throughout his work and only modified a little towards the end of his life.

Foucault's first major publications—*Madness and Civilization* (1961), *The Birth of the Clinic* (1963), *The Order of Things: The Archaeology of the Human Sciences* (1966) and *The Archaeology of Knowledge* (1969)—could be said to be broadly structuralist (although he rejected the structuralist label, as he did all labels). Subsequently, his themes and approach became more Nietzschean, stressing the role of power in society and in knowledge, as in works such as *Discipline and Punish: The Birth of the Prison* (1975) and his massive *History of Sexuality* (three volumes of which were published, beginning in 1976, out of a projected six). In the 1970s Foucault became a major inspiration for radicals, campaigning on behalf of oppressed individuals and groups of various kinds. He was especially interested in sexual **identity**. In later years he spent much time in America, exploring the experimental gay scene in California. He died of AIDS-related illness in 1984.

In his early writings, Foucault seeks to explain the origins of our modern conceptions of madness, of clinical medicine and of the modern conception of 'man' generated by the new social sciences. He develops the concept of 'discursive formation', by which he means the complex of concepts and arguments and techniques and technologies relating to a particular practice, and how these in fact create not only the knowledge of a subject but the object of study itself. Thus, the modern conception of madness was created along with the creation of psychiatry in the early 19th century. This was not, as is generally assumed, the result of improved science and greater humanity, but of the wider needs of society for increased discipline.

The new conceptualizations, including that of 'man', are therefore arbitrary and could be otherwise. Just as psychiatry defines and therefore creates the madness it studies and treats, the human and social sciences create and help to control 'man', in the sense of establishing what is 'normal', what are the rules and norms of normal human functioning and how we understand ourselves. This is the point of Foucault's notorious remark towards the end of *The Order of Things*: 'As the archaeology of our thought easily shows, man is an invention of recent date. And one perhaps nearing its end' (387). What he means by this is that the way we presently tend to understand ourselves can change and perhaps soon will. He may also have in mind the triumph of structuralism, which would fatally undermine current notions of what is human—a view he held at the time, but subsequently abandoned.

These studies of modern forms of thinking are set against a background of a series of 'epistemes', which are broad frameworks that have changed over time from the Renaissance to the Classical period (*c.* 1750–1800), to the modern. These frameworks provide basic assumptions that underpin particular disciplines and practices at a given period. Foucault's epistemes and discursive formations bear a close resemblance to the 'paradigms' described in Thomas Kuhn's *Structure of Scientific Revolutions* (1962), which are similar structures of theory and practice that define 'normal science' for long periods. Like these scientific paradigms, epistemes and discursive formations are arbitrary and do not follow each other in a rational and progressive way that scientists, and the rest of us, generally assume. Foucault's use of the term 'archaeology' rather than 'history' is meant to emphasize the disjunction and discontinuities of historical change in the way we understand the various aspects of reality, especially ourselves. Foucault wanted to emphasize that these pictures are not necessary but contingent and therefore changeable.

After 1970 Foucault abandoned the somewhat rigid and deterministic framework of epistemes and made discursive formations more flexible and, at the same time, more arbitrary by concentrating upon the concept of power and how power and knowledge are deeply interwoven. It is a move signalled by his use (following Nietzsche) of the term 'genealogy' rather than 'archaeology' to describe his method. Genealogy studies the way in which what we count as knowledge and the content of discourse is the outcome of power struggles between groups whose views then count as universally valid truth and knowledge. Power is an integral part of the production of truth and there is a 'politics of truth' in any society whose outcome determines what is deemed true and by what procedures it is legitimately arrived at.

This is expressed above all in what many regarded as Foucault's most important work, *Discipline and Punish* (1973), in which he analyses the origin of the modern prison in the early 19th century. The reason for this development, he insists, was nothing to do with greater rationality or humanity, but with the need of an industrializing society for more efficient techniques of social control, with schedules of activity and systems of surveillance, culminating in Jeremy Bentham's all-seeing 'Panopticon' prison design, with its central tower from which all prisoners can be observed, where those subject to it internalize the system and become docile bodies. The prison, Foucault believes, became the model for all kinds of disciplinary systems in schools, factories, asylums and so on. 'Panopticism' characterizes modern society, where all are scrutinized and made to conform to standards of 'normality'.

Through the social sciences the modern individual, or '**subject**', is created in the sense of defined and given a sense of what is normal. In creating disciplined citizens the social sciences determine the nature of the modern state, which uses its knowledge of these sciences, rather than force or custom, to control society.

Disciplinary power thus pervades the whole modern society. However, Foucault's analysis of power is a subtle one. Power in society can be creative as well as oppressive, disciplines can enable as well as oppress. In modern societies, particularly in democracies, nobody 'owns' power: it moves and flows among different groups and institutions. It inheres in the system rather than in individuals. Society cannot function without it. Nevertheless, individuals are shaped, manipulated, restricted and oppressed by webs of power relations of which they understand little.

Foucault's history of sexuality also gives a picture of the individual hopelessly entangled in a web of power relations, although in this work there is another major theme (especially in volumes two and three, published in the year of Foucault's death): namely, the creation of the subject and how the subject can resist. If there is always power, there is always the possibility of resistance.

The modern subject is created by the human sciences and the systems of disciplines associated with them, and thus the systems of socialization and social expectation that shape us. However, the picture they create of what it is to be human and behave in appropriate ways can be challenged. We can intervene in the process and shape ourselves, although not in the sense of discovering our 'true humanity'. There is no such true humanity to discover; we would only be conforming to some constructed image generated by the system and thereby submitting to the system's disciplinary power. Instead, we must understand the processes involved and be creative in the construction and reconstruction of our individual selves. Therein lies the possibility of freedom.

Foucault has been the subject of criticism from many quarters. Some have criticized his historical methods, arguing that his interpretations are based on sweeping generalizations that rely on selective use and over-interpretation of evidence. Others have pointed to ambiguities in Foucault regarding the truth of theories that claim that truth is relative or that posit thought as socially determined without being determined themselves.

Criticisms have also come from the left. Although an inspiration for radicals in the 1970s, Foucault ultimately disappointed. He gives no indication of what a better world might be like, and so no reason for changing the present one on any more than a local level. Furthermore, given the oppressiveness of modern society as he portrayed it, Foucault's account of the possibilities of resistance seems very feeble. Limited local action and shaping one's life as a work of art hardly answer to the case. Habermas has been among those who have criticized Foucault's ideas as ultimately conservative, since they suggest that radical change is impossible.

Despite undoubted weaknesses, though, Foucault's ideas have been highly influential across the social sciences, especially his concepts of discourse, the operations of power, the relationships between power and knowledge, and—perhaps above all—his perception that what we consider fixed in our concepts and practices (especially reliance on social sciences) could always be different. Through genealogical research we understand the historicity and contingency of our understanding of ourselves and the possibility of freedom.

Ian Adams and R.W. Dyson, 'Michel Foucault (1926–84)', in *Fifty Major Political Thinkers*, first edn (Routledge, 2003). Politics Online. Taylor & Francis. www.routledgepoliticsonline.com:80/Book.aspx?id=w063 (accessed 15 October 2009).

Foundationalism/anti-foundationalism

Foundationalism means an adherence to the belief that systems of knowledge must be secured on fixed, permanent foundations. Foundations are posited to offer reassurance, authority and certainty in the realm of knowledge. Modern philosophical and political thought has displayed a strong desire for, and faith in, foundations. Anti-foundationalism is sceptical about the possibility of, or requirement for, fixed or permanent foundations. At stake is the question of whether and how knowledge claims can be justified.

It was against the backdrop of considerable intellectual and political upheaval in the 16th and 17th centuries that the clash between foundationalism and anti-foundationalism came into sharp relief. This period not only saw transnational religious violence reconfigure the political landscape, but also brought to the surface profoundly unsettling questions about the sources and legitimation of knowledge. This was in no small part due to the Renaissance revival of Pyrrhonian scepticism associated with the thought of Michel de Montaigne (1533–92). The words he had inscribed on a beam in his house nicely capture the gist of his scepticism: 'All that is certain is that nothing is certain'.

Tremendous intellectual and social change transformed Europe during this time. Feudal social relations slowly gave way to modern ones based on sovereignty and territoriality, and closed medieval cosmologies like the 'great chain of being' gradually gave way to post-Copernican world views of an infinite universe. The Scientific Revolution and heliocentric astronomy together with baroque aesthetics and absolutist political ideologies radically altered the order of things by emphasizing order, coherence, harmony among the parts, and a governing centre. It was believed that in the absence of foundations, moral, theological and political disagreement would fuel disorder and destructive violence. The quest for political stability thus ran parallel to the quest for certitude.

It was in this intellectual and social atmosphere that René Descartes (1596–1650) inaugurated the modern quest for certainty. In response to scepticism, Descartes sought to establish firm foundations for knowledge. Driven by this epistemological imperative, Descartes' aim was to reach certainty by casting aside loose sand and arriving at solid bedrock. This ground-clearing exercise to establish unshakable grounds had to be the first move of the modern philosopher. It required demolishing extant philosophical edifices and beginning anew on solid ground.

For Descartes this meant revisiting the foundational assumptions or 'first principles' of philosophy, which were commonly based on errors and falsehoods. To remedy the situation Descartes would withhold his assent from all but the most indubitable beliefs. Doing so momentarily throws the modern thinker into a chaotic vortex. However, the anxiety of being without foundations is, only a fleeting feeling, for the rational thinker quickly realizes that there is one thing that survives radical philosophical doubt: the thinking self. The quest for certitude ends when we discover the indubitable foundation of the philosophical **subject**, the 'I' who thinks and doubts: *cogito ergo sum*, 'I am thinking therefore I am'. This arrests the anxiety generated by doubt and the temporary absence of foundations, returning the philosopher's feet to philosophical bedrock, and providing grounds of justification.

Geological and architectural metaphors were common to 17th century attempts to establish firm foundations. Both Thomas **Hobbes** (1588–1679) and Gottfried Leibniz (1646–1716) joined this Cartesian quest. They wrote, respectively, of the dangers of building states or knowledge on sandy ground, emphasizing the necessity of solid ground. Common to Descartes, Hobbes and Leibniz was the imperative to use reason to question shaky foundations and replace them with surer ones. However, what they do not do is question the grounds of reason.

By questioning the grounds of reason, thinkers such as Friedrich Nietzsche (1844–1900), Max Weber (1864–1920), Michel **Foucault** (1926–84) and Jacques Derrida (1930–2004) challenge the trust placed in foundations by questioning reason's reasonableness. In famously declaring 'God is dead!', Nietzsche cast doubt over the possibility of offering any reassurances about the authority or objective rationality of knowledge claims. Weber similarly questioned the possibility of rationally grounding moral, legal and political norms because reason itself is divided. This returns us to the threat of cognitive and normative chaos.

Because Nietzsche's original declaration that 'God is dead!' comes from the mouth of a madman, it is often assumed that to question foundations or the need for them is to side with irrationality against rationality, or relativism against objectivism. It seems to imply that all knowledge claims are equal, and that 'anything goes'. In short, it seems to lead to a fatalistic, nihilistic anti-foundationalism. When framed in this way, the opposition between foundationalism and anti-foundationalism is simply another form of the time-honoured political dualisms between reason and unreason, order and disorder, sovereignty and anarchy, male and female that feminists and post-structuralists have convincingly deconstructed.

In the post-positivist study of international relations, anti-foundationalism is almost always associated with **poststructuralism**. Poststructuralist theorists of international relations have drawn attention to parallel crises in the realms of knowledge and politics by casting doubt over the faith in sovereign certitude. What is at stake, Ashley and Walker (1990) argue, is the very question of sovereignty. Sovereignty here needs to be understood simultaneously as a predicate of the state and as a symbol of authoritative foundation. Under conditions of globalization the conceptual and political supremacy of sovereignty has been considerably challenged. New social, political and economic dynamics have generated empirical and normative challenges to the sovereign state as the foundation of the political realm and the guarantor of authority and certainty.

Critics have argued that poststructuralism, taken to its logical conclusion, leads to a hazardous epistemological and political anarchy. If we are deprived of foundations, they enquire, how are knowledge claims to be justified? To retreat from the poststructuralist abyss, some international relations theorists have proposed a 'minimal foundationalism'. While sharing the scepticism towards absolute, permanent foundations, the more interpretive or hermeneutic theorists are equally sceptical of anti-foundationalist assertions. They and critical theorists such as Jürgen Habermas have expressed concern about the apparent normative incoherence or relativism in poststructuralism's alleged anti-foundationalism. They seek to move beyond the antithesis of objectivism and relativism.

While it appears that we must choose between foundationalism and anti-foundationalism as if it were a choice between reason and madness, or objectivism and relativism, critical theorists and poststructuralists both suggest that the antithesis is overdrawn, or even false. Indeed, poststructuralists are perhaps better described as *non*-foundationalists than *anti*-foundationalists. They question and critically analyse all foundations, while positing none of their own. They urge us to expose what foundations mean, and how they function to authorize certain forms of rationality and exclude particular types of knowledge. However, the key point for both poststructuralism and **critical**

theory is that there are no foundations that cannot be questioned and contested. There may well be foundations, but the most important thing is to continue questioning them and their status. The certitude afforded by faith in absolute foundations, like that invested in absolute monarchs, has been subjected to an ongoing critical questioning, one consistent with a democratic ethos. (*See also* **Critical theory**; **Deconstruction**; **Feminism**; **Positivism**; **Post-structuralism**.)

Richard Devetak in Martin Griffiths (ed.), *Encyclopedia of International Relations and Global Politics* (Routledge, 2005). Politics Online. Taylor & Francis. www.routledgepoliticsonline.com:80/Book.aspx?id=w163 (accessed 15 October 2009).

G

Gangster/criminal/detective thriller/private-eye film

The gangster film is the one most readily identified as an American genre even though the French film-maker Louis Feuillade's *Fantomas* (1913–14) is one of the earliest prototypes. It is in the contemporary of its discourses that the gangster film has been so widely perceived as an American genre. This genre, which dates from the late 1920s, came into its own with the introduction of sound and fully blossomed with three classics in the early 1930s: *Little Caesar* (Melvyn Le Roy 1930), *The Public Enemy* (William A. Wellman 1931) and *Scarface* (Howard Hawks 1932), during which the manufacture, sale and transportation of alcoholic drinks was forbidden, and there was the Depression (1919–33), when world-wide economic collapse precipitated commercial failure and mass unemployment.

These two major events in the USA's socio-economic history helped to frame the mythical value of the gangster in movies. Prohibition proved impossible to enforce because gangsters far outnumbered the law enforcers. However, prohibition brought gangsters and their lifestyle into the limelight as never before. Gang warfare and criminal acts became part of the popular press's daily diet and soon became transferred onto film. In fact, many of the gangster films of that period were based on real life. Gangsterism viewed from this standpoint was about greed and brutal acts of violence—in summary, about aggression in urban society. However, the gangster movies were not as straightforwardly black and white as that. The male protagonist embodied numerous contradictions that made spectator identification possible. If we look to the second socio-economic factor mentioned above, we can find a possible reason for the complex and nuanced characterization of the gangster-hero.

The Depression exposed the American Dream—which said that success, in the democratic and classless society guaranteed by the American Constitution, was within the reach of everyone—as a myth. How could this be so when the society was so evidently hierarchized into the haves and have-nots—as the effects of Depression made so blatantly clear? According to the American Dream, success meant material wealth and, implicit within that, the assertion of the individual. The gangster was associated with the proletarian class, not the rich and moneyed classes of the USA. Therefore, the only way he [sic] could access wealth and thereby self-assertion—that is success, the

American Dream—was by stealing it. Accruing capital meant accruing power over others. In this respect, the gangster points up these contradictions, his death at the end of the film is an ideological necessity. He must ultimately fail because the American Dream cannot be allowed to show up the Dream's contractions.

The classic gangster film came into its own with the advent of sound, which reinforced the realism of this genre. Warner Brothers was the first studio to launch this genre in a big way with *Little Caesar* and *Public Enemy*, and their films are seen as the precursors to the film noir. By 1928 this production company had finally become vertically integrated and entered into full competition with the other majors. It was the first company to introduce sound (*The Jazz Singer*, Alan Crosland 1927) and was now poised, as a major in its own right, to outstrip the other four. Warner Bros became associated with the genre because its production practices had set a certain house style of low-budget movies and short shooting schedules. Cheap-to-make films influenced the product, and so backstage musical and gangster films were the genres to prevail. Warner Bros was also very much associated with social content films, and indeed after the launch of Roosevelt's New Deal (1933–34) became indentified with the new president's politics of social and economic reform. Social realism and political relevance, combined with a downbeat image, endowed Warner's films with populism that made their products particularly attractive to working-class audiences, a major source of revenue for film companies.

The gangster movie, in its naked exposure of male heroics, has been likened to an urban western. However, unlike the western where rules are observed, the gangster film is the antagonism between the desire for success and social constraint. The gangster will choose to live a shorter life rather than submit to constraints. Hence the aura of fatalism that runs through the film. As far as the spectator is concerned, for the duration of the film, where violence is countered with violence, where there are no rules, she or he is witnessing an urban nightmare as the narrative brings the plot to the brink of a social breakdown.

The gangster film is highly stylized with its recurrent iconography of urban settings, clothes, cars, gun technology and violence. (In recent history, the film that most fulsomely parodies and yet pays homage to this genre and its iconicity is Quentin Tarantino's *Reservoir Dogs*, 1992.) The narrative follows the rise and fall of a gangster; a learning curve that of course has ideological resonances for the spectators—but not before there has been a first pleasure in identifying with the lawlessness of the 'hero'. During his (fated) trajectory towards death, the protagonist's coming to self-awareness—rather than self-assertion, which is what he initially sought through success—functions cathartically for the spectator: we 'learn' from his mistakes. Furthermore, the use of the woman who is romantically involved with the protagonist and in whose arms he (often) dies—as the law enforcers stand menacingly around the prostrate couple, armed to the teeth—positions the spectator like her and therefore as sympathetic, understanding even, to the gangster. Thus, although the 'message' of the film—'the gangster must die for his violent endangering of American society'—intends to provide the spectator with a sense of moral justification, there is none the less an inherent criticism of American society which says that ultimately the 'little guy' must fall.

The classic age of the gangster movie (1930–34) was brought to a swift halt in an ambience of moral panic. Pressure was put on the Hays Office (of the Motion Picture Producers and Distributors of America) to do more than ask the film industry to apply self censorship. In 1934 the Production Code (or Hays Code), which condemned among other things films glorifying gangsters, became mandatory. Given the popularity of the genre, film companies were not going to give up such a lucrative scenario. Forced to water down the violence, they produced a set of sub-genres: private-eyes forms and detective thrillers. That is, without dropping much of the violence, they now foregrounded the side of the law and order resolving disorder. Told to put a stop to the heroization of gangsters and violence, they simply shifted the role of hero from gangster to cop or private eye. Thanks to the Hays Code intervention, the seeds for film noir were sown. The sadism of the gangster became transformed into the guilt and angst of **masculinity** in crisis of the film noir protagonist. Against the ambiguous urban landscape of some modern American city, the hard-boiled detective seeks justice. The ambiguity of the city reflects the ambiguity and complexity of a society where corruption reigns and law cannot easily bring the guilty to justice. Thus the detective is often a private eye, outside 'official' law, a law unto himself. As a marginal, by being outside he can solve the crime and bring the perpetrators to justice. Traditionally the detective is male; however, recent hard-boiled detective novels written by women (such as Sara Paretsky) have introduced the phenomenon of the woman private-eye detective, who has subsequently found her place on the screen—they are 'just as tough' and equally as poor as their male counterparts. Being poor is part of the construct of the private-eye, pointing to the fact that, even though their methods may be outside the law, none the less they are not in the job for the money. Women in the detective thriller, on the whole, have a more central role than in the earlier gangster movies. They are beautiful and dangerous, often the murderess and, therefore, subjected to investigation by the detective.

Susan Hayward, *Cinema Studies: The Key Concepts* (London and New York: Routledge, 2000), 153–56.

Gaze, the

The gaze is a term that derives from psychoanalysis and, in particular, from the account of vision offered by the French psychoanalytic theorist Jacques Lacan. For Lacanian theory the crucial point is the fact that the gaze is reducible to human vision. The gaze precedes and makes possible human vision but remains itself elusive and indefinable: it is on the side of objects rather than viewing subjects, others rather than offering visual mastery. In summary, Lacan presents the gaze as an external force that shapes the **subject**. The term is used in this Lacanian sense by feminist theorists such as Kaja Silverman, Bracha Lichtenberg Ettinger and Joan Copjec.

In feminist theory the term is also used much more loosely as a synonym for the act of looking, and most particularly the act of viewing cultural products such as films, advertising and the visual arts. Feminists contend that these cultural products generally presuppose (or naturalize) a masculine viewpoint and hence a masculine gaze. In other words, the gaze is largely, if not exclusively, a masculine property or phenomenon and thus implicated in the production and maintenance of gendered social positions. Laura Mulvey discusses these **gender** positions in her classic essay of 1975, 'Visual Pleasure and Narrative Cinema'. Her pithy phrase 'woman as image, man as bearer of the look' demonstrates how the polarization of the sexes is refracted through vision and visuality.

In this article, Mulvey identifies two common kinds of viewing pleasure—identification and objectification—both of which presume, or conform to this asymmetrical relation between the sexes. Women, she argues, are usually objectified by mainstream realist cinema; visual pleasure here derives from voyeuristically looking at the women-as-spectacle. Men, on the other hand, are the active figures with whom one should narcissistically identify.

The identification of the gaze with **masculinity** has generated much further discussion in both art history and **film theory**. A central question, pursued by theorists such as Mary Ann Doane, is how to think about the female gaze. Are women spectators obliged to either assume a masochistic identification with the masculine viewpoint, or to accept their role as objects of that gaze? Doane uses the idea of femininity as a masquerade to disrupt this untenable dichotomy. Other lines of inquiry have pursued the diversification of the gaze by looking at such issues as the queer spectator, the black gaze and so forth.

Susan Best in Lorraine Code (ed.), *Encyclopedia of Feminist Theories* (London and New York: Routledge, 2000), 219.

Gender

Gender can take on either of two distinct meanings. The first, and most obvious, is the biological; in that sense, gender is just the description of female or male. The second is the social construction of the concept referring to associations, stereotypes and social patterns concerning the differences between women and men. It is becoming more the norm to refer to biological differences as sex, and to use the word 'gender' to encompass the social and cultural constructions based on differences between the sexes. Often the term 'gender' is used only for the female sex, even though it more properly refers to the social construction based on sex differences (Nelson 1995).

Studying gender can be deceptively difficult. The biological **difference** between men and women is easily defined: sex depends upon the number of X and Y chromosomes present in genetic make-up. Gender differences hinge on a great deal more. The social construction of gender has implications about interests, behaviour, value structure and even communication styles.

Joan Scott suggests that gender is both an element of social relationships and a primary way of signifying relationships of power. An example she cites is the practice by art historians of reading social implications from depictions of women and men. These depictions then signify a primary way to decode meaning and status among forms of human interactions and the reciprocity of gender and society.

Michele Barrett (1988) discusses the ideology of gender in relation to materialism in her book, *Women's Oppression Today*. Gender ideology—the social constructs revolving around **masculinity** and femininity—is not and cannot be completely separated from the historical and class contexts in which it appears. However, placing it solely within the context of economic relations is also stifling. Gender's meaning within society is tied to the particular household structure and **division of labour** as it has evolved historically so that ideology and materialism must intertwine.

The recognition that gender differences have implications for the practice of science has a long, if nearly hidden, history. Even before 1900, voices could be heard saying that research centred on the male experience is inadequate when dealing with issues such as household work or women's experiences in the labour market. Much of this research is sex blind (Ferber and Nelson 1993). For instance, the Marxist theory of deskilling and the theories of segmented and dual labour markets assign technical competency or other work attributes (that keep women in secondary jobs) by gender to explain why women are eliminated from higher paying jobs. However, they underplay the essential interrelationships between gender and the workplace that may account for these differences (Beechey 1988). Another example of this is the analysis of poverty without recognition of the disproportionate representation of women. These analyses also usually reinforce the assumptions of 'natural' gender roles and the existing societal status quo. For example, women are assumed to earn less in the market place because of 'their' household responsibilities and yet, women 'should' spend time in the household because of the lower opportunity cost of their time.

Another example of how the workplace can interact with gender norms is suggested by Wayne Lewchuk (1993). He proposes that, as assembly lines changed the nature of work, firms maximized productivity by accommodating men to tedious work by 'masculinizing' these occupations so that labourers felt a sort of fraternalism between fellow workers. Therefore, they could celebrate their work even though it was monotonous, unchallenging and generally tedious. This plays to the masculine self-image as it relates to occupation and reinforces preferences for occupational segregation.

Researchers explaining the unique experiences of women, in contrast to those of men, must first recognize that there are basic differences between the sexes (Jacobsen 1994). However, there are two disparate views of those differences. **Biological determinism** suggests that gender differences stem, primarily, from biological differences, that is, male dominance is the natural outcome of the greater physical strength and hormonally induced aggression of the male. Biological potentiality, on the other hand, suggests that biology shapes only the potential, not the outcome. For instance, women give birth but are not necessarily a child's primary caretaker.

In essence, this is the nearly insoluble 'nature versus nurture' debate that raged in psychology. The problem is compounded here because gender traits are only loosely associated with sex. For instance, there are some women who are more aggressive than some men and there are some men who are more nurturing than some women. Still, aggression is considered a male trait and nurturing is considered a female trait. Similar problems are faced by all studies involving the litany of **race, ethnicity, gender and class**, where social traits are assigned to groups characterized by such outwardly manifested signals as colour, sex and social standing.

Feminist perspectives on gender differences include two basic views (Ferber and Nelson 1993). One is the maximalist or essentialist view that basic differences between the sexes are so deep, whether they are biologically or culturally based, that there exists a distinct women's culture. According to this concept, women's culture should be valued, studied and, perhaps (in the view of some more radical believers in this paradigm), should replace the existing male culture. Another view is the minimalist or contructivist paradigm which see less of a schism between the sexes. What differences are observed are attributed mainly to the imposition of social structures. Therefore, any inequities based on gender differences can be redressed through manipulation of such social structures.

Gender-related research in the social sciences takes on the form of redefining both the implications of gender for science and how the form of scientific investigation shapes the meaning of gender. For instance, in political science, gender studies involves an analysis of government policy as it affects women differentially than men, and what the state's role should be to rectify this difference. Standard neoclassical economics has at its base the autonomous agent, totally separated from physical and social constraints. This agent's characteristics are generally centred on the male experience. Certain themes in feminist political economy seek to reintegrate this 'homo economicus' (the rational, autonomous, self-interested economic agent) back into a world with interpersonal connections and as a responder to social influences.

In general, academic **feminism** seeks to investigate how gender shapes the course of scientific inquiry and thought. However, how to change the course of science to integrate women's issues and experiences is not always clear. There are three levels at which this debate operates (Nelson 1995). The first approach says that an increase in female representation among practitioners will redress the situation, leaving aside as unassailable both the methods and the objects of inquiry. A second strategy suggests that women's lives offer areas of study previously ignored while continuing to assume it unnecessary to alter current methodology. A third approach recognizes that current methodology in economics may be too rigid and formalistic to enable the integration of the social and cultural relationships essential to the study of gender differences. This last approach suggests that the holistic method, such as argument by metaphor, storytelling and pattern models, should either augment or replace current methods. (*See also* **Butler, Judith**; **Race, ethnicity, gender and class**.)

Nancy J. Burnett in Phillip Anthony O'Hara, *Encyclopedia of Political Economy*, first edn (Routledge, 1999, 2001). Politics Online. Taylor & Francis. www.routledgepoliticsonline.com:80/Book.aspx?id=w026 (accessed 15 October 2009).

Gender and class – *see* **Race, ethnicity, gender and class**

Gender and development

A broad concept used analytically to understand women's role in international development and normatively to promote their effective participation in politics and society. More narrowly the term refers to a particular model of development, grounded in neo-Marxist **feminism**, that emerged in the late 1970s. The notion of women and development has been variously interpreted and applied since becoming part of development discourse in the 1970s, but it has retained its primary focus on the particular experiences and perceived needs of women in developing countries. There is now broad agreement that women should (and do) play an important role in development, although scholars and policy-makers continue to debate the nature, purpose and effects of women's participation.

Pre-colonial societies in much of the non-Western world accorded women a relatively high status. Thus, for example, women could often own property or hold political office. The advent of European **imperialism**, and the institutionalization of Western values in colonized territories, diminished this status. Because colonizers regarded the public sphere (politics and the formal economy) as men's domain and the (home-based) private sphere as women's 'proper place', women lost much of their socioeconomic power and were (further) marginalized as political actors.

The end of the colonial era and the onset of the Cold War brought increased attention to political, economic and social conditions in former colonies. As a result, development programmes were implemented (with mixed success), but few of these paid specific attention to women. This began to change in 1970, when a seminal study by Danish economist Ester Boserup helped place 'women and development' on the international agenda.

Documenting her analysis with data from various countries, Boserup argued that—contrary to prevailing wisdom—development programmes intended to facilitate modernization generally were not benefiting women and, in many cases, were proving detrimental to them. Subsequent studies have both supported and contradicted Boserup's conclusions. This has led to continued debate and the development of contending approaches to issues of women and development. These approaches have been variously categorized, based on their theoretical orientations and/or the nature of their policy prescriptions.

One classification scheme identifies three approaches—'**women in development**' (WID), 'women and development' (WAD), and '**gender** and development' (GAD)—that are distinguished primarily by their theoretical perspectives. The first of these, WID, developed in the early 1970s in the wake of Boserup's study. Solidly grounded in **liberal feminism** and modernization theory, this approach seeks better to integrate women into development programmes and economic systems without challenging existing social structures. Policy prescriptions thus focus on ensuring equality of opportunity, and increasing women's participation in areas such as education and employment. Despite some conceptual shortcomings, WID has legitimated the separate study of women's experiences and perceptions with regard to development. WAD, a second approach that emerged in the late 1970s, is critical of modernization theory and the WID approach. Adopting a neo-Marxist feminist perspective, WAD focuses attention on women's role in society as determined by inequitable social structures. Thus, rather than supporting the integration of women into existing programmes, the WAD approach calls for carefully designed strategies and targeted development policies that can alleviate broader social inequities. WAD arguably represents an improvement over WID. Nevertheless, like the WID approach, it remains preoccupied with women's income-generating ('productive') activities and fails to address their unpaid ('reproductive') household labour. A third approach, GAD, took shape in the 1980s. Based on socialist and postmodernist feminisms, and significantly broader in scope than WID or WAD, GAD takes a holistic approach to women and development issues by focusing on socially constructed gender roles and their implications. Policies consistent with the GAD approach reject the public/private dichotomy, challenge patriarchal social structures, and seek to facilitate women's empowerment. GAD's critical and transformative orientation makes it a relatively less popular, yet potentially more effective, approach than either WID or WAD.

Another typology distinguishes approaches to women and development by a combination of their theoretical orientation and policy prescriptions. Theoretically, a key distinction is made between **liberalism** and structuralism. Liberal approaches advocate capitalist strategies or emphasize modernization processes, while structuralist approaches support socialist strategies or focus on dependency relationships. Policy prescriptions are distinguished by their emphasis on 'practical gender needs' or 'strategic gender needs'. Those that address practical gender needs are defined in terms of women's gendered social position and do not threaten patriarchal structures. Examples include teaching women dressmaking or providing child care facilities at a mother's workplace. Policies that address strategic gender needs, on the other hand, challenge the status quo and seek to erode patriarchal social structures. Examples include teaching women masonry or providing child care facilities in a father's workplace. Those who employ this typology regard approaches to women and development as an ongoing process, by which institutions and policymakers have adopted various theory/policy combinations since the 1970s based on political ideology and lessons learned. Most recently, the 'empowerment' approach has gained currency. Similar to GAD, this approach combines a mix of liberalism and socialism with an emphasis on addressing women's strategic gender needs.

Since the 1990s, as suggested by the development of GAD and empowerment approaches, conceptions of women and development have become increasingly holistic in orientation and more fully integrated into development thinking. For example, the Gender Related Development Index (GDI) and Gender Empowerment Measure (GEM)—indicators created by the UN Development Programme (UNDP) to measure gender inequalities—are increasingly used to help determine the effectiveness of development programmes. The GDI focuses broadly on women's quality of life, as measured by their life expectancy, education and income relative to men. The GEM measures women's formal participation in politics and the economy, their relative decision-making power in society, and their control over economic resources.

Since the 1970s, a focus on the particular needs and experiences of women has done much to improve the effectiveness of international development programmes. Particularly since the 1990s, the concept of 'women and development' has productively expanded to include gender (the socially constructed roles of both men and women) and its implications for development. (*See also* **Capitalism**; **Women in development**.)

Craig Warkentin in Martin Griffiths (ed.), *Encyclopedia of International Relations and Global Politics* (Routledge, 2005). Politics Online. Taylor & Francis. www.routledgepoliticsonline.com:80/Book.aspx?id=w163 (accessed 15 October 2009).

Gender and international relations

A system of power relations governing the interaction of males and females which usually privileges the former over the latter. **Gender** analyses have primarily criticized the absence of or under-representation of women both from international political decision-making, and from traditional narratives in the study of international relations (IR).

Gender analysis in IR is an extension of the broader feminist critique and global activist project that has taken shape since the early 1960s. What binds together diverse feminist approaches to IR is a focus on women as political and historical actors; a foundation in the realm of women's and girls' lived experiences; and a normative contention 'that women and the feminine constitute historically underprivileged, under-represented, and under-recognized social groups and 'standpoints', and that this should change in the direction of greater equality' (Jones 1996, 406).

One cannot pinpoint a foundational text in feminist IR, but there is perhaps a foundational question: Where are the women? It was first posed in the most enduringly popular work of the sub-field, Cynthia Enloe's *Bananas, Beaches and Bases* (1989). Various features of the feminist IR literature in the 1990s were anticipated in Enloe's book: the critique of classical theories of IR that shunted women and other subaltern actors to the sidelines; its global scope, asserting that gender was universally operative in international politics; and its diligent and imaginative location of women as agents in IR. Many of its core themes also received deeper exploration, including by Enloe in subsequent works: women and militarism; women in the international political economy; and gender and nationalism.

At the heart of the feminist IR literature is a critique of the classical (basically realist) tradition, although feminists have also levelled a critical gaze upon liberal internationalist and socialist formulations. Both from a distinctive slant and together with other exponents of **critical theory**, feminist IR scholars have assailed a wide range of basic tenets in the field: the universalization of man/men as the essential 'rational actors'; the idea of a unitary, 'black-boxed' state that obscures divisions of class, **race** and not least gender; and modernist/masculinist epistemologies that underpin the classical tradition and bolster hierarchies of knowledge, power and material privilege. Among the most fervent critics have been poststructuralists, who have sought to undermine dominant epistemologies, notions of stable **identity**, and all generalizing terminologies. The apogee of this trend was perhaps reached with the publication of Christine Sylvester's *Feminist Theory and International Relations in a Postmodern Era* (1994).

As noted, the study of gender and IR has been deeply affected by global trends in feminist theory and **activism**. In the activist sphere, women's mobilizations for equality have scored striking successes over the past four decades. National and international initiatives have addressed the plight of women, and entrenched women's issues at all institutional levels, though women still remain woefully under-represented in leadership positions nearly everywhere. The process was mirrored in academia, where feminist critiques came to constitute both a discipline in their own right and a standard component of introductory courses across the humanities and social sciences. In IR, the feminist sub-field was by the mid-1990s one of the most flourishing in the discipline.

As **feminism** has become global, so has it been globalized. The internationalization of feminist activism, once a Western monopoly, was symbolized by the UN Decade for Women and the UN-sponsored World

Conferences on Women, the fourth and most recent of which was held in Beijing in 1995. Encounters between first world and Third World feminists at these and other gatherings led to a dramatic reappraisal of 'women' as a unitary global category. It prompted greater attention to divisions of culture and class among women; the role of women in female oppression (for example, as employers of female domestic servants and adherents of xenophobic nationalist movements); the relationship of feminism to other emancipatory critiques (such as anti-**imperialism**); and the possibility of men and women working towards common goals, a strategy that Third World women emphasized more than their developed world counterparts.

Mention should be made of a related trend prominent from the early 1970s: the growth of a **women in development** (WID) critique within development studies and international organizations. The critique focused on women's situation in developing countries, quantifying many of the relevant issues for the first time, and arguing that women (and children) should be integrated into national development initiatives and international development work. In the 1990s, WID was increasingly replaced by a **gender and development** (GAD) critique that called for gender relations (rather than women) to be adopted as the primary analytical tenet.

Increasingly in the 1990s, as part of a broader revision of security studies, explorations of gender have revolved around conflict and conceptions of security. Rejecting traditional state-centric formulations, feminist scholarship has pointed to the gendered insecurity that afflicts women (for example, through domestic and sexual violence), even within superficially secure nation-states. Increasing attention has been paid to women's and girls' gender-specific vulnerabilities in wartime. A central theme here is the **rape** of females in wars and **genocides** from Bangladesh to **Bosnia**. As well, the collapse of communism in Eastern Europe has prompted sceptical, feminist-informed appraisals of the impact of democratization processes on gender relations.

Early feminist contributions were sometimes marked by gender essentialism: the notion that women and men held beliefs and attitudes that reflected their differing physiologies. Thus, a greater nurturing orientation for women could result from their capacity to bear children. However, such essentialism was decisively supplanted by a stricter demarcation between biological sex (male versus female) and socially constructed gender (**masculinity** versus femininity), with the overwhelming explanatory weight assigned to the latter. Hence, women were not necessarily predisposed to peace, nor men to violence. Rather, cultural formations shaped embodied men and women to behave according to limited and stereotypical visions of masculinity and femininity.

Is gender thus to be defined as exclusively a cultural formation? Increasingly, gender theorists have come to view sex and gender as mutually constitutive. Among feminists, the shift has perhaps been prompted by the feminist emphasis on damage done to the female **body** under conditions of male dominance, through violence and labour exploitation. The corporeality of gender seemed not only evident, but integral to feminist normative prescriptions.

What of the subject of men and masculinities in the study of gender and IR? It is fair to say that the attitude of most IR gender theorists has tended to be sceptical or hostile. Males have figured primarily as leaders, powerbrokers and gender oppressors. Attention to diverse, especially subaltern masculinities has, until recently, been extremely limited. (*See also* **Feminism and international relations**; **Liberalism**; **Poststructuralism**.)

Adam Jones in Martin Griffiths (ed.), *Encyclopedia of International Relations and Global Politics* (Routledge, 2005). Politics Online. Taylor & Francis. www.routledgepoliticsonline.com:80/Book.aspx?id=w163 (accessed 15 October 2009).

Gender mainstreaming

This is the process of evaluating the implication for women and men of any programmatic action, including legislation, policies or activities, in all fields and at all levels. It is a strategy for making women's and men's concerns and experiences an integral part of the plan, implementation, monitoring and assessment of policies and programmes in all political, economic and societal domains so that women and men benefit equally and inequality is not perpetuated. The fundamental goal is to achieve **gender** equality.

Gender performativity – *see* Butler, Judith

Gender and international political economy – *see* International political economy and gender

Gendercide

Gender-selective mass killing: whether directed against females or males, the practice is as old as history. In the politico-military context, sources from the Old Testament to Thucydides record instances of mass killing of 'battle-age' males, usually accompanied by the enslavement and forced **concubinage** of women and girls. In a sociological context, gendercidal institutions such as female infanticide, witch-hunts, forced labour, and military conscription are among the more venerable and destructive human institutions.

The term 'gendercide' derives etymologically from the term **genocide**, referring to the destruction of human groups. Under the 1948 Genocide Convention, genocide was limited to national, ethnic, racial and religious groups. Groups structured according to political **identity** or socio-economic class were excluded. So, too, were gender groups. However, it is notable that the most far-reaching internal reappraisal of the Genocide Convention, the United Nations Whitaker Report of 1985, *did* recommend incorporating 'sexual groups' as targets of genocide. These included both biological men and women, and minorities (such as homosexuals) defined by actual or imputed sexual identity (Charny 1999).

Even a cursory glance at the numerous genocides of the 20th century, along with other campaigns of mass killing, attests to gender's prominence as a variable. The most common trend has been the gendercidal targeting of battle-age but non-combatant males, between roughly 15 and 55 years old. Such gendercidal massacres constitute an end in themselves and/or a means of weakening and undermining the broader target community.

Quite commonly, as in Bangladesh in 1971 and **Bosnia**-Herzegovina in the 1990s, the selective killing of adult men and adolescent boys has been seen as both a necessary and a *sufficient* genocidal strategy, at least as far as mass killing is concerned. Accompanied by other measures that are usually not fatal, such as mass expulsion and widespread war **rape** of women, the gendercidal killings of males serves to terrify a target population and reduce its will and capacity to resist. Gendercide of community males alone is genocidal, though it falls short of full-scale genocide. As Daniel Goldhagen argues in his study of the Jewish holocaust, '[t]he killing of the adult males of a community is nothing less than the destruction of that community' (Goldhagen 1997, 153).

For contemporary observers, the paradigmatic case of gendercidal killing was the July 1995 slaughter at Srebrenica in Bosnia. In the worst massacre in Europe since the Second World War, over 7,000 Muslim men and boys were separated from the community and executed *en masse*, or hunted down by Serbian regular and irregular forces in the surrounding hills (Honig and Both 1996). Children and women, along with some elderly men, were allowed to flee to safety in Muslim-controlled territory.

In many other cases, including (by definition) the most extreme genocidal campaigns in history, gendercide against 'battle age' men has served as a tripwire or precursor of genocide against all members of the target population. The three classic genocides of the 20th century (against Turkish Armenians in 1915–16, European Jews in 1939–45, and Rwandan Tutsis in 1994) all share this feature. Armenian men, including eventually the very old and the youngest boys, were mercilessly eliminated from the population prior to, or at the start of, the death marches to the Syrian Desert for which the genocide is notorious. These forced expulsions eventually killed hundreds of thousands of Armenian children and women. Hitler's campaign against the Jews began with round-ups and selective killings of Jewish men, such as after the *Kristallnacht* (Night of Broken Glass) in 1938. It was overwhelmingly gendercidal in the earliest stages of mass killing by *Einsatzgruppen* death squads in occupied Poland and Russia during summer 1941, at a time when millions of male Soviet prisoners of war were also being corralled for subsequent extermination. The killing of Jews rapidly expanded beyond traditional gender parameters, and was extended to children and women. As many scholars have noted, this caused considerable psychological stress to the Nazi killers, and led to the development of distancing technologies such as gas vans and eventually gas chambers to implement the extermination orders. As for Rwanda, although mass slaughters of Tutsis took place almost from the start of the 1994 genocide, they were accompanied by an especially merciless targeting of Tutsi males, who appear overall to have constituted a substantial majority of those killed. Tutsi women were exposed to systematic war rape and sexual slavery that can itself be considered both genocidal and gendercidal.

Gendercide may be old, and a regular feature of contemporary conflicts world-wide, but the term itself is a product of modern feminist social inquiry. It was coined by the American scholar Mary Anne Warren in her book *Gendercide: The Implications of Sex Selection* (1985). Warren drew an analogy between extermination on racial grounds (genocide) and gendercide, the deliberate extermination of persons of a particular sex (or gender). A particular advantage, in her view, was the gender-inclusive character of gendercide, compared with female-specific terms like gynocide and femicide. (*See also* **Feminism**.)

Adam Jones in Martin Griffiths (ed.), *Encyclopedia of International Relations and Global Politics* (Routledge, 2005). Politics Online. Taylor & Francis. www.routledgepoliticsonline.com:80/Book.aspx?id=w163 (accessed 15 October 2009).

Geneva Conventions

These are four treaties signed between states parties to the UN in Geneva, Switzerland in 1949. Each concerns how states parties are obligated to conduct themselves during an armed conflict (both national and international) and how they are to treat prisoners of war. The treaties have received much attention within the last few years as the USA, under President George W. Bush, attempted to limit the applicability of Common Article 3 concerning the humane treatment of detainees, pursuant to his Administration's 'war on terror'.

Tawia Ansah

Genocide

The term 'genocide' was coined in 1944 by Rafael Lemkin most immediately in reaction to the Nazi 'Final Solution' directed against the Jews, but it was also meant to identify that crime more generally as the annihilation or attempted annihilation of the members of a group (*genos*) solely because of group association. Lemkin, a lawyer and himself a Polish-Jewish refugee, had previously (at the Fifth International Conference for the Unification of Penal Law, Madrid, 1933) unsuccessfully proposed international recognition of the crime of 'barbarity'—'oppressive and destructive actions directed against individuals as members of a national, religious, or racial group'. In his book, Lemkin expanded the concept of genocide to include attacks on political, economic and cultural groups; in addition to the Nazi campaign of annihilation, he cites among earlier instances of such attacks the Roman destruction of Carthage (146 BCE), the conquest of Jerusalem by Titus (72 CE), and the Turkish massacre of the Armenians (1915–17). The crime of genocide, he claims, extends beyond the attacks on civilian populations in 'occupied' territory that had been addressed and in some measure guarded against in international law by the Hague Conventions. The designation of *groups* as targets for destruction, in Lemkin's view, expands the possible rationale (and thus the threat) of systematic killing.

In December 1948, the UN General Assembly adopted the Convention on the Prevention and Punishment of the Crime of Genocide, which defines genocide as an act or acts 'committed with intent to destroy, in whole or in part, a national, ethical, racial or religious group'. (After considerable controversy, political groups were removed from the Convention's protection.) The means of genocide cited by the UN Convention include, in addition to actual killing, less direct methods such as the prevention of births within a group and the transfer of children to other groups. Under the UN Convention, charges of genocide are to be heard before the International Court of Justice. (Legislation for implementing ratification of the Convention by the USA, an original signatory in 1948, was not passed by the US Congress until 1988 and then with the attachment of substantive—to an extent, disabling—reservations. Some 97 other members of the UN had previously ratified the Convention.)

Although the term 'genocide' is often used as a synonym for 'mass killing', the element of intention implied in genocide—the destruction of a group *as* a group—distinguishes the two terms. This distinction also suggests an intrinsic role in genocide for premeditation, insofar as determination of the group which is the object of genocide must be made before, and then maintained during, the act. In certain respects, the issue of what defines a group as the object of genocide remains a matter of convention—for example, whether a minimum number of people is required to constitute the group or, as victims, to warrant the charge of genocide, and what the connections among members of a group must or may be. Here, as in other 'paradigm' cases, such methodological difficulties need not affect claims of the occurrence of the phenomenon. The deliberateness and systematic extent of the Nazi 'Final Solution' have led some writers (e.g. Bauer) to view it as historically unique and thus to distinguish the category of 'Holocaust' from other putative instances of genocide; but it is arguably the distinctive features of genocide itself that mark the various occurrences.

Attempts have been made to develop a typology of genocide—for example, as ranging from physical destruction of a group carried out irrespective of individual considerations (e.g. geographical boundaries), as in the Nazi genocide against the Jews; to physical destruction, which is restricted by such boundaries or by other distinctions within the group itself (as in the differences ascribed by the Nazis to the several 'tribes' of Gypsies); to 'ethnocide' or cultural destruction, as in the transfer or dispersion of populations or the repression of cultural and linguistic traditions. (A recent extension of the concept of genocide associates the prospect of nuclear destruction with the threat of 'omnicide' or 'anthropocide'—the killing of *all* groups and individuals.) Attempts are also ongoing to formulate an 'early-warning' system to anticipate the occurrence of genocide; but inasmuch as the most flagrant 20th-century instances of genocide were directed against a populace under the cover of a more general war, the difficulty of identifying such predictive features is significant.

It has sometimes been claimed that genocide is essentially a modern phenomenon, related to the historical development of totalitarianism and the technological society; but these, although contributory, seem to be neither necessary nor sufficient conditions. (The technology employed in the occurrences of genocide on record did not involve 'advanced' means.) The question of how to determine responsibility for genocide is complicated by the fact that genocide characteristically occurs as a group or corporate act (in addition to being directed *against* a group). The UN Convention conceives of **accountability** for genocide as primarily individual, attributing liability to anyone who takes part in the act, 'whether they are constitutionally responsible rulers, public officials or private individuals'. Legal and moral questions of how such a standard can be applied do not differ, it seems, from the issues of how and to what extent individual responsibility can be assigned for other corporate acts. Beginning in 1998 and continuing, international tribunals considering charges of genocide against individuals for acts committed in **Bosnia** and Rwanda strengthen the possibility of enforcing the UN Convention.

The relatively recent recognition and designation of genocide suggest that social awareness of wrong-doing is itself subject to development—that it has a history. However, if this new consciousness is construed (as it sometimes has been) as evidence of moral progress the events which impel that history—in this instance, the occurrence of genocide itself—provide stark counter-evidence.

Michael Freeman in Lawrence C. Becker and Charlotte B. Becker (eds), *Encyclopedia of Ethics*, second edn (New York: Routledge, 2001).

Genocide, Rwanda

The **genocide** that turned world attention to Rwanda in 1994 has been described in graphic terms. The May edition of *Time* magazine offered a damning headline: 'There are no devils left in Hell. They are all in Rwanda.'

To understand the role of religion in the Rwanda genocide, a historical overview is necessary. Prior to 6 April 1994, Rwanda was an obscure and distant country in Africa except to people who had seen the film *Gorillas in the Mist*. Rwanda is the smallest country south of the Sahara Desert. It covers only 25,000 sq km, but 8m. people live in that area. They have been inaccurately identified in ethnic groups. The three major groupings share a culture and language (Kinyarwanda) but differ in physical appearance, social status and occupation.

Myths of origin were fabricated to prove a separate origin for the Tutsi people of Rwanda. The Hamitic (relating to Noah's son Ham) 'theory of origin' was introduced by John Hanning Speke in 1863 in his *Journal of the Discovery of the Source of the Nile*. According to his pseudoscientific theory, the Tutsis came from India. The main thesis of the Hamitic theory came from the first British administrator of the Ugandan protectorate, Sir Harry Johnston, who claimed that the Tutsis originated in Ethiopia. One variation of the Hamitic theory follows the biblical myth of the three sons of Noah. The variation emphasizes that although blacks are descendants of Ham, not all blacks, such as the Tutsis, are included.

The 'superior race' concept and practices of the Germans in 1914 were limited but were formalized as colonial policy by the Belgian colonialists. Rwanda was colonized by the Germans in 1890 and by the Belgians from 1916 to 1946 under the League of Nations mandate along with Burubdi, and from 1946 to 1961 under the UN as a trust territory. During the 1930s the Belgians would ensure that most Hutu chiefs were deposed. They promoted Tutsis in every sphere of public life and expressed confidence in Tutsi leadership and their ability to succeed in education. Missionaries expressed the same bias in their evangelization methods. They first converted the Tutsi chiefs and sent them to Catholic schools and seminaries to prepare for leadership roles in the church. From 1931 to 1933 the Belgian colonialists introduced ethnic identity cards that would radically alter and solidify an otherwise fluid ethnic **identity**. After all, Tutsis and Hutus intermarried, and there was no untameable animosity between the two. The Roman Catholic Church was everything to the Tutsis in terms of ambition and substance. Its education system, including a school for children of kings, was the gateway to social position, based on its acceptance of the dangerous mythology.

The year 1959 was a major turning point because the anticolonial movement arose in European colonies, and the UN forced the Belgians to change their policies to accommodate the majority population's interests in representational government. This meant the replacement of the ruling Tutsi elite for the sake of democracy. Leading up to 1959 the first public expression of animosity against the Tutsis, which was shared by Catholic Church leaders, was the 'Bahutu Manifesto' in 1957. The Parmehutu (the political party promoting Hutu emancipation) received strong support from Monsignor Perraudin, a Swiss bishop. By the time independence came in 1962 20,000 Tutsis were massacred, and over 200,000 Tutsis would seek refuge in neighbouring Burundi and Uganda. Since independence the country has been greatly favoured for development by non-governmental organizations and Western nations because it has been generally perceived to be stable under the power of the Hutus. In order to justify and maintain that image, the new leaders have employed two types of rhetoric: one for a domestic audience and the other for external aid donors. In order to maintain the urgent need for the Hutu power structure, 'social revolution' was utilized to remind the majority Hutus that Hutus are the true and legitimate inhabitants of Rwanda and deserve to be freed from the oppressive rule of the outsider (the Tutsis). The second type of rhetoric focused on the need for 'development', even though most of the money was diverted to bolster the president's power base.

After Gregoire Kayibanda took power as the first president, he established close ties with President Mobutu of Zaire with the hope that Mobutu would help demobilize the Tutsi rebel camps in eastern Zaire. President Kayibanda appeared to be an ineffective leader to his top military leaders. In 1973 General Juvenal Habyariman overthrew Kayibanda in a bloodless coup. Habyariman soon gained support from militant anti-Tutsi Hutu elements. The 1980s were difficult years for Rwanda. One reason was the growing strength of the Rwanda Patriotic Front (RPF), which was made up of mostly exiled Tutsi refugees in Uganda. They had been arming themselves in preparation for a return to their country. Their first attempt was made in 1990. The defeat of the group gave President Habyariman an excuse to make scapegoats of it and summon support for its elimination. Habyariman found ready support in President François Mitterrand of France.

International pressure from Western donor countries, the International Monetary Fund and the World Bank for greater democratization put the president in a tight spot. The Arusha Accords, to which he had been a signatory, required the inclusion of Tutsis in power sharing and the elimination of ethnic identifications on documents. For Habyariman and his northern Hutu clique (the Akazu), these accords spelled disaster because their implementation would mean the loss of power and privilege. In order to enforce the terms of the Arusha Accords, the **UN peace-keeping** force would supervise. Unfortunately, the Arusha Accords could not be carried out because of the growing suspicion and hatred of the Tutsis.

Hell broke loose in Rwanda on 6 April 1994. Although the Hutus had identified moderate Hutus and Tutsis to be massacred, the immediate pretext came when the presidential plane carrying President Cyprien Ntaryamira of Burundi and Habyariman from a regional conference in Dar-es-Salaam, Tanzania, was struck by missiles. The purpose of the conference was to break the impasse over the accords. The killing and maiming of civilians by soldiers and Hutu extremists were methodical and brutal, using machetes, spiked tools and weapons. Neighbours turned in neighbours, and some Tutsis paid killers to spare them the anguish of gruesome death as they pleaded to be shot. The two groups responsible for much of the carnage were the Interhamwe and the Impuzamugambi. All the while the state-run Hutu radio station, Radio Mille Collines (thousand hills) encouraged Hutu audiences with hate-filled messages to kill the Tutsis.

Church leaders stood by while the propaganda filled the airwaves. The Rwandan conversion to Christianity, when put to the test, turned out to be superficial and opportunistic. The example of Bishop Rwankeri illustrates the abandonment of Christianity among the Roman Catholic hierarchy. During the crisis he encouraged Christians to support the interim government and its policies. Also, in August 1994 some 29 Hutu priests wrote a letter to the Pope denying Hutu responsibility for the massacres. Protestant clergy and other Christians also gave in to evil. Pastor Elizaphan Ntakirutima of the Seventh-Day Adventist Church in Kibuye and his son, Dr Gerard Ntakirutimana, a physician, gave public support to Hutu power and helped to assemble fleeing Tutsis in a church, where they were killed by Hutu militia. However, a few ordinary Christians and priests risked their lives for their fellow men and women.

The conspiracy of silence and the condoning of carnage have exposed the hypocrisy of Rwandan Christianity in the proceedings of the **International Criminal Tribunal for Rwanda** in Arusha, Tanzania. The tribunal was set up in November 1994 to try war criminals of the Rwanda massacre. Approximately 125,000 people have been jailed for war-related killings. Out of the 1,200 tried in 1999, over 100 were sentenced to death, and 22 have been executed. Convictions of prominent Church leaders followed, including those of Bishop Samuel Musabyimana, the Anglican bishop who, prior to the killings, declared that the situation for the Tutsis was dire and that their end had come; Father Athanase Seromba, who escaped to the diocese of Florence, Italy in 1997; and two Rwandan nuns who were tried in Belgium and convicted. Upon the conviction of Sister Maria Kisito Mukabutera and Sister Gertrude Mukangango for handing over 7,000 Tutsis who were being sheltered in a southern Rwanda convent, the Vatican distanced itself from their actions. Perhaps the most notorious Church accomplice to be recently convicted is Pastor Ntakirutimana and his son, who fled to the USA.

The conviction of top Rwandan politicians, professionals and religious authorities signals a new awareness about the gravity of crimes against humanity and the need for immediate action by the international community. The success of the Rwanda criminal tribunal brings hope and closure to those who long for justice and **accountability** in regions of the world that hardly make headlines in Western capitals. However, the success of the tribunal will be complete when the UN and Western nations, which had a stake in Rwanda but did nothing to stop the massacres, are made to own up to their complicity as well.

Victor Wan-Tatah in Gabriel Palmer-Fernandez (ed.), *Encyclopedia of Religion and War* (Routledge, 2004). Religion Online. Taylor & Francis. www.routledgereligiononline.com:80/Book.aspx?id=w029 (accessed 15 October 2009).

Geographies, feminist

One of the central tenets of the **feminism** that informs our scholarship and political action as feminist geographers is the need to challenge the construction of authority. Putting into print a list of key works canonizes particular contributions, something that supports the construction of *a* feminist authority. However, a dilemma arises when feminist geographers must legitimate feminist geographies within the discipline of geography in order to establish them as acceptable fields of study. As a provisional resolution, we suggest the accompanying list of readings as an introductory survey of the range of feminist geographies in the English language-based academy. We want to emphasize, though, that despite the history of struggle over establishing feminist geographies, the amount of literature available today compared with only a decade ago is remarkable. We encourage readers to review this list of readings and the bibliographies in these works to see for themselves the scope of feminist geographies today.

Like many of the social sciences in the English language-based academy, feminism—as both scholarship and a political movement—has been influential in informing theories and practices in the discipline of geography. Yet feminist geography is not an integrated or cohesive academic subdiscipline itself. Although not singular in definition or practice, feminist geographies do place central sexual **difference** as a primary organizing element of society. For nearly three decades, feminists through their scholarship have been forging innovative analyses of spatial phenomena that frame both the understanding and explanation of themselves as scholars, their topics of study, the methods through which they collect and convey information, and their everyday politics. These analyses have drastically changed the way key traditional geographic constructs are conceptualized—space, place, home, community, city, environment and work.

From early on, feminist geographers demonstrated that being a woman matters. Zelinsky (1973) makes a persuasive argument that women are missing from the discipline of geography not only in topical terms, but also as geographers themselves. He argues that geography is diminished by their absence, intellectually and substantively. Although the specifics of his arguments may appear to be dated, the intention of his 'outburst' (Ibid., 104) is much like a feminist call to political action today. In other words, geography needs more research with/by/for/on women. His piercing remarks sparked a rash of publications over the next decade to justify the existence of either a feminist perspective in geography or a feminist analysis of spatial phenomena. For example, Monk and Hanson (1982) suggest that building a geography that is feminist entails critically analysing issues involving women and sorting out sexist bias in geographic research. They usefully point out ways that feminism can inform a spatial analysis and list multiple ways that conventional research neglects and excludes women and women's interests.

Until the late 1980s, feminist geographers tended to focus on women's labour primarily because initial feminist approaches in geography were closely tied to the newly emerging Marxist analyses. Massey (1984) exemplifies this best with her work in locality studies. She details how global restructuring processes transformed the industrial organization of Great Britain. Using the concept of spatial **division of labour**, she demonstrates that the production process is concentrated in specific geographical areas according to sectors and tasks. She also shows that these spatial divisions of labour are *gendered*. Suzanne Mackenzie (1989), working at the scales of the household and community, investigates the paid and unpaid labour of women in post-war Brighton. The scrutiny of these women's daily lives shows that these women, through engaging in the mundane, actively participate in the construction of their environments. Her work showcases the predominance of socialist analyses in feminist geography at the time while still taking seriously the notion that women shape and are shaped by their experiences of place. Linda McDowell provides a detailed review of the types of English language-based feminist geographical studies throughout this period.

Feminism effected change by creating new analyses in a variety of geographical subdisciplines, including, for example, political, cultural, social, urban, historical and economic geography. For example, Massey's (1994) collection of essays, written over a period of two decades, illustrates how feminism transformed her own economic analysis. She begins in the late 1970s as a Marxist industrial geographer with no attention given to **gender**. however, by the mid-1990s she uses gender relations as a central construct in explaining the impact of globalization.

As feminist analyses in geography in the English language-based academy matured throughout the late 1980s and early 1990s, sexual difference and gender became merely two sets of power relations constituting social space. With the introduction of poststructuralist thought into feminist geography, issues of difference and representation became paramount. Drawing on these poststructuralist notions, Jones, Nast and Roberts (1997) edited a volume that demonstrates how some feminist geographers are elaborating the spatial aspects of these issues. One goal these articles collectively achieve is the recasting of 'woman'—moving away from a monolithic ensemble acting and reacting in unison with all other women toward a complex being constituted by a multiplicity of power relations—a task that signifies the integration of diversity as an organizing theme for feminist geographies. Another example of the way poststructuralist thought is informing feminist critiques is Gibson-Graham's (1996) reworking of **capitalism**. They take to heart this notion of diversity and apply it to concepts in Marxist political economy. The result is an innovative look at how multiple capitalisms structure women's lives locally and globally and the ways in which they resist.

McDowell (1999), in her review of feminist geographies, systematically demonstrates how feminist geographers have critically assessed and extended notions of conventional geographic constructs such as home, workplace, community, city, public/private space and nation-state. She also includes more recent feminist work on new geographical terms such as **identity**, **body**, sexuality and **travel writing**. Her work highlights the attention feminist geographers are giving to the multiple sets of power relations including, but not limited to, those arranged around ability, age, citizenship, class, ethnicity, nationality, **race** and sexuality.

Another area that interests feminists is the construction of geographical knowledge. Rose sets up a framework to explain how geographical thought and practice, including, for example, time geography, fieldwork and humanistic, is 'masculinist'. Masculinism in a discipline means that men alone define what counts as knowledge, control the processes of creating knowledge, and deem men's experiences as the only worthwhile subject matter. Together, these epistemological claims support the notion that the existing (masculinist) geographical knowledge is exhaustive, and assumes there can be no other contributions. She goes on to demonstrate how masculinist thinking dominates several spheres of geographical thought and practice as, for example, time geography, fieldwork and humanistic conceptions of home. In an attempt to reposition the subject of masculinism, she proposes 'paradoxical space' as a way of thinking about how to access entry points into the construction of knowledge. In this way she is able to offer feminism as a new way of knowing geography—one that 'refuses the exclusions of the old' (Rose 1993, 1,410).

Part of the rejection of a masculinist geography involved reworking approaches to research. The collection of articles edited by Moss (1993) initiated a discussion about what makes research feminist. Although perspectives varied, contributors did agree that what makes research feminist is how research methods are used within and outside the academy. North American feminist geographers at that time held that researching with numbers had a dehumanizing effect on women's lives and many turned to qualitative methods of data collection. However, the wholehearted acceptance of qualitative methods raised issues around experience, representation, critical reflection and subjectivity. Feminists began recognizing the exploitation inherent in such interactive research methods. Nast introduced a collection of articles that addresses these specific issues in relation to being 'women in the field'. Each contributor problematizes the relationships among participants in research projects. Rather than promoting a set of ethical guidelines for undertaking feminist research (which would be a masculinist solution), these articles challenge feminist geographers to further reflect upon their roles as feminist researchers.

Mattingly and Falconer al-Hindi (1995) edit a set of essays wherein each contributor reflects on how numbers can or should be used in feminist research. Coming from a different location within debates over experience, representation, critical reflection and subjectivity, these feminists assume that feminist geographers can undertake research without necessarily exploiting participants. The edited collection by Moss (1995) takes up this issue of exploitation by offering sets of reflections on specific research projects. Each contributor discusses various sets of politics shaping their experiences of conducting academic research and their ways of coming to interim solutions to the dilemmas they encountered in the field. Methodological concerns continue to be central in feminist geographies and remain open for critique, discussion and debate.

There are few explicitly feminist geography courses in the English language-based academy and even fewer required feminist courses (although there are some). Most books are pitched at graduate students and professional academics with only a few textbooks for undergraduates. Women and Geography Study Group of the Institute of British Geographers 1984 was the only available text from its publication until the early 1990s. It includes the central theme of gendering spatial phenomena. This book is still useful because of its place in the history of feminist geographies. A more recent publication by the Women and Geography Study Group of the Royal Geographical Society with the Institute of British Geographers 1997 attempts to capture the shifts in feminist geographies over the years. The group sought to produce a textbook for undergraduates that includes the 'breadth, diversity, intellectual vibrancy, debate and difference currently to be found in feminist geography' (Ibid., 1). This is really the only textbook explicitly for undergraduate students in their second or third year of study.

Rose's (1993) book would be useful as an upper-level undergraduate text if the focus were feminist geography as a field of study. She provides an historical overview of the development of feminism in geography, a feminist critique of the construction of geographical knowledge, and a psychoanalytical conception of paradoxical space. Jones, Nast and Roberts, too, could serve as an upper-level text. It would be most useful if the course were oriented toward women in the city or toward the undertaking of feminist research in specific settings. The book is packed with concrete examples of how feminists conduct research and analyse data. Spain (1992), who extends the concept of gendering to discursive spaces, and Matrix (1984), who exposes the masculinist assumptions underlying the construction of spaces, are examples of good companion texts for courses in urban studies. McDowell (1999) is probably the best text for an overview of feminist geography—inside the classroom and as background reading. It can be used as a resource manual, a survey text, or a primary discussion text for upper-level courses. Depending on the organization of the course, it might also be used for second- and third-year students. All these books could be used in postgraduate studies.

Much feminist geographical scholarship is sandwiched between other critical or progressive analyses in mainstream journals or edited collections. Yet gaining some understanding or seeking an introduction to various feminist geographies is not a formidable task. Perusing the pages of journals such as *Gender, Place & Culture, Environment & Planning D: Society & Space*, and *Antipode*, is useful in finding current examples of feminist research and analysis. Reviews of the feminist literature in geography appear regularly in *Progress in Human Geography*. It might also be useful to look specifically for feminist geography links through, for example, Geographical Perspectives on Women Speciality Group (GPOW).

Pamela Moss and Kathleen Gabelmann in Jonathan Michie (ed.), *Reader's Guide to the Social Sciences*, first edn (Fitzroy Dearborn Publishers, 2001). Politics Online. Taylor & Francis. www.routledgepoliticsonline.com:80/Book.aspx?id=w065 (accessed 15 October 2009).

Green consumerism

Green consumerism is a discourse that places the onus on individual consumers to take responsibility for reducing the environmentally harmful impacts of contemporary life. Governments and environmentalists place emphasis on the role of individuals as consumers to tackle **climate change** by conserving energy, taking public transit, recycling waste, growing food and foregoing flights. One example is the UK Government's 'Act on CO_2' campaign which gives helpful hints for reducing individuals' 'carbon footprint' when 'in the home', 'on the move' and 'out shopping' (campaigns.direct.gov.uk/actonco2/home).

Sherilyn MacGregor

Greening of hate

This phrase was coined by environmental justice and feminist scholars to describe the dovetailing of environmentalist, populationist and far right anti-immigration agendas. Population and immigration controls have recently been advocated as means to solve environmental problems such as **climate change**. The critique is that there is a disproportionate emphasis on controlling the fertility of poor, racialized women in the Global South (whose own ecological footprints, and that of their children, are miniscule compared to people in the affluent North) and insufficient attention to the real causes of environmental destruction (e.g. mass consumption, concentration of extreme wealth, militarism) (Hynes 1999; Urban 2007).

Sherilyn MacGregor

H

Harem

This Turkish word is derived from the Arabic *haraam*. It refers to the private and exclusive domain of women in polygynous households in countries now encompassed by the term Middle East. Harem literature refers to travel narratives written by Western women, who were granted unique access to women's quarters, and whose accounts are deemed less eroticized and rather less sensationalist than those of men.

Corinne Fowler

Hegemony

A term used to explain the relative stability of political authority and control in and between capitalist democracies, such as the US and many Western European countries, despite economic crises and World Wars. The original theorist in this vein is Antonio Gramsci (1891–1937), an Italian who wrote his major work as a political prisoner in the 1920s and 1930s. Gramsci began with the insight that, most of the time, power in liberal democracies is exercised not by overt coercion (such as imprisoning political dissenters), but through a dominant world view, or ideology. This common set of ideas and symbols legitimates existing rulers, helping them to win the citizens' consent, or at least acquiescence. Thus, in a medieval feudal economy, where serfs (agricultural labourers in bondage to the lords who owned the land they worked) were ruled over by an aristocracy, and the aristocracy by a monarch, a whole set of political structures and ideas had to be invented to legitimate and perpetuate the aristocracy and monarch's exclusive control of property. The notion that kings had a divine right to rule, given to them by God, is a good example of an idea that seems archaic to us today, but which served to support centuries of rulers.

Martin Griffiths (ed.), *Encyclopedia of International Relations and Global Politics* (Routledge, 2005). Politics Online. Taylor & Francis. www.routledgepoliticsonline.com:80/Book.aspx?id=w163 (accessed 15 October 2009).

Hobbes, Thomas

Describing himself as the 'twin of fear' because his premature birth was supposedly caused by rumours of the Spanish Armada off the coast of England, Thomas Hobbes (1588—1679) made fear of death the centrepiece of his political and moral theorizing. In a series of works culminating in *Leviathan* (1651), Hobbes argued for the institution of an absolute sovereign as a way to further the peace of the community and thereby promote the preservation and comforts of its citizens. To reach this conclusion he forged a social contract argument, a type of argument popular among some intellectuals of his day but which he revolutionized in a way that powerfully influenced the political thinking of subsequent contractarians such as John Locke (1632–1704), Jean-Jacques Rousseau (1712–78) and Immanuel Kant (1724–1804). He also published other works tackling issues in **metaphysics** and epistemology (including ontology, scientific method and free will), and topics in science and mathematics (including optics, geometry and human physiology).

In his political writings, Hobbes asks us to imagine a world devoid of government, filled with people who are 'sprung up like mushrooms' (*De Cive*, chapter 8, section 1), free of either political or social influences. In this 'state of nature' people act to satisfy their desires, chief of which is the desire to preserve their lives. In the process they come into conflict with one another: first, because of their inevitable competition for objects that each of them takes to be necessary to satisfy this desire; second, because of their fear of the aggressive motives of others; and third, because certain passions, among them the desire for glory, impel them to fight. The state of nature thus becomes a state of 'war of every one against every one', so that every person's life in that state is 'solitary, poor, nasty, brutish, and short' (*Leviathan*, chapter 13).

Hobbes argues that in order to end the war threatening their self-preservation, people in this state would agree to create the only genuine remedy: a commonwealth headed by an absolute sovereign. Hobbes is not arguing for a tyranny: sovereignty, in his view, can be invested in one person (a monarchy), several persons (an oligarchy), or all the people (a democracy). However, he argues against dividing sovereignty between or among different jurisdictional units or branches of government (advocated by contemporary English parliamentarians) on the grounds that these sovereign-parts will inevitably dissolve into factions violently competing for power, precipitating a return to the **state of war**. He also argues against attempts to limit the sovereign's power by a contract or constitution enforced by the people, claiming that if the people have the right to judge the sovereign's performance, disagreements among themselves or between them and the sovereign about how well he was governing could be adjudicated only by violent conflict, once again precipitating civil strife. Finally, he argues that the state must have the power to intervene in any area of human life in order to prevent internal strife.

While the structure and conclusion of Hobbes's social contract argument are well known, his moral theory is less famous and is the subject of many interpretive controversies. In virtue of the chaos in the state of nature, for which Hobbes proposed a political rather than a moral remedy, it may seem that he did not take seriously the idea that there could be moral imperatives which all human beings ought to follow and which, if followed, could ensure stable co-operation. In all of his political works he gives a detailed list of what he calls 'laws of nature'—moral imperatives dictating various kinds of co-operative behaviour. Although Hobbes claims these laws have a kind of validity in the state of nature, he none the less argues that they can neither stop the violence in that state nor provide the content of a constitution limiting the sovereign's rule or governing the sovereign's legislation. What are these laws? What is their justification? Why don't they provide a remedy for conflict in the state of nature?

The Taylor-Warrender Interpretation: A group of critics, headed in this century by A. E. Taylor and Howard Warrender, has insisted that Hobbes was an ethical objectivist and deontologist propounding laws of nature establishing rights and obligations. Whereas Taylor sees Hobbes as a proto-Kantian, Warrender believes he is still tied to the medieval natural law tradition and offers an interpretation that is probably the most influential of the recent deontological approaches to Hobbes. Warrender finds textual evidence for interpreting Hobbes as a deontologist in those passages in which Hobbes discusses rights and obligations. When, for example, Hobbes speaks of the rights of a sovereign, Warrender argues that he is using this term in a traditional way, to designate that to which the sovereign is morally entitled. Such a right is a correlate of a duty which others are obliged to respect. However, Warrender also argues that Hobbes sets out a different and original notion of right in passages such as the following:

> For though they that speak of this subject, use to confound *jus* and *lex, right* and *law*: yet they ought to be distinguished; because RIGHT, consisteth in liberty to do, or to forbear; whereas LAW, determineth, and bindeth to one of them: so that law, and right, differ as much, as obligation, and liberty; which in one and the same matter are inconsistent.
>
> (*Leviathan*, Chapter 14)

Warrender argues that this passage contrasts a duty, i.e. that which a person is obliged to do, with a right in Hobbes's new sense, which is what a person cannot be obliged to renounce. He maintains that Hobbes believes all obligations have validating conditions which define when an obligation holds, and that the Hobbesian study of rights is the study of when these validating conditions do not hold, and thus when one is freed from an obligation.

Warrender claims that the principal validating conditions for Hobbes are defined by applying certain principles: a) possessing the capacity to comprehend the law setting out one's obligation; and b) having a sufficient motive to perform the action that is obliged. For example, even though the third law of nature insists that contract-makers have an obligation to keep their promises,

Hobbes argues that when either promisor realizes that reneging will be to his advantage, he lacks sufficient motive to keep the promise, so that the second validating condition does not hold and the obligation is removed.

Warrender also argues that Hobbes rejects the idea that the mere knowledge of one's obligation could give one a reason to perform it. According to Warrender, Hobbes's psychological theory insists that human beings can only act so as to secure that which they take to be good for themselves. Hence, if Hobbes's moral theory is to be consistent with his psychological and motivational pronouncements, morality must be understood to supply us merely with a descriptive account of our duties and rights, and not to provide any additional motivational elements. Warrender argues that this way of interpreting Hobbes's moral theory makes that theory consistent with his psychological pronouncements, although it also means that the two views have no logical connection with one another. Hobbes's ethical theory, in Warrender's view, merely answers the question: 'How ought we to act?', whereas his psychological theory answers the question 'What motivates us to act?'

Why are the laws authoritative pronouncements of our duty? Warrender contends, on the basis of passages in which Hobbes claims the laws of nature are the commands of God (e.g. *Leviathan*, Chapter 15), that—like the Medieval natural law theorists—Hobbes regards these laws as morally obliging because our supreme divine authority commanded them. However, Warrender also argues that these divine commands are not followed in the state of nature because the validating conditions for the obligations set forth in these laws are not met.

Critics have raised a number of problems with Warrender's view. First, as we shall see below, they have criticized Warrender's objectivist reading of the text. Second, they have argued that Warrender's objectivist view amounts to attributing to certain actions a non-material and non-natural quality of 'rightness' that cannot be reduced to any material object or physiological feature of human beings. Moreover, Hobbes's persistent tendency to shy away from all property talk (discussed by Watkins), and his rejection of non-material moral objects proposed by Aristotelian moral theorists (e.g. *Leviathan*, Chapters 11 and 15) are inconsistent with an advocacy of the objectivist moral theory Warrender attributes to him.

Third, critics complain that Warrender's attribution to Hobbes of an inert notion of obligation is tantamount to denying that he has any ethical theory at all. If our obligations cannot move us, morality becomes merely an intellectual activity, a way of evaluating events and actions in the world that can have no motivational effect on human action. So we lose the idea that, as Kant would put it, one can do one's duty for duty's sake. It is arguable whether or not the notions of duty and obligation survive if they are interpreted as motivationally powerless, and whether what remains is a genuinely deontological moral theory.

Fourth, Warrender's interpretation of Hobbes's ethical views would also make his moral theory largely irrelevant to his political argument. That argument is concerned to persuade people to institute and maintain a sovereign. Given Hobbes's psychological theory, people will do this only if they believe it is in their self-interest. Hence, self-interest is all that can yield obedience to the laws of nature and political obedience to the sovereign. However, if this is so, then why would Hobbes talk about these actions in moral terminology lacking motivational force? Given his goals, there seems to be as little point in his describing them in moral terms as in aesthetic terms—either description would have no role to play in an argument designed to motivate people to institute a sovereign.

In part because of the religious language in Hobbes's writing, many theorists since Warrender have persisted in developing objectivist interpretations of Hobbes's views. Note, though, that any objectivist interpretation of Hobbes's ethical theory faces one or more of the problems besetting Warrender's interpretation. Either such a moral theory will be made consistent with Hobbes's psychological pronouncements by an interpretation that renders it motivationally powerless, in which case it loses its normative power and becomes irrelevant to Hobbes's political argument, or it will be understood to be motivationally efficacious, in which case it is inconsistent with Hobbes's psychological pronouncements and his nominalist and materialist metaphysics.

The Systematic Interpretation: Those who object to an objectivist interpretation of Hobbes's ethical theory tend to endorse a subjectivist interpretation that they claim is not only consistent with but reducible to Hobbes's psychological views. It has been called by Warrender the 'systematic' interpretation insofar as it purports to take seriously Hobbes's claim that his ethical views are derived from his psychological and physiological theories of human behaviour, and by Gauthier the 'contractarian' interpretation insofar as it captures a sense in which Hobbes's moral laws could be contracted on by everyone.

On this interpretation, some or all of Hobbes's normative language (including his use of terms such as 'obligation' and 'right') is to be given both authoritative and motivational force, but in a purely prudential way, and not in any morally objectivist way. 'Good' and 'bad' objects or states of affairs are defined by a person's desires and aversions, and 'right' or 'rational' actions are understood instrumentally, as those actions that are the most effective ways of attaining good (i.e. desired) objects or states of affairs. The resulting theory, exemplifying what is called naturalism in ethics, makes no attempt to refer to non-material objects or qualities, attributes no strange powers to human reason, and is entirely consistent with a physicalist metaphysics. Hobbes appears to endorse this kind of theory when he dismisses the existence of any *summum bonum*—the prescriptive entity that his academic contemporaries were most likely to embrace.

The cornerstone of this moral theory, according to the systematic interpreters, is a subjectivist conception of value, which they argue is put forward in passages such as the following:

> […] whatsoever is the object of any mans Appetite or Desire; that is it, which he for his part calleth *Good*: And the object of his Hate, and Aversion, *Evill*; And of his Contempt, *Vile* and *Inconsiderable*. For these words of Good, Evill, and Contemptible, are ever used with relation to the person that useth them: There being nothing simply and absolutely so; nor any common Rule of Good and Evill, to be taken from the nature of the objects themselves.
>
> (*Leviathan*, Chapter 6)

A subjectivist position on value is consistent with Hobbes's nominalist and materialist metaphysical tendencies. On this view, that which is responsible for our evaluations and which moves us to act is not some kind of intrinsic good 'out there', but our desires, the satisfaction of which is accomplished by securing objects which, in virtue of their power to satisfy desire, we call good. Hobbes does not explicitly put forward a projection theory of value, i.e. one in which a positive or negative evaluation is projected onto objects in virtue of our perception of them as either satisfying or impeding desire, but systematic interpreters regard his remarks in the passage above as highly suggestive of and consistent with such a theory.

After taking this naturalistic approach to value, Hobbes explains the sense in which moral philosophy is a 'Science of what is *Good* and *Evill*, in the conversation, and society of mankind' (*Leviathan*, Chapter 15). He points out that, although men differ greatly in what they desire, 'all men agree on this, that Peace is good' (*Leviathan*, Chapter 15). Hobbes calls this type of good (i.e. one that all human beings want and that they can all share) a common good (*De Homine*, Chapter 11).

However, Hobbes also distinguishes between two sorts of desired goods: real and apparent. The former is what a person would desire if he had true beliefs as well as a rightly functioning reason and desire-formation system in his **body**; the latter is what a person actually desires given the beliefs that he has and the physiological state he is in. Therefore, systematic interpreters claim that when Hobbes speaks about moral philosophy as the science of what is good in the conversation of mankind, he is not interested merely in what people actually seek, given their desires, as means to achieving those desires; he is also interested in what they *should* seek as means to achieving them (i.e. as the correct or most effective way to realize the object they are pursuing). Peace is, in his eyes, a 'real' common good insofar as it actually does lead to the furtherance of what people desire most—their self-preservation. Moreover, he also believes that peace is actually perceived by all as a good—the apparent and the real coincide in this case. What is not so manifest to everyone is that if peace is good, then also, 'the way, or means of Peace, which (as I have shewed before) are *Justice, Gratitude, Modesty, Equity, Mercy*, and the rest of the Laws of Nature, are good' (*Leviathan*, Chapter 15).

Systematic interpreters take this passage to present the laws of nature as prudential imperatives (what Kant would call 'hypothetical imperatives') asserting a causal connection between co-operative forms of behaviour and self-preservation because these forms of behaviour effect peace and thereby help to effect longer life. As further evidence they cite Hobbes's description of them as 'Conclusions, or Theorems concerning what conduceth to the conservation and defence of [people]' (*Leviathan*, Chapter 15). Moreover, they are mutually beneficial laws, and thus laws that people 'could agree to', because everyone benefits when they are followed.

Why aren't these laws followed in the state of nature? Hobbes answers this question by maintaining that:

> The Laws of Nature oblige *in foro interno*; that is to say, they bind; to a desire they should take place: but *in foro externo*; that is, to the putting them in act, not always. For he that should be modest, and tractable, and performe all he promises, in such time and place, where no man else should do so, should but make himself a prey to others, and procure his own certain ruin, contrary to the ground of all laws of nature, which tend to nature's preservation.
>
> (*Leviathan*, Chapter 15)

Two systematic interpreters (Kavka, Hampton) take this passage to say that every law has a rider attached to it, requiring that the action dictated in the law be performed only if others are willing to do so, too. This means the laws have the following structure: If you seek peace (which is a means to your preservation), provided that others are willing to do x, then do action x. So interpreted, the laws can be taken to be accurate axioms of prudence, and not directives that generate obligations that are opposed to self-interest.

What about a situation in which others are willing to do action x, and action x is conducive to peace, but I am able to attain more for myself if I refrain from doing x and 'free ride' on others' performance of the action? Does Hobbes believe I should follow the rule anyway? Kavka argues that he does, interpreting Hobbes as a 'rule egoist' who accepts that following the laws of nature will be contrary to self-interest in this one case. Hampton claims he does not, on the grounds that such a position is inconsistent with his psychological view that each of us always acts to procure some good for ourselves.

The systematic interpreters disagree over exactly how to interpret Hobbes's use of the term 'right'. However, all of them tend to think that the term is primarily used to denote an individual's liberty rather than a claim, and all of them want to interpret his use of that term to make it consistent with what they take to be his subjectivist moral theory and his largely egoistic psychology.

The Aristotelian Interpretation: The systematic interpreters believe that the key to Hobbes's moral theory is his subjectivist theory of value; it is this theory that provides the non-normative building blocks for the theory that enables it to be derivable from his psychology and consistent with his metaphysics. However, what if, upon examination, Hobbes's selection or definition of what counts as value-defining turns out to presuppose a norm either directly (i.e. that which does the selection or defining is a norm) or indirectly (i.e. a norm motivates the requirement that does the selecting or defining)? If so, Hobbes's moral theory presupposes the existence of the kind of inherently prescriptive moral object that the systematic interpreters claimed he was supposed to be explaining. The Aristotelian critics of Hobbes's texts argue that we can often see him relying on this kind of moral object, making him a neo-Aristotelian thinker rather than a proto-Humean one.

Aristotle (384–322 BCE) posited the existence of an objective value, commonly called the *summum bonum* by subsequent philosophers (including Hobbes), which is good for its own sake and which one ought to desire. The *summum bonum* is supposed to be accessible through the use of reason, so that reason has an end of its own, which may conflict with the ends of our desires. When someone pursues an end set by desire that is in conflict with the end set by reason, he is rightly criticized as irrational. Hobbes's Aristotelian interpreters note that when Hobbes labels as irrational those people who do not pursue their self-preservation (especially the glory-prone among us), he sounds quite Aristotelian. How can a subjectivist say that the effective pursuit of any good defined by desire is contrary to reason? A true subjectivist is barred from attributing to reason a goal that is taken to be objectively valuable and which can be opposed to the goals of desire.

Therefore, these critics argue that such passages show that Hobbes embraces a very Aristotelian theory about the role of reason in defining value. However, systematic and Aristotelian interpreters quarrel over whether or not these passages really do rely on an objective conception of value. Suppose, for example, that we distinguish between 'basic' and 'motivated' desires: whereas the object of a basic desire is desired for its own sake, the object of a motivated desire is desired, at least in part, as the means to the satisfaction of some other desire. One can be a thoroughgoing subjectivist and still criticize desires as irrational, if those desires are motivated rather than basic. Insofar as motivated desires are informed by reasoning about how to achieve either a basic desire or another motivated desire, then whenever the object of the motivated desire is not a means to the satisfaction of a more basic desire, criticizing that motivated desire as irrational amounts to a criticism of that person's reason, which has motivated a desire for the wrong object and which has therefore failed to be an effective servant to the more basic desire. One might argue that Hobbes's distinction between 'real' and 'seeming' goods is prompted by his wish to make this kind of criticism of motivated desires.

It is unclear that all of Hobbes's criticisms of certain passions, such as the passion for glory, can fit this analysis. He often criticizes as irrational a person who acts to achieve one basic desire, the desire for glory, because this pursuit impedes the satisfaction of another basic desire, the desire for self-preservation. Such criticism appears to be motivated by the Aristotelian thought that self-preservation is the 'right' good for people to pursue, regardless of what they actually desire. Hobbes also calls vain-glorious people 'insane' (e.g.

Leviathan, Chapter 8, and *De Homine*, Chapter xii), a pejorative label that appears to convict them of falling short of a norm of reasoning and choice.

If these passages cannot be rendered consistent with Hobbes's subjectivism, we can either see them as consciously Aristotelian (an interpretation that sits uneasily with Hobbes's constant disparaging of Aristotelian ethics in his writings), or as regrettable Aristotelian 'slips' from the austere moral science that may not have enough content to generate the results Hobbes needs to justify his political conclusions.

The latter conclusion holds attractions for those who wonder if it is possible to construct a synthesis of these three interpretive approaches to Hobbes's texts, one which admits that proto-Humean, Aristotelian and deontological language all surface in the Hobbesian texts. This view takes it that Hobbes, despite setting out to construct what we may describe as a proto-Humean conception of reason and morality, was forced to resort to Aristotelian and deontological ideas in order to generate a reasonable, plausible moral system. If this interpretation is right, it may show that it is difficult, perhaps impossible, for anyone, including Hobbes, to sustain a purely naturalistic moral theory.

Jean Hampton in Lawrence C. Becker and Charlotte B. Becker (eds), *Encyclopedia of Ethics*, second edn (New York: Routledge, 2001).

Human rights

Human rights are defined as the inalienable rights of any individual that are derived from being human. These rights are considered to provide the moral foundations for a life in human dignity. Human rights encompass two notions of 'right'; first, they refer to moral righteousness since they are based on assumptions about how human beings should be conceived of and treated, 'offering a framework for debate over basic values and conceptions of a good society' (Charlesworth and Chinkin 2000, 210). Second, as rights they also are entitlements, rights to something, e.g. bodily integrity, not to be enslaved or totally dispossessed. It is the combination of their moral, legal and political nature that establishes the specific character of human rights as social practice. According to Donnelly (1985, 12), 'having a right' means to be 'in a special, and controlling, relation to certain persons (who are bound by correlative obligations) and "objects" (to which one is entitled)'. Human rights as a social practice are constituted through compliance as well as through claims and negotiation. Human rights exist independently of their observation by particular governments or acts of law. However, without any legal recognition they only exist as legitimate moral claims, but not as rights. The content, practices and universality of human rights are subject to continuous discussion, and human rights are far from being universally accepted. None the less, there is general agreement as to the fundamental scope of human rights as enshrined in the **International Bill of Rights**. These documents define human rights predominately as the relationship between individuals (citizens) and their sovereign states. They acknowledge the universality of human rights and thus the 'inherency of rights' for all individuals and their mutual relations. Governments are obliged not to deprive individuals of their human rights, and they also have the duty to protect human rights and to prohibit abuse and **discrimination** in the private and public sphere on their territory. Consequently the state serves 'as both the guardian of basic rights and as the behemoth against which one's rights need to be defended' (Ishay 2004, 363).

Lena J. Kruckenberg

Human trafficking

Article 3(a) of the UN Palermo Protocol to prevent and suppress trafficking states: '"Trafficking in persons" shall mean the recruitment, transportation, transfer, harbouring or receipt of persons, by means of the threat or use of force or other forms of coercion, of abduction, of fraud or deception, of the abuse of power or of a position of vulnerability or of the giving or receiving of payments or benefits to achieve the consent of a person having control over another person, for the purpose of exploitation. Exploitation shall include, at a minimum, the exploitation of the prostitution of others or other forms of sexual exploitation, forced labour or services, slavery or practices similar to slavery, servitude or the removal of organs."

Sharron FitzGerald

I

Identity

As a term, this can assume different meanings and connotations depending on the context within which it is used (such as an academic discipline or theoretical framework). Current debates on **gender** politics often start from a critique of essentialist conceptualizations of a fixed, stable and integral individual identity, e.g. as a personal characteristic. Instead, identities are predominantly understood as socially constructed and therefore as procedural and positional. From such an explicitly inter-subjective perspective, identities are informed by societal structures and constructed through social practices. They are in constant flux, never singular but multiple, and constituted across different social positions, practices and discourses. Gender, **race** or class are framed as attributes that emerge in some interactions and fade in others, thereby shaping some social relationships, but not others. A white mother, for example, appears in some situations predominantly as a white person, in some she is primarily a woman, in some she is recognized as a mother and nothing else, and in many others she enacts all of the listed properties, and many more.

Lena J. Kruckenberg

Imperial feminism

This is a term usually linked to British **feminism** that came out of the black feminist critique of 'white feminist theory' of the 1980s. It is a term that seeks to encapsulate what black feminists regarded as the refusal by 'white feminist theory' to take full cognizance of the consequence of white ethnic power within feminist discourse. It proposes that conventional feminist discourse is a reproduction of the knowledge and experiences of white women as that of all women, which is rooted in normative ideas of modernity and rational universal humanism. The accusation of **imperialism** rests on the premise that conventional feminist discourse unfolded within a power structure of white dominance without acknowledging the problematic of **race** and as such erased the experience of the black female and black women's contributions to feminism and the advancement of women. Black feminists who have contributed to the imperial feminist debate include Valerie Amos, Hazel Carby and Pratibha Parmar.

Corinne Fowler

Imperialism

A policy aimed at conquering or controlling foreign people and territory. The essence of an imperial state is that it seeks to derive a benefit of some sort from those states and peoples unable to defend themselves against its superior military and/or economic force. This benefit may take the form of power, prestige, strategic advantage, cheap labour, natural resources, or access to new markets. Imperial states have achieved their goals in a number of ways. The most common method is through conquest and occupation, but the transportation of settlers and missionaries as well as market domination have also played a part in maintaining effective control over an empire.

Martin Griffiths and Terry O'Callaghan, *International Relations: The Key Concepts*, second edn (Routledge, 2002). Politics Online. Taylor & Francis. www.routledgepoliticsonline.com:80/Book.aspx?id=w043 (accessed 15 October 2009).

Intergovernmental Panel on Climate Change (IPCC)

The international organization that collects, analyses and synthesizes research on **climate change** conducted by scientists and other researchers from all over the world. Its reports, such as the recently released Fourth Assessment Report Climate Change 2007, are the main sources of research upon which governments around the world rely when setting national targets and making policies to address climate change.

Sherilyn MacGregor

International Bill of Rights

The first piecemeal steps towards the establishment of a body of international law on **human rights** were taken in the late 19th and early 20th centuries by a number of multinational agreements designed to correct specific abuses. The long struggle to eliminate slavery had produced the *Berlin Conference Resolutions, 26 February 1885*, the *Brussels Act, 2 July 1890*, and the *Convention to Suppress the Slave Trade and Slavery, 25 September 1926*. International agreements designed to suppress 'White Slavery', signed in 1904 and 1910, were followed by a *Convention for the Suppression of the Traffic in Women and Children, 30 September 1921*, and another on the same subject on *11 October 1933* and by a *Convention Concerning Forced Labor, 28 June 1930*.

However, it was not until the post-war period that serious steps began towards the creation of a comprehensive international regime of human rights. The fundamental document was the *Universal Declaration of Human Rights, 10 December 1948*. This drew on many sources, containing echoes, for example, of the English *Bill of Rights* (December 1689), the French *Declaration of the Rights of Man and Citizen* (26 August 1789), and the American

Bill of Rights (15 December 1791). The Declaration was adopted by the UN General Assembly without dissent—though six communist-bloc countries, Saudi Arabia and South Africa abstained. The document recognized not only civil and judicial but also economic, social and cultural rights, including the rights to own property, to work and to social security. It sought to establish a 'common standard of achievement for all peoples and all nations'. Although it accrued considerable moral authority, the Declaration lacked the full force of a treaty since its provisions were not legally enforceable. Another 18 years of detailed discussion were required before the *Covenant on Civil and Political Rights, 16 December 1966*, added legal substance to the provisions of the Declaration. The Covenant, which entered into force only on 23 March 1976 upon ratification by 35 states, added certain rights not included in the Declaration, notably rights of peoples to self-determination and cultural rights of minorities. A proposal to include the latter in the Declaration had been rejected in 1948, although a note was attached to the Declaration stating that 'the United Nations cannot remain indifferent to the fate [note "fate", not "rights"] of minorities'. The Covenant also established a Human Rights Committee to monitor compliance by signatory states. An accompanying *Covenant on Economic, Social and Cultural Rights, 16 December 1966*, which entered into force on 30 September 1976, upon ratification by 35 states, remained, like the Declaration, more an expression of goals than an enforceable treaty, although certain provisions, notably those prohibiting racial, sexual, linguistic, religious, political or other forms of **discrimination**, were prohibited outright.

Two *Optional Protocols* were added to the Covenant on Civil and Political Rights. The first, adopted by the UN General Assembly on 16 December 1966, enabled the Human Rights Committee, set up under the Covenant, to receive and consider complaints from individuals claiming to be victims of violations of rights. This Protocol entered into force simultaneously with the Covenant on 28 March 1979. The second Protocol, which was adopted by the General Assembly on 15 December 1989, aimed at the abolition of the death penalty and stated in its first article that no one within the jurisdiction of a state party to the Protocol might be executed. The second Protocol entered into force on 11 July 1991, upon ratification by 10 states. By 1995 28 states were parties to the second Protocol. Since all these had already abolished the death penalty (save, in some cases, in exceptional circumstances) in internal legislation, the effect of the Protocol was more symbolic than practical. Taken together, the Universal Declaration, the two Covenants and the two Optional Protocols were held to constitute an International Bill of Human Rights.

J.A.S. Grenville and Bernard Wasserstein (eds), 'The International Bill of Human Rights', in *The Major International Treaties of the Twentieth Century*, first edn (Routledge, 2001). Politics Online. Taylor & Francis. www.routledgepoliticsonline.com:80/Book.aspx?id=w044 (accessed 15 October 2009).

International Covenant on Civil and Political Rights (ICCPR)

This covenant is, together with the UN Universal Declaration on Human Rights (UDHR) and the International Covenant on Economic, Social and Cultural Rights (ICESCR), one of the key constituents of the **International Bill of Rights**. Unlike the UDHR, the two Covenants were each designed as legally binding treaties. The need for treaties, and indeed for treaties that enumerated two very different categories of rights, stemmed from the ideological cleavage that divided states after the Second World War and that was associated with the Cold War. The UDHR had the handicap of listing rights that were privileged by the 'free world', or those on the capitalist side of the ideological cleavage, along with rights that were privileged by those on the socialist and communist side of the ideological cleavage. The antagonism between these sides was such that those on the right did not consider economic, social and cultural rights to be rights as such, while those on the left felt that civil and political rights meant nothing if the subject of these rights had neither food nor shelter. The political way forward was the creation of two covenants, each a UN Treaty. They were drafted by the Human Rights Commission of the UN's Economic and Social Council.

The distinctive contribution of the ICCPR in the protection of **human rights** may be found in articles 6 to 27. The earlier articles and preamble share with the ICESCR a commitment to a broad programme in which the protection of human rights are made central to the purpose and conduct of states. Then the ICCPR goes on to delineate its specific rights agenda. Commencing with the right to life, the covenant requires that no one be subject to **torture**; cruel, inhuman or degrading treatment or punishment; slavery or compulsory labour, arbitrary arrest or detention. Those deprived of their liberty must be treated humanely, and people must not be imprisoned just because they cannot fulfil a contractual obligation. Further, there are provisions for freedom of movement, the lawful presence of aliens, equality of persons before courts and the law, and guarantees in criminal and civil proceedings. There are prohibitions on retroactive criminal law, and on arbitrary or unlawful interference in an individual's affairs, reputation and standing. The Covenant protects the rights to freedom of thought, conscience and religion, and the freedom to have and express opinions. Propaganda for war is prohibited, as is the advocacy of hatred (national, racial, or religious) that would constitute incitement to **discrimination**, hostility or violence. It protects the rights to peaceful assembly, freedom of association, marriage, and the founding of a family (including the equality of rights and responsibilities within and in the event of the dissolution of marriage). Children's rights also are affirmed. Citizens are recognized to have the right to partake in public affairs, to vote, be elected, and have equal access to public service. All are equal before the law and are entitled to equal protection from the law. Ethnic, religious and linguistic minorities are to be protected. The final article of the Covenant provides for a Human Rights Committee to be established in order to supervise the implementation of these rights.

Certain limitations and exceptions were granted in the provision of these rights. In emergency situations that threatened the life of the nation rights might be suspended, but only as strictly necessary and in accordance with the nature of the emergency. Such suspensions may not involve discrimination on the basis of **race**, colour, sex, language, religion or social origin. Suspensions must be reported to the UN. Some rights may never be abrogated; these are the rights to life and to equal recognition before the law; the freedoms of thought, conscience and religion; and freedoms from torture, slavery, imprisonment for debt and retroactive penal laws.

There were, in addition, two optional protocols. These protocols were optional in the sense they were in addition to the basic Covenant, and were to be subscribed to independently. The first optional Protocol is symbolic of the radical changes to be wrought in international politics by the human rights movement. Through this Protocol, individuals are given legal standing in international relations—in contrast to the prevailing doctrine that only states had legal personality. Under the Protocol, states that are signatories recognize that the Human Rights Committee established by the Covenant has the competence to receive communications from individuals who claim to be victims of rights violations, and who have exhausted all domestic means available to them for redress. The Committee, if it determines the communication to be admissible, brings the communication to the attention of the state party, which must then offer an explanation within a certain time frame, indicating what steps if any have been taken to ameliorate the complaint.

The second Optional Protocol to the ICCPR aimed at the abolition of the death penalty. No individual within the jurisdiction of states parties to the protocol may be executed; states must offer information about the steps they have taken to ensure that this is the case; and individuals may use the procedures of the first protocol in relation to the subject of the second (unless the state in question withdrew this option at the time of ratification or accession).

It took the Human Rights Commission from 1948 until 1966 to develop the two Covenants, in large measure because of the ideological divide expressed in international affairs as the Cold War, noted above. The express wish of the General Assembly of the UN was that the two Covenants, once tabled, would be ratified and come into effect without delay. However, another decade was to transpire before this was to be fulfilled. The delay was caused by the requirement that 35 states become parties to the Covenant before it could enter into force. Once the target number was eventually reached, the Covenant continued to be dogged by lack of support from the superpowers and many other key states.

The implementation of the ICCPR is facilitated by the Human Rights Committee (HRC). The Committee has two roles. The first is the operation of the state reporting mechanism. The Covenant requires that states parties place periodic reports about their compliance to the terms of the treaty. The Committee reviews these reports, returns comments to the reporting states and produces its own report. This report notes any relevant problems with state behaviour and reiterates the terms of the treaty. Second, the Committee accepts and acts on communications delivered via the Optional protocols, as discussed above.

The expectation is that states parties, in response to these reports, will make changes to their domestic legal systems and political practices in order to conform to the requirements of the treaty. The HRC often holds states to be in violation of their treaty requirements; how efficacious the committee's reports are in changing the conduct of offending states is another matter. Even in cases of significant positive change in state behaviour, other causal factors (such as domestic politics) apply. Further, the Covenant has no influence over states who are not parties to the treaty, or to the protocols. By 2003, the ICPCR had 149 member states, but fewer than 100 of these signatories had ratified the optional protocols. (*See also* **Capitalism**.)

Anthony J. Langlois in Martin Griffiths (ed.), *Encyclopedia of International Relations and Global Politics* (Routledge, 2005). Politics Online. Taylor & Francis. www.routledgepoliticsonline.com:80/Book.aspx?id=w163 (accessed 15 October 2009).

International Criminal Tribunal for the former Yugoslavia (ICTY)

In May 1993 the **UN Security Council**, acting under Article VII of the UN Charter, adopted Resolution 827, which established an ad hoc 'war crimes' tribunal. The so-called International Tribunal for the Prosecution of Persons Responsible for Serious Violations of International Humanitarian Law Committed in the Territory of the Former Yugoslavia (also referred to as the International Criminal Tribunal for the former Yugoslavia—ICTY) was inaugurated in The Hague, Netherlands, in November. The ICTY consists of a Chief Prosecutor's Office and 16 permanent judges, of whom 11 sit in three trial chambers and five sit in a seven-member appeals chamber (with the remaining two appeals chamber members representing the **International Criminal Tribunal for Rwanda—ICTR**). In addition, a maximum at any one time of nine *ad litem* judges, drawn from a pool of 27, serve as required. Public hearings were initiated in November 1994. The first trial proceedings commenced in May 1996, and the first sentence was imposed by the Tribunal in November. In July and November 1995 the Tribunal formally charged the Bosnian Serb political and military leaders Radovan Karadžić and Gen. Ratko Mladić, on two separate indictments, with **genocide**, crimes against humanity, violation of the laws and customs of war and serious breaches of the **Geneva Conventions**. In July 1996 the Tribunal issued international warrants for their arrest. Amended indictments, confirmed in May 2000, and announced in October and November, respectively, included the withdrawal of the fourth charge against Mladić. Karadžić was eventually detained in July 2008; however, Mladić remained at large in March 2009. In April 2000 Momčilo Krajišnik, a senior associate of Karadžić, was detained by the ICTY, charged with genocide, war crimes and crimes against humanity. Biljana Plavšić, a further former Bosnian Serb political leader, surrendered to the Tribunal in January 2001, also indicted on charges of genocide, war crimes and crimes against humanity. In the following month three Bosnian Serb former soldiers were convicted by the ICTY of utilizing mass **rape** and the sexual enslavement of women as instruments of terror in wartime. In February 2003 Plavšić was sentenced to 11 years' imprisonment, having pleaded guilty in October 2002 to one of the charges against her (persecutions: a **crime against humanity**). (Under a plea agreement reached with the Tribunal the remaining charges had been withdrawn.) In mid-1998 the ICTY began investigating reported acts of violence against civilians committed by both sides in the conflict in the southern Serbian province of Kosovo and Metohija. In early 1999 there were reports of large-scale organized killings, rape and expulsion of the local Albanian population by Serbian forces. In April ICTY personnel visited refugee camps in neighbouring countries in order to compile evidence of the atrocities, and obtained intelligence information from NATO members regarding those responsible for the incidents. In May the President of the then Federal Republic of Yugoslavia (FRY, which was renamed Serbia and Montenegro in February 2003, and divided into separate states of Montenegro and Serbia in 2006), Slobodan Milošević, was indicted, along with three senior government ministers and the chief of staff of the army, charged with crimes against humanity and violations of the customs of war committed in Kosovo since 1 January 1999; international warrants were issued for their arrest. In June, following the establishment of an international force to secure peace in Kosovo, the ICTY established teams of experts to investigate alleged atrocities at 529 identified grave sites. The new FRY administration, which had assumed power following legislative and presidential elections in late 2000, contested the impartiality of the ICTY, proposing that Milošević and other members of the former regime should be tried before a national court. In April 2001 Milošević was arrested by the local authorities in Belgrade. Under increasing international pressure, the Federal Government approved his extradition in June, and he was immediately transferred to the ICTY, where he was formally charged with crimes against humanity committed in Kosovo in 1999. A further indictment of crimes against humanity committed in Croatia during 1991–92 was confirmed in October 2001, and a third indictment, which included charges of genocide committed in **Bosnia** and Herzegovina in 1991–95, was confirmed in November 2001. In February 2002 the Appeals Chamber ordered that the three indictments be considered in a single trial. The trial commenced later in that month. However, Milošević continued to protest at the alleged illegality of his arrest and refused to recognize the jurisdiction of the Court. The case was delayed repeatedly owing to the ill health of the defendant, and in March 2006 he died in captivity. In August 2001 the ICTY passed its first sentence of genocide, convicting a former Bosnian Serb military commander, Gen. Radislav Kristić, for his role in the deaths of up to 8,000 Bosnian Muslim men and boys in Srebrenica in July 1995. In January 2003 Fatmir Limaj, an ethnic Albanian deputy in the Kosovo parliament and former commander of the Kosovo Liberation Army (KLA), was indicted by the ICTY on several counts of crimes against humanity and war crimes that were allegedly committed in mid-1998 against Serb and Albanian detainees at the KLA's Lapusnik prison camp. Limaj was arrested in Slovenia in February 2003 and transferred to ICTY custody in early March. At March 2009 the ICTY had indicted a total of 161 people. Of those who had appeared in proceedings before the Tribunal, 10 had been acquitted, 58 had received a final guilty sentence, and 13 had been referred to national jurisdictions. Some 26 people had completed their sentences. At that time 44 people remained accused by the Tribunal, including some 21 who were on trial, nine at the appeals stage, and two at large. It was envisaged that the Tribunal's trial activities would be terminated in 2010 and that appeals procedures would be completed by 2012. The Tribunal assisted with the establishment of the War Crimes Chamber within the Bosnia and Herzegovina state court, which became operational in March 2005, and also helped Croatia to strengthen its national judicial capacity to enable war crimes to be prosecuted within that country.

The ICTY is supported by teams of investigators and **human rights** experts working in the field to collect forensic and other evidence in order to uphold indictments. Evidence of mass graves resulting from large-scale unlawful killings has been uncovered in the region.

International Criminal Tribunal for the Former Yugoslavia (ICTY), in Europa World online (London: Routledge), www.europaworld.com/entry/io.un.sc.472.1 (accessed 1 June 2009).

International Criminal Tribunal for Rwanda (ICTR)

In November 1994 the **UN Security Council** adopted Resolution 955, establishing the ICTR to prosecute persons responsible for **genocide** and other serious violations of humanitarian law that had been committed in Rwanda and by Rwandans in neighbouring states. Its temporal jurisdiction was limited to the period 1 January to 31 December 1994. The Tribunal consists of 11 permanent judges, of whom nine sit in four trial chambers (based in Arusha, Tanzania) and two sit in the seven-member appeals chamber that is shared with the ICTY and based at The Hague. In August 2002 the UN Security Council endorsed a proposal by the ICTR President to elect a pool of 18 *ad litem* judges to the Tribunal with a view to accelerating its activities. In October 2003 the Security Council increased the number of *ad litem* judges who may serve on the Tribunal at any one time from four to nine. A high security detention facility had been built within the compound of the prison in Arusha. The first plenary session of the Tribunal was held in The Hague in June 1995; formal proceedings at its permanent headquarters in Arusha were initiated in November. The first trial of persons charged by the Tribunal commenced in January 1997, and sentences were imposed in July. In

September 1998 the former Rwandan Prime Minister, Jean Kambanda, and a former mayor of Taba, Jean-Paul **Akayesu**, both Hutu extremists, were found guilty of genocide and crimes against humanity; Kambanda subsequently became the first person ever to be sentenced under the 1948 Convention on the Prevention and Punishment of the Crime of Genocide. In October 2000 the Tribunal rejected an appeal by Kambanda. In November 1999 the Rwandan Government temporarily suspended co-operation with the Tribunal in protest at a decision of the appeals chamber to release an indicted former government official owing to procedural delays. (The appeals chamber subsequently reversed this decision.) In 2001 two ICTR investigators employed on defence teams were arrested and charged with genocide, having been found to be working at the Tribunal under assumed identities. Relations between the Rwandan Government and the ICTR deteriorated again in 2002, with the then Chief Prosecutor accusing the Rwandan authorities of failing to facilitate the travel of witnesses to the Tribunal and withholding access to documentary materials, and counter accusations by the Rwandan Government that the Tribunal's progress was too slow, that further suspected perpetrators of genocide had been inadvertently employed by the Tribunal and that Rwandan witnesses attending the Tribunal had not received sufficient protection. Reporting to the UN Security Council in July, the then Chief Prosecutor alleged that the Rwandan non-co-operation ensued from her recent decision to indict former members of the Tutsi-dominated Rwanda Patriotic Army for **human rights** violations committed against Hutus in 1994. In January 2004 a former minister of culture and education, Jean de Dieu Kamuhanda, was found guilty on two counts of genocide and extermination as a **crime against humanity**. In the following month Samuel Imanishimwe, a former military commander, was convicted on seven counts of genocide, crimes against humanity and serious violations of the **Geneva Conventions**. In early May 2004 Yussufu Munyakazi, accused of directing mass killings by the Interahamwe militia in Cyangugu and Kibuye Provinces, was arrested in the Democratic Republic of the Congo. In December 2008 Théoneste Bagosora, Aloys Ntabakuze and Anatole Nsengiyumva, former high-ranking military commanders, were found guilty of genocide, crimes against humanity and war crimes, and were each sentenced to life imprisonment. By March 2009 the Tribunal had delivered judgments against 41 accused, of whom six were acquitted. A further 24 people were on trial at that time, while eight indictees were awaiting trial and nine others were awaiting transfer. It was envisaged that the first trial activities of the Tribunal would be completed by the end of 2009. Some 13 of those accused remained at large. By early 2009 the ICTR had established 10 small information and documentation centres in Rwanda.

The ICTR is supported by teams of investigators and human rights experts working in the field to collect forensic and other evidence in order to uphold indictments. Evidence of mass graves resulting from large-scale unlawful killings has been uncovered in the region.

International Criminal Tribunal for Rwanda (ICTR), in Europa World online (London: Routledge), www.europaworld.com/entry/io.un.sc.472.2 (accessed 1 June 2009).

International political economy and gender

Feminists are introducing **gender** analysis into the field of international political economy (IPE) to explain why women are underrepresented in positions of political and economic power in most states and why the world economy differentially affects and rewards women and men. They draw on data, disaggregated by sex, which are exposing systematic political, social and economic inequalities between women and men on a global basis. Feminists claim that these persistent inequalities cannot be understood without including gender as a category of analysis. While IPE has focused on the behaviour of states and markets, gender analysis emphasizes the global gendered **division of labour** as an explanation for women's subordination.

Feminists claim that gender ideologies, as defined above, have the effect of assigning to women work that is unremunerated, such as the management of households, the care of children and subsistence agriculture. When women work in the labour force, they are disproportionately located in poorly paying jobs, such as service industries and the caring professions, which are typically defined as 'women's work'. Feminists believe that these gender ideologies naturalize this gendered division of labour, making it necessary to look beyond legal forms of **discrimination** and examine how deeply embedded patriarchal structures act to reinforce women's oppression.

Outside the field of IPE as conventionally defined, there is a considerable literature on women and development that uses gender as a category of analysis. This literature demonstrates that the development process has often led to a deterioration of women's status relative to men's. Feminists claim that liberal strategies that advocate economic growth through market forces and integration into the global economy have a differential impact on women and men. The negative effects of structural adjustment programmes on women have been an important focus of this literature on women and development.

Ann J. Tickner in R.J. Barry Jones (ed.), *Routledge Encyclopedia of International Political Economy*, first edn (Routledge, 2001). Politics Online. Taylor & Francis. www.routledgepoliticsonline.com:80/Book.aspx?id=w053 (accessed 15 October 2009).

International relations – *see* Feminism and international relations; Gender and international relations

Intersectionality

This is a concept to describe intersecting, interdependent and mutually constitutive relationships between categorical dimensions of **identity** such as **race,**

ethnicity, gender, class or sexual orientation. Intersectionality implies that these and other categorizations cannot be understood as separate or essentialist, since they are always experienced simultaneously and derive their particular meaning in relation to one another. Intersectionality, however, does not only define a concept, it is also closely connected to the research programme and theoretical work of Kimberlé Crenshaw. Building on a tradition of Black Women's Studies, Crenshaw coined this term and emphasized the ways in which *subordinate devalued* identities intersect, focussing on the complex patterns of multiple **discrimination** experienced by those with two or more devalued facets of identity, such as black women. The concept of intersectionality received substantial scholarly attention during the 1990s and thoroughly transformed the discussion of gender issues. More recently, Patricia Hill Collins's concept of intersectionality as 'matrix of domination' became influential (Collins 2000). West and Fenstermaker's concept of '**doing difference**' is part of the discourse on intersectionality. The intersectionality perspective still poses significant challenges for empirical research, not the least due to the innumerable categories which need to be taken into account. Further difficulties arise from different levels involved, from the individual to the interpersonal and structural level, and the complexity of the contextualized accomplishment of different facets of identity in social interaction. Approaches to intersectionality vary widely, with some of them defining social categories such as race and gender as given but having interdependent relationships, others understanding them as given but interdependent in themselves, while some seek to deconstruct the categories as such. Identities are related in a synergistic way when intersecting at the 'crossroads' of race, gender, class and other categorizations. As Crenshaw focused on 'overlapping systems of subordination' (1991, 1,265), such as racism and **sexism**, intersectionality is now often understood as mainly referring to these two.

Lena J. Kruckenberg

Intertextuality

A central tenet of poststructuralist literary studies, intertextuality refers to the presence of one text in another and considers the way in which particular narratives are shaped by anterior writing. The concept of intertextuality is also routinely used in the study of journalism and literature, or literature and film.

Corinne Fowler

Irregular migration

This is a broader and less normative term than 'illegal migration'. De Haas notes that 'illegal migration is primarily a legal term, which does not necessarily reflect the actual experiences of migrants' (De Haas 2007, 4).

Sharron FitzGerald

L

***Leviathan* – *see* Hobbes, Thomas**

Liberal feminism

Liberal **feminism** advocates equal rights for women. It was most prominently advocated in the 19th century by John Stuart Mill in *The Subjection of Women.* The dominant form of feminism in the 19th and early 20th centuries, for second-wave feminists it has primarily been an object of criticism. This is despite the fact that most gains by the women's movement in the West have been made on terms of liberal rights. Socialist feminists have criticized the empty formalism of rights that do not enable women to achieve substantive equality, and the way the public/private distinction central to **liberalism** obscures women's subordination in the domestic sphere, ruling it outside legitimate intervention. Other feminists have argued against the 'malestream' assumptions of liberalism, with particular reference to the supposedly **gender**-neutral individual. However, alongside this critique there is also a feminist revision of liberalism, which argues that it is not *necessarily* hostile to feminist claims, even if liberal arguments have generally been used to support the status quo.

Mill's *The Subjection of Women* (1869), a statement of liberal feminism, was enormously influential in the first-wave feminist movement. From the initial premise that men and women have essentially the same capacity for reason and, therefore, self-determination, he argues that women should not be excluded from exercising that capacity in professional work and political life by being confined to the domestic sphere under the direction of their husbands. They should have equal rights with men to education and access to training and work, to the representation of their political interests by means of the vote, and to personal autonomy with rights over property, divorce and so on. However, he does not attack the sexual **division of labour** as such; on the contrary, he supposes that only exceptional women would choose to compete in the workplace and that most would be principally wives and mothers, dependent on husbands who would support them and their children in the home. The liberal acceptance of socio-economic relations that contribute to the inequality of the sexes is the basis of more radical feminist criticism of liberal feminism.

Kate Nash in Lorraine Code (ed.), *Encyclopedia of Feminist Theories* (London and New York: Routledge, 2000), 303.

Liberalism

Liberalism can mean either a particular party creed in a particular time period, especially the late 19th century (the heyday of Liberalism), or a general social and political attitude and orientation. Historically, Liberalism was a middle-class or bourgeois movement for freedom from remaining feudal and monarchical control, and was associated, *inter alia*, with freedoms both legalistic, such as the economic theory of laissez-faire, and individual. From this position of supporting basic civil liberties or **human rights**, liberalism has developed a modern political creed in which the independence of the ordinary citizen against any powerful body, whether the state or, for example, organized labour, is taken as vital. Modern liberal parties, and they exist in most democratic states, although not necessarily under that title, tend to argue that traditionally organized class politics, with an apparently insoluble conflict between **capitalism** and some form of socialism or Marxism, is misplaced, and that a greater concentration on the talents, capacities and needs of actual individuals rather than systems of social composites is possible and desirable. Liberalism is one of the best reasons for doubting the suitability of the standard left/right model of politics because it contains the commitment both to equality by the left and to approval of individual human effort and freedom by the right. In this sense it is often seen as being in the middle of the political spectrum, but most Liberals would argue that, far from being 'centre' or 'moderate', they are in fact radical, wishing to change much in society. Their opposition to class politics is illustrated by the example of recent British general elections, in which the Liberal Party/Liberal Democrats have gained almost exactly the same percentage from *all* social classes. A similar pattern tends to be found in most other Western countries.

David Robertson, *A Dictionary of Modern Politics*, third edn (Europa Publications, 2002). Politics Online. Taylor & Francis. www.routledgepoliticsonline.com:80/Book.aspx?id=w007 (accessed 15 October 2009).

Literary canon

This refers to those literary works that are considered to have had a determining role in shaping Western culture. However, this canon, and the cultural processes that surrounded its formation, began to come under critical scrutiny around the 1960s. In recent years, further debates have attended the selection of texts for university courses on postcolonial writing.

Corinne Fowler

Lowbrow and middlebrow writing

Originally intended as terms of abuse, the categories of lowbrow (an older term) and middlebrow have been used, often interchangeably, to denote accessible and popular literature of inferior literary quality. The term middlebrow was first coined by *Punch* magazine in 1925 to bridge the concepts of 'highbrow' and 'lowbrow'. Both terms cover popular fiction and low-status genres such as **travel writing**, romance or crime writing. However, such writing is now commonly judged to be of real and enduring interest to literary and cultural studies. The theory surrounding these categories is still evolving. In his 2005 study, *From Lowbrow to Nobrow*, Peter Swirski convincingly argues that, even since the early 20th century, the intellectual distinction between highbrow and lowbrow has been founded on cultural prejudice rather than on reality, since the two have long since been intertwined and interdependent.

Corinne Fowler

M

Mainstreaming gender – *see* **Gender mainstreaming**

Man/woman – *see* **Categories and dichotomies**

Marriage, institution of – *see* **Beauvoir, Simone de**

Masculinity

This refers to the range of physical, behavioural and attitudinal qualities that characterize what it means to be a 'man' in any given historical or cultural context. Masculinity increasingly has come under critical scrutiny largely as a result of feminist theory and **activism**. First-wave feminist challenges to **patriarchy** initially focus on issues such as women's equal access to the political arena and educational institutions. Yet entrenched assumptions about what women were and were not capable of doing and, indeed, what they ought to be allowed to do, stood in the way of these challenges. At issue were certain reigning beliefs about the nature of 'femininity'. In the 1970s, feminist scholarship and activism began to analyse and rethink **gender** more directly, focusing on the social meaning and influence of the idea of femininity itself. In such critiques, masculinity was necessary implicated—at first implicitly then, increasingly, explicitly—as the gender norm against which femininity is both defined and diminished. Such critiques have spawned a proliferation of **critical theory** about masculinity.

Traditional thinking about gender is often essentialist. Many people presume that masculinity and femininity are universal and natural expressions of being male or female. However, essentialist assumptions are not confined to ordinary thinking about gender. Some theorists argue, for example, that possession of the Y chromosome, and the subsequent production of the hormone testosterone during puberty, naturally result in the normal gender expression known as masculinity. Others offer sociobiological arguments that, like males of other species, the human male is subject to the laws of natural selection; on this view, masculinity—which is presumed naturally to include sexual aggressiveness, competition and dominance—is simply a function of the evolutionary drive toward maximizing 'inclusive fitness'. However, not all essentialist arguments are biological. Drawing on Jungian psychology, Robert Bly (1990) maintains that masculinity is the manifestation of universal primordial archetypes such as the Wild Man and the Warrior.

In general, essentialist explanations hold that the outward, socially apparent qualities of maleness derive from certain elements inherent to the male human being. Often, such explanations do not question the universality or appropriateness of male dominance in modern society. Indeed, some appeal to essentialist theories to argue that patriarchy reflects a natural order between men and women. However, essentialists need not be apologists for male dominance and violence. For example, Bly argues that contemporary society has alienated men from their true, balanced masculine selves.

By contrast, social constructionist approaches to gender are suspicious of the presupposition that prevailing social relations and structure express a natural biological or psychological order. Instead, masculinity and femininity are regarded as powerful social concepts that are produced, shaped, and maintained by social conventions and institutionalized practices.

Constructionist theory focuses on how the ideal of masculinity is exemplified in cultural productions, such as literature, film and art, and in the conventions and practices of male-dominated activities, such as sports and the military. Constructionist arguments show how the predominant ideals of masculinity and femininity constitute a rigid gender binarism that are not only complementary in structure, but establish a social and political asymmetry. Those qualities deemed masculine—for example, independence, rationality, assertiveness, physical strength and protectiveness—place men in a relation of power over men who are construed as dependent, emotional, passive, weak and in need of protection. The construction of the masculine/feminine dyad is also inherently heterosexual and therefore integral to sustaining heterosexism. Social constructionists further argue that, in patriarchal contexts, masculinity functions as the normative gender. The qualities associated with masculinity tacitly structure the key institutions of modern society, including law, science and medicine. Feminists such as Anne Fausto-Sterling and Carol Gilligan have shown how scientific explanation has uncritically presumed the **normativity** of masculinity. Others have argued that the standards for such things as 'reasonable behaviour' in the law and 'merit' in the workplace reflect masculinist values.

Constructionists maintain that masculinity and femininity not only structure social relations and institutions, they also mediate the psycho-sexual development of individual persons. The development of boy-children into adults occurs under the force of social sanction; those who conform to the prevailing standards of masculinity are socially accepted, while those who do not are punished. Masculinity is unevenly exhibited in the behaviours and practices of actual men. While all persons identified as 'men' tend to benefit socially and politically, to some degree, from their inclusion in the ambit of masculinity, many (perhaps most) men do not and cannot embody the gender ideal. Indeed, while early feminist theory often implicated a monolithic and reductive notion of masculinity, theorists have come to recognize that, even as a gender ideal, masculinity is neither monolithic nor universal. It varies along lines of culture, **race** and class. For example, some theorists have argued that the prevailing norm of masculinity in the West is ineluctably white, affluent and non-disabled. If this is the case, men who are dark-skinned, poor and/or disabled are systematically less able to conform to its demands and, thus, are subject to greater diminishment.

Finally, masculinity does not only have to do with men. From the woman who is simply assertive, strong and self-confident to the woman who explicitly identifies as butch, to defy the boundaries imposed by femininity is to be deemed 'masculine'. Excursion into 'masculine' by women would not be socially frowned upon were they not recognized to be claims on social power.

Initial feminist resistance to focusing energy on masculinity has given way to the recognition that transformation of oppressive gender relations requires critical scrutiny of the predominant ideals of masculinity no less than of femininity. (*See also* **Biological determinism**; **Categories and dichotomies**.)

Joan Mason-Grant in Lorraine Code (ed.), *Encyclopedia of Feminist Theories* (London and New York: Routledge, 2000), 322–24.

Metaphysics

Metaphysics is a broad area of philosophy marked out by two types of inquiry. The first aims to be the most general investigation possible into the nature of reality: are there principles applying to everything that is real, to all that is?—if we abstract from the particular nature of existing things that which distinguishes them from each other, what can we know about them merely in virtue of the fact that they exist? The second type of inquiry seeks to uncover what is ultimately real, frequently offering answers in sharp contrast to our everyday experience of the world. Understood in terms of these two questions, metaphysics is very closely related to ontology, which is usually taken to involve both 'what is existence (being)?' and 'what (fundamentally distinct) types of thing exist?'

The two questions are not the same, since someone quite unworried by the possibility that the world might really be otherwise than it appears (and therefore regarding the second investigation as a completely trivial one) might still be engaged by the question of whether there were any general truths applicable to all existing things. However, although different, the questions are related: one might well expect a philosopher's answer to the first to provide at least the underpinnings of their answer to the second. **Aristotle** proposed the first of these investigations. He called it 'first philosophy', sometimes also 'the science of being' (more or less what 'ontology' means); but at some point in antiquity his writings on the topic came to be known as the 'metaphysics'—from the Greek for 'after natural things', that is, what comes after the study of nature. This is as much as we know of the origin of the word. However, it would be quite wrong to think of metaphysics as a uniquely 'Western' phenomenon. Classical Indian philosophy, and especially Buddhism, is also a very rich source.

Edward Craig (ed.), 'Metaphysics', in *Routledge Encyclopedia of Philosophy* (London: Routledge, 1998), www.rep.routledge.com/article/N095.

Method acting

A style of acting that was adopted by the Actors Studio, founded in 1947 in the USA and which was derived from the Soviet actor and director Konstantion Stanislavsk. The method was to act completely naturally, to so infuse one's own self with the thoughts, emotion and personality of the character that one became that character. Simultaneously the actor must draw on her or his own experiences to understand what motivates the character she or he is to play. Often the performance is understated—certainly never in excess. The method is seen as totally realistic and is best exemplified by the actors Marlon Brando, Montgomery Clift, James Dean, Rod Steiger and Julie Harris (see, for example, *A Streetcar Named Desire*, Elia Kazan 1951; *From Here to Eternity*, Fred Zinnemann 1953; *On the Waterfront,* Kazan 1954).

Susan Hayward, *Cinema Studies: The Key Concepts* (London and New York: Routledge, 2000), 227–28.

Motherhood – *see* **Beauvoir, Simone de**

Mutually exclusive – *see* **Categories and dichotomies**

N

Neoliberalism

This is a political philosophy that asserts that the global economy is driven 'by free choices of customers, which dictate a radical opening of economies. The task of government is to end protectionism and to ensure that its people have a good life by ensuring stable access to the best and cheapest goods and services from anywhere in the world' (Ohmae 1991, 12). Western nation-states have used this rhetoric to 'combine trade and security' (Douzinas 2007, 192). This notion of openness is also put side by side with the spread of Western democracy world-wide.

Sharron FitzGerald

Neo-Malthusian population discourse

Invoking the dismal predictions of 19th-century thinker Thomas Malthus, some environmentalists warn that the world's population is surpassing the planet's carrying capacity. They predict that lack of resources will bring certain conflict, mass starvation and other miseries (Hardin 1968). Comparing population growth with a cancer, some early environmentalists advocated authoritarian government, 'voluntary human extinction', and compulsory sterilization in developing countries as possible solutions. By the mid-1980s, this particular brand of environmentalism had been rejected on the grounds of questionable ethico-politics as well as faulty demographics and computer modelling. This Hobbesian view of humanity, in which conflict over scarce resources is inevitable, has recently re-emerged as a theme in debates about **climate change**, as exemplified by the United Kingdom-based Optimum Population Trust.

Sherilyn MacGregor

Normativity

This is any theory that seeks to establish the values or norms that best fit the overall requirements of society. However, I deploy the term as a pejorative term in the service of those who govern through prescriptive and exclusionary rules and regulations (Douzinas 2007).

Sharron FitzGerald

O

Other

The French philosopher Simone de **Beauvoir** is credited with publishing the first extensive analysis of the way in which women are defined as the 'other'—the ways in which women are invisible in culture and in which female experience is ignored. Beauvoir argued that man was the **subject**, the norm, and that women were *defined* as the 'other'. Men are defined positively, and women negatively. Woman is not just different from man, but the antithesis of man. To be like a man is to be not like a woman.

Women are defined from a male perspective. Men are seen as cultured, as rational, as possessors of scientific knowledge. Women are seen as 'natural' and as the object of scientific knowledge rather than its producers. They are seen as controlled by their bodies and their hormones. Scientific knowledge, of which men are the possessors, is concerned with controlling by 'cultured' man. Beauvoir thought that by accepting themselves as defined by men—as other to men—women allowed themselves to be dominated by men.

Feminists have challenged man-made 'scientific' definitions of women and men's dominance as definers of reality. While some feminists have argued for androgyny, others have stressed the need for discourses that offer self-definitions for women that value their unique characteristics as much as those of men—or more. (*See also* **Beauvoir, Simone de.**)

Cheris Kramarae and Dale Spender (eds), *Routledge International Encyclopedia of Women: Global Women's Issues and Knowledge* (New York and London: Routledge, 2000), 1,482–83.

P

Patriarchy

The literal and historical meaning of 'patriarchy' is 'the rule of fathers'. Although there is no consensus on the contemporary definition of the term, many feminists have extended it beyond the realm of the family to include the rule of men over women. Patriarchy so defined encompasses all systems of male dominance. Most feminists outside the liberal tradition claim that men dominate women through a variety of political, economic and social structures that vary across time and place. Since patriarchy is not seen as a universal state, but as a social construction, feminists believe that it can be overcome. Therefore, revealing and critically analysing patriarchy's various manifestations can contribute to ending it.

Liberal feminists have argued that forms of **discrimination** against women have decreased in many societies as legal barriers to voting and participation in the labour force have broken down. However, since forms of **gender** discrimination remain even after legal barriers are eliminated, other feminists believe that they can only be explained by the existence of a patriarchy that is sustained by patriarchal ways of thinking.

Patriarchal thinking is based on socially constructed gendered dichotomies such as reason/emotion, culture/nature, independent/dependent and public/private. The first term in each of these pairs is typically associated with men, the second with women. Such thinking has the effect of assigning women to reproductive and maintenance tasks in the private sphere. Based on a belief that women are closer to nature than men, these roles come to be seen as natural ones for women. Through the association of reason and culture with **masculinity**, men are able to transcend biological categories, thus legitimizing their predominance in activities in the public sphere.

This public/private distinction is integral to feminist definitions of patriarchy. Feminists assert that the separation of public and private spheres began in the 17th century, concurrent with the birth of modern states and **capitalism**. Modern state-formation marked a shift to relatively independent household units, legally headed by men; women became vulnerable and dependent on fathers and husbands. As the workplace became separated from the household, women were consigned to the role of housewife, and men assumed the role of breadwinner. Even though many women have always worked, the notion of the family wage has meant that their wages have been perceived as supplementing men's, thus justifying women's lower pay. Since men's labour moved into the market, production has been seen as more important than reproduction. Feminists claim that understanding these public/private distinctions can help to explain the persistence of gender inequalities.

The privileging of the public over the private has led to the use of market models to explain the economic and political behaviour of individuals and states. Suggesting that the concept of rational economic man assumes characteristics that correspond to a socially constructed masculinity, feminists assert that women in their reproductive roles as providers of basic needs for families do not conform to the behaviour of instrumental rationality. Feminists claim that this has implications for liberal and nationalist theories of international political economy (IPE), both of which draw on rationalist models to explain and prescribe the behaviour of states and markets. Feminists also criticize Marxian theories for focusing on public-sphere activities and neglecting the private sphere of reproduction.

Feminists in the field of IPE believe that, if the reasons for the continued under-representation of women in positions of political/economic power and their disproportionately high numbers among the world's poor are to be adequately understood, the patriarchal underpinnings of states/markets and the theories that are used to explain their behaviour must be revealed, analysed and transcended.

Ann J. Tickner in R.J. Barry Jones (ed.), *Routledge Encyclopedia of International Political Economy*, first edn (Routledge, 2001). Politics Online. Taylor & Francis. www.routledgepoliticsonline.com:80/Book.aspx?id=w053 (accessed 15 October 2009).

Performativity – *see* **Butler, Judith; Body, the**

Phallocentrism

This is the belief that the phallus or penis is a major symbol of male power, and feminists' use of the term extends the analysis provided through the term **sexism**. Phallocentrism includes the view that assumptions about the supposed innateness of heterosexuality are crucial to behavioural, attitudinal and institutional sexism. It goes further than the analysis of heterosexism by focusing on sexual power and its relationship to heterosexuality. In this analysis, 'heterosexuality' in its institutional form indicates considerably more than sexual preference, for the use of phallocentric definitions of 'sex' positions women as subordinate and passive objects to male sexual subjectivity in its narrowly penile and penetrational form. For example, an indication of the prevalence of this view lies in the range of expletives and insults based on the penetrational use or misuse of women's genitals by men. Phallocentrism also makes 'the lesbian' vanish from epistemological sight except as rejection by men, for 'the lesbian' is outside symbolic economy based on the penis and what it penetrates.

A critique of phallocentrism has been central to a wide range of feminist discussions of heterosexual sex from the early days of the present women's movement. It has also underpinned lesbian feminist theorizations of institutional heterosexuality and lesbian oppression. In addition, it has been crucial to analyses of power within patriarchal society that have seen **rape** as the archetypal instance of the 'force and the threat of force' that constitutes women's oppression as distinct from women's inequality and exploitation (Brownmiller 1975). Lacanian ideas about 'the phallus' see this as having a *symbolic* role in signifying binary **gender**, particularly in language, whereas analyses of phallocentrism are concerned with phallocentric *behaviour*, specifically the threatened and actual use of the penis as a weapon against women.

Liz Stanley in Cheris Kramarae and Dale Spender (eds), *Routledge International Encyclopedia of Women: Global Women's Issues and Knowledge* (New York and London: Routledge, 2000), 1,535.

Philosophy of science, feminist

The notion of science as an objective enterprise—value free, politically neutral and **gender** blind—is being contested on a variety of fronts. The feminist critique reveals that modern science, with its ideals of detachment and domination, displays a distinctly androcentric bias. There are three main categories of feminist philosophy of science that emerge from this critique: feminist empiricism, feminist standpoint epistemology and postmodern feminist epistemology. All start from the observation that science is a socially constructed activity: the social location, status and gender of scientists and scientific communities all play a significant role in determining the methods and practices of science. Throughout much of the history of Western science, women and people of colour have not been part of scientific communities, and this exclusion has influenced which questions are deemed appropriate for scientific inquiry, what types of research methods are employed and what evidence counts in evaluating hypotheses.

Feminist empiricism has its origins in the work of feminist scholars in biology and the social sciences who recognized that the answers to many questions involving sex and gender reflected a distinct androcentric and/or sexist bias (see Tuana 1989). Moreover, many questions concerning women's lives and bodies either were not answerable within mainstream theory, or had received inadequate attention from the mainstream scientific establishment. They believed that the scientific method was not the problem; the problem was that researchers were not following it. **Sexism** and **androcentrism** can be eliminated from science if researchers rigorously follow the existing norms of scientific methodology. Many feminist practitioners in the social and natural sciences today subscribe to this view

Feminist standpoint epistemology calls for a more radical change in the conception of good science. In this view, all knowledge is socially situated; there is no one position that is value neutral and objective. It has its roots in Marx's dialectical materialism which holds that material conditions structure the way we apprehend the world. Hartsock (1983) argues that if material life is structured in fundamentally opposing ways for two different groups, then the vision of each will represent an inversion of the **other**, and in systems of domination the vision of the rulers will be partial and distorted. The institutionalized gender **division of labour** structures men and women's lives differently and forms the basis of a feminist standpoint. A feminist standpoint is neither objective nor ethically neutral. Since it is rooted in the material conditions of women's lives, it is epistemologically privileged and offers a more humane vision of the relationships between people and between people and the natural world.

Object-relations theory posits a psychoanalytic explanation for the **difference** in male and female apprehensions of the world, and the way that those apprehensions have influenced scientific practice. Keller (1985) argues that the equation of knowledge and power is linked to the masculine developmental process. In learning to become males, boys must learn that they are fundamentally different from their mothers, and in learning this difference boys come to perceive that the self and the object world are separate and distinct. Pervasive in the masculinist world view is the construction of the self in opposition to another. It is this psychological construction of the self that lies behind the persona of the scientist as an autonomous, disinterested observer, and the notion that the purpose of science is to gain control over the object of study.

Harding (1995) argues that feminist standpoint epistemology provides the foundation for the notion of 'strong objectivity'. Ideally, the concept of impartial, unbiased, value-free research should eliminate social values and prejudices from science. However, in practice it eliminates only those values that differ among researchers. Shared values within the scientific community will not be questioned. To the extent that the community excludes women and people of colour, implicit assumptions about **race, ethnicity, gender and class** will not be apparent. Strong objectivity rejects the ideal of value neutrality, and extends the notion of the scientific method to include an examination of hidden cultural assumptions that remain invisible from the standpoint of the dominant groups. Strong objectivity is necessary if science is to escape containment by the interests and values of the powerful.

A similar position is found in a philosophically grounded version of feminist empiricism (see Longino 1990; Nelson 1990). They hold that knowledge is not constructed by individuals using the scientific method, but rather by individuals in dialogic communities. It is within the context of these communities that scientists' observations, theories, hypotheses and patterns of reasoning are shaped and modified. Background assumptions and values partially define what it is to be a member of that community. Thus, effective criticism and, by extension, good science require alternative points of view. The crucial difference between these feminist empiricists and standpoint theorists is that the empiricists reject the claim that the standpoint of oppressed groups is epistemologically privileged.

Postmodern ***feminism*** argues that both feminist standpoint epistemology and feminist empiricism are flawed because they require an uncritical appropriation of Enlightenment ideals (see Flax 1992; Haraway 1990). **Postmodernism** contests the central assumptions of the Enlightenment metanarrative, and rejects the notion of innocent knowledge: that is, knowledge that is separate from power and works for the benefit of all. Truth, reason, universality and objectivity are seen as mere artefacts created by humans rather than transcendental truths. Truth is an effect of discourse rather than an apprehension of the real. Thus any transcendent authorization of meaning is lost, and with it the ontology that grounds Western epistemology.

Order is imposed on the world through the use of binary oppositions—nature/culture, female/male, reason/emotion—which create the necessary boundaries between order and chaos. Haraway argues that the boundary maintaining images generated by the modern episteme are inadequate for today's realities. She suggests that a feminist science must begin with a new episteme and politics. It must address the fact that our world is characterized by massive insecurity and a common failure of subsistence networks for the most vulnerable, and these conditions are inseparable from the social relations of science and technology. She suggests that feminist science look to the possibilities of new unities based on affinities rather than essential characteristics.

Drucilla K. Barker in Phillip Anthony O'Hara (ed.), *Encyclopedia of Political Economy*, first edn (Routledge, 1999, 2001). Politics Online. Taylor & Francis. www.routledgepoliticsonline.com:80/Book.aspx?id=w026 (accessed 15 October 2009).

Population control – *see* **Contraception**

Positioning/positionality

Linda Alcoff (1988) uses the concept of positionality as a way to avoid essentializing the category of Woman while making it possible to take up gender as a political position from which to make change. Positionality includes an understanding both that the concept Woman is in relation to a constantly changing context and that positions in which women find themselves are locations in which to construct, rather than discover, meaning and values. Recognizing that women are situated in relation to others in a constantly shifting capillary of relations and that women are subject of and subjected to social construction, positionality both de-universalizes and de-essentializes the category Woman. Rather than posit female gender as essential and universal, positionality makes it possible to account for an interaction between social forces and individual agency in the production of the subject.

Gloria Filax in Lorraine Code (ed.), *Encyclopedia of Feminist Theories* (London and New York: Routledge, 2000), 394–95.

Positivism

Positivism is a term found generally in the social sciences to indicate a particular approach to the methodology of study. Broadly it indicates a 'scientific' approach in which human behaviour is to be treated as an objective phenomenon to be studied in conditions of value freedom. At its crudest this means that beliefs, attitudes and values of human actors are to be dismissed as insufficiently concrete or objective to become data for scientific study. Thus Durkheim, the leading exponent of positivist social science, would not accept that what an actor thought he was doing was a relevant part of any social science description. Even so personal an act as suicide could only be measured 'externally', and suicide rates, as statistics, rather than the accounts of would-be suicides, were the appropriate subject matter. Although there is no logical necessity, positivism tends to go hand in hand with a preference for statistical and mathematical techniques, and with theories that stress the 'system' rather than the individual in explaining political phenomena. Positivism sees as its enemy those who would study political values, either as political philosophers or as, say, political psychologists, the first because their approach is '**metaphysical**', the second because they are concerned with individuals and their perceptions, rather than with systems and the externally measurable. Though very popular in the immediate post-war development of political science, few today hold to such an extreme position, and the label is increasingly a vague and general way of indicating the main thrust, rather than the detailed methodology, of a social scientist. This is partly because the naïve view of what it is to be a 'scientist', or the attraction of being one, has declined considerably with the development of more subtle philosophies and sociologies of scientific activity, and partly because anyone interested in empiricism and its related theoretical and research techniques has had a more obvious refuge in identifying with Marxists in the fundamental split with non-Marxists that at one time seemed to dominate social science.

David Robertson, 'Positivism', in *A Dictionary of Modern Politics*, third edn (Europa Publications, 2002). Politics Online. Taylor & Francis. www.routledgepoliticsonline.com:80/Book.aspx?id=w007 (accessed 15 October 2009).

Postcolonialism and women – *see* **Women and postcolonialism**

Postmodern feminism – *see* **Philosophy of science, feminist**

Postmodernism

Often called by its detractors 'PoMo', postmodernism is one of the widest ranging intellectual fashions seen in the last 150 years. Like many of its predecessors this self-consciously radical generalized social commentary and theory originated in France and has recruited extensively amongst the American intelligentsia. Its range of influence is remarkable, because postmodernism is seen not only in academic and cultural activities from sociological theory to art criticism, but also in architectural style. It is so often and viciously derided by those who have not fallen to its fashionable influence, that those who are simply neutral tend to an instinctive sympathy, even if they neither pretend to understand or apply postmodern theory.

Postmodernism's core conception can be guessed from its name. By describing the mode of thought as 'post' and 'modern', its advocates are claiming to reject the consensus in Western thought that sees the Enlightenment, with its break with medieval thought and its celebration of rationality, as heralding an unstoppable progress in human life, understanding and experience. It is not conservative in claiming that things were somehow better before the Enlightenment, simply asserting that Enlightenment rationality is as time-bound and relativist in its truths as any preceding period. The universalistic claims of rationality, and the hubris of thinking society on the way to real truth rather than local belief, are the objects of postmodern scorn. A central concept of postmodern thinking, the idea of a 'meta-narrative' can best display what it is all about. The modern age, postmodernists claim, thinks in terms of a great narrative or explanatory story which applies to us all, and gives us universally valid truth. Postmodernists argue that we are all prisoners of our conditions, our characteristics, our communities, and that only local narratives, no longer presented as 'meta' can tell us our own partial truths. Clearly, this can produce what seem like very reactionary arguments. Postmodernists have very little time for grand debates about **human rights**, for example, and positively loathe Marxism, because both are based on claims to absolute and unvarying truths. At the same time, the typical causes espoused by postmodernism are, by those old Enlightenment standards, rather radical. **Gender** and sexual **identity**, and racial and ecological concerns all figure strongly in approval and explication by postmodern thought, its proponents suggesting, with some merit, that classic **liberalism** and classic Marxism thought alike, were deficient in their concern for gays, transsexuals, blacks and those who set the global ecological status higher than scientific progress, whether capitalist or Marxist.

Whether postmodernism, which is already ageing at the beginning of the 21st century, will fade away and be no more important in the long term of intellectual progress than, for example, Dadaism, is yet to be seen. However, as postmodernists do not believe in the concept of progress, it may matter less to them.

David Robertson, 'Post-Modernism', in *A Dictionary of Modern Politics*, third edn (Europa Publications, 2002). Politics Online. Taylor & Francis. www.routledgepolitics online.com:80/Book.aspx?id=w007 (accessed 15 October 2009).

Poststructuralism

Poststructuralism inherits from structuralism the conviction that our knowledge of ourselves and the world is the effect of structures that are not obvious or even perceptible to us. However, it breaks with structuralism's assumption that these structures are both universal and binary, insisting on **difference** as the key to such understanding as is available.

Poststructuralism begins from an account of meaning. On the basis that the distinctions it is possible to make in one language do not always find precise equivalents in another, the work of Ferdinand de Saussure indicated that meaning was not the origin of language, but its effect (de Saussure 1988). Meaning is not given either in the world or in our heads, but is produced by the differences between signifiers (words and phrases, or their symbolic analogues, visual, logical, mathematical or scientific). The implication is that we have no access to free-standing concepts, pure intelligibility or foundational, **metaphysical** truths. On the contrary, what we know exists in its inscription, as a result of the signifying differences we learn in culture. Language is thus not transparent to a fact or an idea, understood as having an independent existence behind or beyond it, and knowledges are in consequence culturally and historically relative. This should not, of course, be taken to imply that the world does not exist, but only that it does not determine what we (think we) know about it.

The structuralist **anthropology** of Claude Lévi-Strauss, in quest of the universal values linking diverse cultures, shared a tendency of Western thought to cement difference as opposition between antithetical alternatives (*Totemism* 1962). On the basis that binary oppositions are seductive, but ultimately reductionist, and drawing on de Saussure's account of meaning, poststructuralism reinscribes difference as the critical term. Jacques Derrida's *Of Grammatology* (1976) still stands as the classic poststructuralist analysis of the impossibility of **foundationalism** and the unsustainable character of binary oppositions. There Derrida demonstrated the inevitable intrusion of the defining opposite into the self-same. If with de Saussure we understand a meaning by reference to its differentiating **other**, the trace of that alterity necessarily enters into our understanding of the meaning itself. He included in his analysis Lévi-Strauss's own inadvertent ethnocentrism, displayed in the idealization of the tribal innocence that the anthropologist opposed directly to Western corruption. *Of Grammatology* also draws attention to de Saussure's phonocentrism, his reproduction, in spite of himself, of the traditional privilege accorded to speech, conventionally treated as pure, immediate and direct, in contrast to writing, which is commonly regarded as secondary, fallen. Derrida shows that neither Lévi-Strauss nor de Saussure manages to sustain the purity each attributes to the privileged term of the antithesis.

Opinion varies on whether it is legitimate to classify as poststructuralist other influential figures who constituted Derrida's immediate context, including Roland Barthes, Jacques Lacan, Louis Althusser and Michel **Foucault**. The conclusion we reach on this issue probably depends on the features of their work we choose to emphasize.

Roland Barthes, for instance, began as a structuralist literary critic, but registered increasing unease about the irony that de Saussure's theory of meaning as differential had been appropriated on behalf of a quest for universal patterns. His *Mythologies* (1957) acknowledged in its title the eminence of Lévi-Strauss as the structuralist mythographer of the day, but repudiated the concept of universality, wittily unveiling the cultural specificity of our own Western myths, where history masquerades as timeless nature and the changeable human being is misrepresented as 'Eternal Man'. In Barthe's *S/Z* (1970), silently registering Derrida's intervention, he parodied structuralist criticism, subjecting a short story to detailed analysis in the light of five codes designed to comprehend its meaning. However, the textual analysis repeatedly gives way to apparently anarchic digressions, isolated on the page and independently numbered. These interpolations offer wide-ranging reflections prompted by the text but are quite unable to be contained within the self-imposed restriction of the codes.

Lacan reread Freud for a post-Saussurean generation, arguing in his *Écrits* (1967), as well as a succession of annual seminars from 1953–54 onwards (and still in the process of publication), that the unconscious was best understood in terms of an absent imperative, refused admission into language and culture. The existence of this other motivating force, independent of the conscious mind and often in conflict with it, has the effect of differentiating human beings from themselves. Insofar as we are not what we believe ourselves to be, and are driven by desires for which we have no name, our self-image as subjects in possession of the objects of our knowledge, or in quest of the objects of our conscious wishes, is always and inevitably a misrecognition.

Althusser, meanwhile, reread Marx in the light of both de Saussure and Lacan to differentiate between the levels that interact in the social formation. In Althusser's *For Marx* (1965), his Marxism plays down the economic determinism of previous accounts. Instead, politics, on the one hand, and ideology, on the other, also exert their own pressures from a position of 'relative autonomy', and may even conflict with the economy and with each other. His account of ideology stresses its unconscious character: what we think we know, the 'obvious', is ideologically constructed and historically produced. Ideology has the effect of constituting subjects who misrecognize their place in the process of production and 'work by themselves' to further its interests, mistakenly regarded as their own (Althusser, *Lenin and Philosophy and Other Essays*, 1971).

Foucault, conversely, developed a history of knowledge-as-power that took direct issue with Marxism. If there is a poststructuralist element in his work, it resides above all in the differential account of history as crucially discontinuous. What might be read as progress or increasing liberalization is reinterpreted as a series of refinements of power that have, Foucault argues, the effect of subjecting us still further to disciplines that mask the power relations they construct.

Calling into question, as it does, the sovereignty of the human **subject** as the origin of meaning and history, poststructuralism is easily misread as a form of determinism. If we are not absolutely self-defining and self-determining, so the argument goes, we must be incapable of **agency**. However, poststructuralism characteristically deconstructs such binary oppositions: in practice, the stress on difference always presents the subject with alternative perspectives, knowledges and political options. Moreover, poststructuralists commonly align themselves with resistance to oppression. Lacan argues that psychoanalytic health depends on the pursuit of unconscious desire (if only we could identify it), in defiance of the conventional ethics imposed by the tyrannical demands of civilization; Althusser takes for granted a framework of class struggle and the possibility of a heroic refusal to obey the ruling ideology; Foucault's protagonists are a succession of criminals, suicides and sexual misfits who resist their own subjection (*Discipline and Punish*, 1975; *The History of Sexuality*, 1976).

The term 'resistance' in post-war French culture carries the full force of the underground struggle against the German occupation. This was a generation that had seen at first hand the consequences of illegitimate power, and the difficulty, as well as the necessity, of opposing it. At the same time, these poststructuralists were also motivated to find ways of accounting for French wartime collaboration and, beyond that, for the susceptibility of civilized Germany to genocidal values.

In view of its emphasis on resistance, poststructuralism was not incompatible with contemporary political movements in the 1960s and 1970s. Derrida has written sympathetically of **feminism** (*Spurs: Nietzsche's Styles*, 1976; 'Choreographies', *Diacritics* 12: 66–76) and of anti-racism. Foucault famously distinguished between homoerotic practices and the construction of the homosexual as a perverse personality. This identification, he argued, took place in the 18th century, and thus belonged to history and culture, not nature (1976). The **politics of difference** promotes sexual and racial diversity.

Poststructuralist analysis has repercussions in Jean-François Lyotard's postmodern emphasis on dissension as the motor of thought (Lyotard 1979), as well as his account of the differend, an incommensurability between positions that means no shared framework is available to resolve the differences between them. The differend implies a politics of struggle for a justice that cannot be definitively identified or attained (Lyotard 1988). Ironically, Lyotard's work also reverts (with a difference, of course) to the starting-point of poststructuralism in its emphasis on social interaction as ultimately a matter of language. To speak, he insists, is to fight, even when our opponent is no more than the language itself (Lyotard 1979).

Catherine Belsey in Adam Kuper and Jessica Kuper (eds), *The Social Science Encyclopedia* (Routledge, 2004). Politics Online. Taylor & Francis. www.routledgepoliticsonline.com:80/Book.aspx?id=w104 (accessed 15 October 2009).

Power – *see* **Foucault, Michel**

Private sphere – *see* **Patriarchy**

Protofeminism

This refers to women who anticipated key feminist concepts and ideals before **feminism** emerged as a coherent concept.

Corinne Fowler

Psychoanalysis, feminism – *see* **Film theory, feminism**

Public/private distinction – *see* **Patriarchy**

R

Race, ethnicity, gender and class

Race and ethnicity: Human social **identity** is constructed historically in complex inter-relationships of criteria of similarity and **difference**. Among the most ubiquitous modes of collective identification is ethnicity. The word derives from the ancient Greek *ethnos*, which referred to a collectivity of humans living and acting together; it is usually translated as 'people' or 'nation'. In the 20th century, the linked concepts of ethnicity and ethnic group have been taken in many directions, academically and politically. In everyday discourse they are central to the politics of group differentiation and advantage in the social democracies of Europe, North America and Australasia. With notions of 'race' in a degree of public and scientific disrepute since 1945, ethnicity has stepped into the ideological breach to become central to the often bloody reorganization of the post-Cold War world: '**ethnic cleansing**' stands beside earlier euphemisms such as 'racial hygiene' and 'the final solution'. Nor is 'race' a uniformly disreputable notion. In the USA, which experienced the Holocaust at a greater distance than Europe, it still finds a place in much public debate.

The inverted commas around 'race' indicate the concept's contested and problematic status. Following a co-ordinated programme of research and education by UNESCO in the decade following the Second World War, it has become widely accepted that 'racial' differences are historically and socially constructed, rather than being biologically fundamental. Differences of genotype between human populations are now held to be relatively trivial in degree and in substance. Phenotypical difference—visible 'racial' differences that are a product of the interaction of genetic endowment and environmental influences—are real enough, but their consequences for individuals and groups are the product of social processes of categorization and racialization rather than biological determination. However, the school of thought that has argued that there are significant genetically-determined 'racial' differences in intelligence and achievement—and which explains systematic disadvantage on this basis—has never gone away; it is, indeed, exhibiting signs of resurgence at the time of writing (e.g. Fraser 1995).

Gender: The other globally ubiquitous social identification is arguably **gender**. The female-male polarity is one of the most basic, if not the most basic, model of similarity and difference that is available to human beings as a principle of classification and social organization. Here also biologically-rooted ideas are germane: sexual differentiation, grounded in the physiology of reproduction and nurture, is distinguished by social science from gender differences, which are socially and culturally constructed institutionalizations of female and male. The distinction between sex and gender is well understood, and decades of feminist debate and scholarship have established the arbitrary nature of much gender differentiation. However, there is some disagreement, within **feminism** and elsewhere, about whether the biology of reproduction and of male and female bodies has determining consequences and about whether men and women are, in any sense, fundamentally different (e.g., Richardson 1993).

Gender and ethnicity: Gender and ethnicity differ fundamentally in some important respects. Ethnicity is a collective social identity; it cannot exist in the absence of collective social organization based on ethnic attachments. However, gender has historically required political work and mobilization in order to materialize—and then only recently—as a potential basis for collective identification and mobilization (although it has always been a potent dimension of individual social identity). Another major difference is that gender identification is in all human societies internalized during the formation of selfhood in primary socialization. This is not necessarily true for ethnicity.

Class: Class can be defined in two basic ways. First, it can be seen as a shared relationship to the means of production (following Marx). Second, it can be seen as social class, as a market position with respect to production, distribution and consumption (following Weber). In both models, private property is important; in both, economic differentiation and hierarchy are central; in both, there is no necessary political or organized awareness of class membership. Marx recognized this in his distinction between a class-in-itself and a class-for-itself; Lenin further acknowledged it in emphasizing the need for 'vanguard' political activity. Even on the broadest of interpretations, class is not ubiquitous in the sense that gender and ethnicity are. Whether one accepts the arguments for the existence of classes before **capitalism** (e.g. Anderson 1974), class as a core principle structuring social relations is essentially a phenomenon of capitalism.

Relationship between categories: Ethnicity, 'race', gender and class have historically all exhibited considerable potential as principles of stigmatization, exclusion and domination. Depending on local circumstances, each may or may not reinforce the other(s). From the point of view of the classic Marxist tradition, for example, the problem has been how to bring gender and ethnicity/'race' into the same analytical framework as the basic dynamo of history, class struggle. To frame this in terms of practical politics, the labour movements of the USA and Europe have long histories of unresolved conflict in which workplace hierarchies of skill, ethnicity and 'race', and gender have intertwined to the detriment of successful collective organization. As subsequent events have dramatically demonstrated, during the period of state socialist **hegemony** in Eastern Europe and Central Asia, the 'nationalities question' was institutionally and coercively contained rather than politically resolved.

For 20th-century feminism the issues may, perhaps, be brought into focus by questioning whether one women's movement, uniting women of all 'colours', cultures and classes in the pursuit of putative interests in common, is a realistic goal towards which to strive. In Europe, for example, one of the issues around which this question has crystallized concerns the stance of (white, often middle-class) feminism with respect to Islamic or African-Caribbean women, and vice versa.

Ethnicity and its ideological allotropes, racism and nationalism, have been among the most enduring, supposedly 'primordial', identifications to conflict with the development of modern state structures and political systems, social democratic and socialist movements, trade unions, labour organizations and movements for the improvement of the social place of women. This can be seen in the often fraught relationship between national liberation, ethnic politics and left-of-centre social reform in the ex-colonial world.

Something similar has also often been observable in Europe, where 19th-century nationalist movements—grounded as they typically were in local bourgeois interests—tended, certainly once the nationalist project had been realized, to be unsympathetic if not utterly hostile to socialism. In the industrialized states of the north and west, the relationship between the struggles of the women's movement, seeking to reform or overthrow 'traditional' **patriarchy**, and the struggles of the labour movement, seeking to reform or overthrow capitalism, has not always been easy. It has not been self evident that the two objectives are co-terminous.

Marxist and feminist writing: Political considerations such as these have produced a steady stream of theorizing. Marxists and feminists, for example, have found it no easier to reconcile their positions in academic print than in other political arenas. Among the earliest examples, and perhaps the most famous, is Engels's *The Origin of the Family, Private Property, and the State* (1884), in which he argued that class antagonism and female-male antagonism developed historically together, alongside and as a consequence of private property and structured social hierarchy, the end result being capitalism. The working through of this approach in the context of 20th-century feminism can be seen, for example, in Delphy's argument that women and men constitute different classes, the one subordinate to the other in the domestic mode of production.

Despite Lenin's writings on **imperialism**, Oliver Cromwell Cox's *Caste, Class and Race* was the first serious attempt at a Marxist analysis of 'race', arguing that 'race prejudice' was a product of capitalism. Cox's work has been subject to a wide range of criticism, as poor scholarship and, by radical Afro-American scholars, as bad politics. Subsequent attempts to square the circle of 'race' and class (such as Bonacich 1980) have never been more than partly successful. Similar comment could, perhaps, be made about recent attempts to integrate 'race' and gender into a unified theoretical framework (such as Anthias 1992).

Yet the problem is not that these things are necessarily unrelated. It is, for example, uncontroversial to argue that imperialism, colonialism and racism are intimately bound up with each other; as indeed colonialism and imperialism are bound up with the development of capitalism and the management of class conflict in the metropolitan centres. Similarly, as already suggested, there are many cases of ethnic attachments and/or racism being deployed to subvert the organization of labour. It is equally clearly the case that class and ethnicity can each be influential in frustrating the political organization and mobilization of women. Once again, to refer back to an earlier point, the women's movement historically has had specific ethnic and class roots, and black women's critiques of feminism are well-known.

Concluding comment: There are relationships between ethnicity, 'race', gender and class, but what are they, and what is the problem? These questions converge on the same answer(s) (see Jenkins 1996, 1997). The first point is that ethnicity, 'race', gender and class—even if there were consensual social science definitions of these terms (and at the moment there are not)—are not the same kinds of things. Ethnicity and gender, different as they are, appear to be ubiquitous features of human social life, in the historical long term, unlike 'race' or class. 'Race' is very much more a matter of imposed, external categorization than ethnicity. Despite its pervasiveness as an index of lifestyle differentiation, class is a social identity the organizing capacity of which in everyday life is, on the historical evidence to date, weak.

The second point is that where there are relationships between these modes of social identification, those relationships have been historically contingent rather than theoretically specifiable. Much of the scholarship that has sought to establish such relationships has attempted to fit the contingencies of history into theory, rather than the other way around (and, although history may not be at an end, it is proving relatively impervious to anything other than *ex post facto* theorization). Since ethnicity, 'race', gender and class are historical and local phenomena, we should be concerned with ethnicities and racisms, and with the local specificities of gender and class relations. Nor does the post-modern celebration of 'difference'—which is arguably a species of historicism, masquerading under another sign—seem likely to provide a solution to the problem. We can only begin to understand better the complexities of these issues through the careful specification of concepts, in the light of the exploration of particular historical and contemporary instances and their interrelations, and in the absence of preconceptions about those relationships. (*See also* **Class and feminism**.)

Richard Jenkins in Phillip Anthony O'Hara (ed.), *Encyclopedia of Political Economy*, first edn (Routledge, 1999, 2001). Politics Online. Taylor & Francis. www.routledge politicsonline.com:80/Book.aspx?id=w026 (accessed 15 October 2009).

Radical feminism

This is an approach to feminist thinking and action that maintains that the sex/**gender** system is the fundamental cause of women's oppression. According to Jagger and Paula Rothenberg, this claim may mean that women's oppression is the longest existing, most widespread, deepest, or worst form of human oppression. It can also mean 'that women's oppression [...] provides a concept model for understanding all other forms of oppression' (Jagger and Rothenberg 1984, 186). However, just because radical feminists agree about the pernicious nature and functions of **sexism** does not mean they agree about ways to eliminate it. On the contrary, with the emergence of so-called essentialism in feminist thought, radical feminists have divided into two camps: radical-libertarian feminists and radical cultural feminists. (*See also* **Philosophy of science, feminist**.)

Katherine M. Mack in Lorraine Code (ed.), *Encyclopedia of Feminist Theories* (London and New York: Routledge, 2000), 419–20.

Rape, war

The rape of women and girls has been a ubiquitous feature of warfare from primitive times. Females are often viewed as 'spoils of war' to be sexually assaulted at will, often as a prelude to their enslavement or murder. Rape reflects broader patterns of misogyny (hatred of women) that have pervaded male-dominant societies throughout history, and are especially pronounced in the hypermasculine context of war (Price 2001).

Particularly in the contemporary era, rape of women and girls has been used as a form of terrorism to coerce broader communities or populations into surrendering or fleeing. It has been adopted as a means of humiliating community males by emphasizing their inability to protect 'their' women from sexual attack. It has also been used as a form of **torture** by security forces around the world—a practice that reached a zenith during the era of 'national security states' in Latin America and elsewhere during the 1970s and 1980s.

A number of 20th-century cases have become paradigmatic of war rape. First was the 'Rape of Nanking' in 1937–38, in which tens of thousands of Chinese women and girls were raped by occupying Japanese troops, with the attacks often followed by savage mutilation and murder. Gendercidal massacres of some 200,000 Chinese men took place during this period; an unknown number of men were also raped. In Berlin in 1945, victorious Soviet forces staged a weeks-long rampage of mass rape against German women and girls. When the Pakistani army staged its genocidal crackdown on ethnic Bengalis in East Pakistan (present-day Bangladesh) in 1971, gendercidal massacres of Bengali men and boys were accompanied by the rape of some 200,000–400,000 Bengali women. Many of these women were forcibly impregnated and experienced severe social isolation after giving birth. According to Brownmiller, the Bangladeshi atrocities marked the first time that serious international attention was devoted to wartime rapes of women. This came about in part because of efforts by the independent Bangladeshi Government to publicize women's suffering, and in part because of 'a new feminist consciousness that encompassed rape as a political issue' (Brownmiller 1975, 84).

This developing feminist consciousness was poised to confront war rape in a far more concerted fashion during the war that consumed the former Yugoslavia in the first half of the 1990s. Rape has long been illegal under international law, but until recently, it has tended to be 'characterized as an outrage upon personal dignity, or as a crime against honor', separate from 'crimes of violence, including murder, mutilation, cruel treatment, and torture' (Copelon 1998, 65). This changed in the wake of the concerted attention that Western media and feminist activists paid to the systematic rape of females as a strategy of war and **ethnic cleansing** in **Bosnia**-Herzegovina. The **International Criminal Tribunal for the Former Yugoslavia** (ICTY) was the first to define rape as a **crime against humanity**, and conduct its prosecutions accordingly. The even greater prevalence of rape as an atrocity during the Rwanda **genocide** of 1994 led to the 1998 conviction of Jean-Paul **Akayesu**, a former mayor, by the **International Criminal Tribunal for Rwanda** (ICTR). This marked the first time that a sentence was handed down for rape in a context of civil war. Rape was also declared to fall within the mandate of the new International Criminal Court. However, despite these advances new accounts and **human rights** reports from around the world attest to the continued prominence of rape in contemporary warfare.

Rape of women and girls in war should be set alongside related phenomena such as sexual torture, forced prostitution and sexual slavery. All these crimes feature rape as an essential tool of violence and subjugation. Perhaps the most dramatic symbols of the wartime sexual victimization and exploitation of women are the so-called 'comfort women', victims of the regimen of forced prostitution instituted by Japanese occupiers in East and South-east Asia during the Second World War. At least 100,000 women, mostly Korean, Indonesian and Philippine, were tricked and/or coerced into providing sexual services for Japanese troops. To the extent that this was deceptive and/or coercive, it can be considered under the rubric of war rape. In recent years, surviving 'comfort women' have mobilized to seek an apology and compensation from the Japanese Government.

Considerable debate has taken place over the concept of genocidal rape, prompted in large part by the war in the Balkans and the genocide in Rwanda during the 1990s. The concept was deployed by Beverly Allen (1996) to designate a military policy of rape for the purpose of genocide. Allen isolated three main forms of genocidal rape: those inflicted in the presence of family and community members to terrorize the targeted community into fleeing; random rapes in detention centres and concentration camps, often as part of torture; and rape in specially-designated rape camps, in which women are often killed following the assault, or forcibly impregnated in the hope that the victim will bear the rapist's child.

Some scholars have criticized the conflation of rape with genocide. Rhonda Copelon (1998, 64) contends that 'rape and genocide are separate atrocities'. However, many such critiques place excessive emphasis on death and physical destruction as features of genocide. Under international law, genocidal strategies can also include the infliction of physical and psychological damage on members of a targeted group, as well as the prevention of births within a group. These commonly result (the latter through forced impregnation) from systematic campaigns of rape in war, genocide and ethnic cleansing. The validity of the concept of genocidal rape was further bolstered by events in Rwanda between April and July 1994. Not only was rape a standard component of the assault against Tutsi communities, but the high rates of HIV infection among Hutu troops and militias meant that their victims confronted the fear and possible reality of death by AIDS as a consequence of the attack(s). It is notable that the conviction of Jean-Paul Akayesu by the ICTR included a judgment that rape had been used as a form and strategy of genocide.

In both domestic and international contexts, rape has overwhelmingly been viewed as a crime inflicted upon women by men, although male family and community members are often part of the designated audience of the assault. Only very recently has attention begun to be paid to the rape of males, both in conflict situations and in domestic society. According to Joshua Goldstein (2001), the rape of male soldiers and non-combatants is intimately bound up with other **gender**-specific atrocities by victorious against defeated males, including castration and large scale gendercidal killing. The intent is not only to inflict physical suffering, but to 'feminize' enemy males by exposing them to violations typically directed against females. (*See also* **Feminism**; **Gendercide**; **Genocide**.)

Adam Jones in Martin Griffiths (ed.), *Encyclopedia of International Relations and Global Politics* (Routledge, 2005). Politics Online. Taylor & Francis. www.routledgepoliticsonline.com:80/Book.aspx?id=w163 (accessed 15 October 2009).

Rationalism/rationalist

This may be broadly defined as a commitment to the standards of rationality. The etymological root from which both words originate is the Latin noun *ratio,* meaning 'proportion'. A 'rationalist', in this broad sense, then, is someone who thinks in terms of proportions or who looks at everything in the world in harmonious proportion to everything else. That is, someone who places special emphasis on the human capacity to think, with precision and coherence, about the manner in which the parts fit together in the whole. However, there is another, more technical definition of the word 'rationalism'. **Aristotle**, for example, used the adjective 'rational' to distinguish humans from animals, but this did not make him a 'rationalist' in the technical sense. Likewise, many philosophers today are firm advocates of the value of reason, yet few are 'rationalists' in the strict sense.

In the technical sense, rationalism is the claim that all knowledge comes ultimately from reason, a philosophy which is generally contrasted with empiricism and the claim that all our knowledge is based on sensory experience. An extreme version of this rationalism rejects the senses as inherently misleading and unreliable. However, some rationalists acknowledge that some use of the senses is necessary for the full development of knowledge. All rationalists claim that it is possible to have *a priori* knowledge—that is, knowledge prior to experience. Empiricists admit that some propositions can be known *a priori,* but insist that such propositions are strictly tautological (for example, 'all bachelors are unmarried') and thus give us no knowledge about the world.

Rationalists, on the other hand, are convinced that we can know some substantive truths about reality through the use of reason independently of sensory experience. Thus, although we do not know the cause of any particular event without knowledge of the facts, we do know *a priori* that every particular event has a cause.

Rationalism is often associated with the work of the great 17th-century philosophers René Descartes (1596–1650), Benedictus Spinoza (1632–77) and Gottfried W. Leibniz (1646–1716). However, these thinkers owe a great deal to the classical Greek tradition, particularly to Plato's philosophy that one can achieve insight into eternal, absolute truths about beauty and justice through the intellect alone. However, it was Immanuel Kant (1724–1804) who produced a systematic account of the nature and limits of *a priori* knowledge, distinguishing two kinds of *a priori* propositions: analytical *a priori* and synthetic *a priori.* Analytical propositions are tautological propositions in which the predicate is contained in the subject ('all bachelors are unmarried'). These are the necessary and universal truths of logic, and to deny such propositions is self-contradictory. However, synthetic *a priori* judgements are judgements that give us knowledge about the world, although they too are not arrived at on the basis of observation, but are concepts imposed by the mind on the impressions received by the senses.

Kant thus argues that, in experiencing the world, the mind necessarily structures it in terms of what he calls 'concepts of the understanding'. Here Kant is following the traditional rationalist principle of innate ideas, which is the idea that the mind is equipped from birth with certain fundamental concepts. Two concepts of this understanding are the categories of causality and of substance. To give an example, knowing that every event has a cause requires an *a priori* concept of causality, since all that our senses can observe is a certain repetition and regularity in events or a series of correlations between events of type A and events of type B. There is no justification, on the basis of sensory experience alone, for saying that every event must have a cause. It is the *a priori* concept of causality which allows us to know that objects are necessarily and universally related to each other by causality. However, Kant rejects the earlier formulation of innatism (advanced by such philosophers as Descartes and Leibniz) according to which innate ideas are completely independent of experience. Kant, instead, concludes that innate ideas (or the categories of the understanding) must be deduced as the very precondition by which we are enabled to make intelligible sense of our experiences.

In the first half of the 20th century, with the rise of a new, more rigorous form of empiricism (commonly known as 'logical **positivism**'), rationalism came under serious, unrelenting criticism. The positivists, including Bertrand

Russell (1872–1970) and the early Ludwig Wittgenstein (1889–1951) argued that all *a priori* propositions are meaningless in so far as they are not truths about the world, but about a 'reality' beyond all possible sense-experience. Unless propositions could be translated into 'observation statements' that one knows how to verify experimentally or observationally, such propositions must be discarded as **metaphysical** illusions. Although the propositions of logic and mathematics are independent of experience, these propositions are truths by definition, in accordance with the meanings of the symbols involved.

However, it was not long before logical positivism itself came under attack. A key point of criticism was the philosophical status of the claim that only statements which can be verified are meaningful. For how does one verify the claim that only verifiable principles are valid? The empirical criterion of truth is not itself a factual statement, so it seems clear that it is a metaphysical assumption, as abstract or general as the rationalist concepts of 'substance' and 'causation'. An additional problem encountered by logical positivism is that, in the higher levels of science, structures and entities are discussed that are not in any way directly testable against experience, but are really theoretical or universal constructs.

With the decline of positivism (or at least the more extreme version of it), rationalism gradually came back into the scene. A central figure in this contemporary revival has been the linguist Noam Chomsky (1928–). According to Chomsky, all humans are born with an innate knowledge of the principles of universal grammar. The role of experience is merely to activate the abstract linguistic structure genetically programmed in the brains of humans. He argues that the child's awareness of the principles of universal grammar is *a priori*, rather than acquired through the habitual repetition of words. However, it would be a mistake to conclude that Chomsky's rationalism excludes empirical research, since his theory is set up as an empirical hypothesis to be evaluated according to the evidence collected on the ways in which children actually learn language, as well as on the evidence about the physiological structure of the brain. Chomsky's linguistics, then, offers us an excellent bridge between rationalism and empiricism.

Ricardo Duchesne in R.J. Barry Jones (ed.), *Routledge Encyclopedia of International Political Economy*, first edn (Routledge, 2001). Politics Online. Taylor & Francis. www.routledgepoliticsonline.com:80/Book.aspx?id=w053 (accessed 15 October 2009).

Reproductive health and rights

The UN Population Fund (UNFPA) recognizes that improving reproductive health is an essential requirement for improving the general welfare of the population and the basis for empowering women and achieving sustainable social and economic development. The International Conference on Population and Development (ICPD) succeeded in raising the political prominence of reproductive health issues and stimulating consideration by governments of measures to strengthen and restructure their health services and policies. In October 2007 the UN General Assembly officially incorporated the aim of achieving, by 2015, universal access to reproductive health into the target for Goal 5 of the Millennium Development Goals. UNFPA encourages the integration of family planning into all maternal, child and other reproductive health care. Its efforts to improve the quality of these services include support for the training of health care personnel and promoting greater accessibility to education and services. Many reproductive health projects focus on the reduction of maternal mortality (i.e. those related to pregnancy), which was included as a central objective of the ICPD Programme, and recognized as a legitimate element of international **human rights** instruments concerning the right to life/survival. Projects to reduce maternal deaths, which amount to about 500,000 each year, have focused on improving accessibility to essential obstetric care and ensuring the provision of skilled attendance to women in labour. The ICPD reported that a major cause of maternal deaths was unsafe **abortions**, and urged governments to confront the issue as a major public health concern. UNFPA is concerned with reducing the use of abortion (i.e. its use as a means of family planning). UNFPA is an active member of a core planning group of international organizations and partnerships that organized the first Women Deliver conference, held in London, United Kingdom, in October 2007. Participants, including government ministers and representatives of organizations, private-sector foundations and non-government bodies, endorsed a final commitment to increase investment in women's health and to make improving maternal health a development priority. In February 2008 UNFPA appealed for donations to a new fund, with a target figure of US $465m. during 2008–11, to support efforts in 75 developing countries to improve maternal health care. In addition to maternal deaths, an estimated 10m.–15m. women suffer serious or long-lasting illnesses or disabilities as a result of inadequate care in pregnancy and childbirth. In 2003 UNFPA launched a Global Campaign to End Fistula, which aims to improve the prevention and treatment of this obstetric condition in 30 countries in Africa and Asia and to achieve its elimination by 2015. UNFPA supports research into contraceptives and training in contraceptive technology; it organizes in-depth studies on national contraceptive requirements and aims to ensure an adequate supply of contraceptives and reproductive health supplies to developing countries. In the early 2000s the Fund and other partners developed a Reproductive Health Commodity Strategy (RHCS), which aimed to improve developing countries' self-sufficiency in the management and provision of reproductive health commodities. UNFPA encourages partnerships between private-sector interests and the governments of developing nations, with a view to making affordable commercial contraceptive products more easily available to consumers and thereby enabling governments to direct subsidies at the poorest sectors of society.

UNFPA is a co-sponsor of the Joint UN Programme on HIV/AIDS (UNAIDS), and is the UNAIDS convening agency with responsibility for young people and for condom programming, as well as taking a leading role in the UNAIDS inter-agency task team on **gender** and HIV/AIDS. The Fund, in co-operation with the other participants in UNAIDS, aims to strengthen the global response to the HIV/AIDS epidemic, and is also concerned to reduce levels of other sexually-transmitted infections (STIs) and reproductive tract infections (RTIs), and of infertility. UNFPA gives special attention to the specific needs of adolescents, for example through education and counselling initiatives, and to women in emergency situations. The Fund maintains that meeting the reproductive health needs of adolescents is an urgent priority in combating poverty and HIV/AIDS. Through the joint Adolescent Girls Initiative, UNFPA, UNICEF and WHO promote policy dialogues in 10 countries.

UNFPA takes a lead role in an emergency situation, following natural disaster or conflict, in providing supplies and services to protect reproductive health, in particular in the most vulnerable groups, i.e. young girls and pregnant women. It also supports counselling, education and training activities, and the construction of clinics and other health facilities, following humanitarian crises. At the end of 2008 UNFPA appealed for $51m. through the UN Consolidated Appeals Process (CAP) for 2009; the largest share of the Fund's assistance through the CAP (32%) was allocated to humanitarian activities in Sudan. During late December 2008–mid-January 2009, in response to the intensive bombardment of the Gaza Strip by Israeli forces (with the stated aim of ending rocket attacks launched by Hamas on Israeli targets), UNFPA provided medical equipment and supplies to hospitals in Gaza. In early 2009 UNFPA, in co-operation with district health authorities, was supporting pregnant women displaced by ongoing violent unrest in northern Sri Lanka through the provision of emergency transport for life-saving obstetric care; the supply of surgical instruments to hospitals for Caesarean operations; the distribution of personal hygiene items and clean delivery kits to women in Internally Displaced Persons (IDP) camps; and the provision of mobile reproductive health clinics. UNFPA was also working with local health authorities to ensure that IDP camps were maintaining separate bathing and toilet facilities for men and women and well-lit paths for the safety of women and girls.

'Reproductive Health and Rights' (United Nations Population Fund—UNFPA), in Europa World online (London: Routledge), www.europaworld.com/entry/io-un.1633 (accessed 1 June 2009).

Resistance – *see* **Poststructuralism**

Retreat from Kabul

One of the worst disasters in British military history, the First Anglo-Afghan War, featured a botched invasion that culminated in the retreat of British

India's invading force from its military encampment in Kabul. Although Great Britain restored supremacy of arms at the war's close, nearly 16,000 troops, soldiers' wives, servants and children perished during the retreat. The events surrounding the retreat helped shape an enduring belief in Afghan treachery in the popular imagination thereafter.

Corinne Fowler

Roma or 'Romani people'

Roma or 'Romani people' is an umbrella term for an archipelago of minority groups living all over the world. The largest, concentrated populations can be found in Europe, in particular Central and Eastern Europe, where the majority of the approximately 10m. European Roma live. In contrast to the stereotype of nomads living in wagons, only a few groups of the Central and Eastern European Roma are still nomadic. Roma supposedly are descended from nomads who were displaced from India in the 10th century and who, moving westwards, discovered Europe mainly in the 14th and 15th centuries. They did not leave early written evidence of their arrival in Europe of their own; however, they are described in documents produced by the majority populations that they encountered. Until the late 18th century the Indian heritage of the Roma was not known to most Europeans, who identified them as different, often giving them names expressing a misunderstood origin such as Turks or Egyptians (the term 'Gypsy' is derived from 'Egyptian'). Albeit sharing a widely similar Indian heritage in terms of language and cultural practices, Roma are divided into several groups and subgroups from which they derive their **identity**. Romani identity and culture developed and diversified according to their experiences within the broader pattern of European history. Roma partially adapted to cultural practices and adopted the religion of the populations surrounding them; they also retained their distinct identity, though to a varying extent in different parts of Europe. Romani people, therefore, can be perceived as an integral part of the societies in which they live. Nevertheless, they remain regarded as distinct and 'alien'. As a consequence, much of Romani history is defined by **discrimination**, oppression and violence from which Roma suffered. This history of exclusion and victimization culminated in the 'Porajmos' in which between 200,000 and 1,500,000 Roma were murdered during the Holocaust. For more information on the Roma and their history see, e.g. Hancock (1987, 2002), Marushiakova and Popov (2001), Liégeois (2007) and Fraser (1992). See also Guy (2001), Pogány (2004), Stewart (1997), Gheorge and Acton (2001), and Petrova (2009).

Lena J. Kruckenberg

Rwanda – *see* Genocide, Rwanda

S

The Second Sex – *see* **Beauvoir, Simone de**

Second-wave feminism – *see* **Beauvoir, Simone de**

Sex – *see* **Gender**

Sexism

This refers to the inferiorization of women because of their 'sex' or, more properly, their **gender**, i.e. the socially constructed view of who they are as women. It might be argued that males also can be inferiorized by gender, and so the term should include both males and females. However, when this happens it is generally a spill-over from the inferiorization of females. Males experience hostile verbal assaults that suggest that they are not fully 'masculine', but 'sissies' (sisters) or 'effeminate'. Such assault is based on the assumption that men become 'masculine' by negating identification with qualities and roles assigned to females.

Inferiorization of homosexuals because of their sexual orientation is often assumed to relate to their lack of '**masculinity**' in the case of males, or 'femininity' in the case of females. Thus homophobia is related to sexism. However, this is properly a separate subject, namely 'heterosexism', and so it will not be treated here. This entry will focus on the central meaning of sexism, namely the inferiorization of women as women.

Sexism is expressed in personal psychology and inter-personal relations, but it cannot be defined in only individualist terms. It is part of a social and cultural system with a long history. This system is generally called **patriarchy**. It has shaped the legal, economic and political systems, as well as the cultures, of most of the societies of the world. Patriarchy is a broader concept than sexism. It refers to social systems in which the male head of the family is seen as exercising collective sovereignty over wives, children, slaves and servants, animals and land. Public political power is held by the collectivity of male heads of family: *patres familiae.*

Rosemary Radford Ruether in Paul Barry Clarke and Andrew Linzey (eds), *Dictionary of Ethics, Theology and Society* (Routledge, 1996). Religion Online. Taylor & Francis. www.routledgereligiononline.com:80/Book.aspx?id=w004 (accessed 15 October 2009).

Sexual difference – *see* **Difference**

Sexuality – *see* **Foucault, Michel**

Social boundaries

Social boundaries correspond to distinctions made by social actors in order to set one social system or group apart from others. They shape social interaction in important ways as they allow the distinction between in-groups and out-groups, between 'us' and 'them', thereby generating and reflecting feelings of similarity and **difference**. Social boundaries translate into identifiable patterns of exclusion and inclusion with respect to resources and opportunities. They become visible in many different ways, e.g. through signs separating different neighbourhoods, clothing indicating **gender** or occupation, or social practices referring to class membership. Boundaries may vary with regard to their clarity, permeability, temporal stability, the degree of their institutionalization as well as the range of contexts in which they matter. A female, dark-skinned bus driver, for example, experiences and enacts different social boundaries in her bus, in the depot, at home or in a theatre.

Lena J. Kruckenberg

Social categorization – *see* **Categories and dichotomies**

Sovereign – *see* **Hobbes, Thomas**

Standpoint, feminist – *see* **Feminism and international relations; Philosophy of science, feminist**

State of nature – *see* **Hobbes, Thomas**

State of war

On Hobbes's definition, 'war' refers not only to active armed hostilities between individuals or collective bodies, but also to conditions in which one has no reasonable expectation that one will not suffer sudden and violent assault. (*See also* **Hobbes, Thomas.**)

Glen Newey

Structuralism – *see* **Poststructuralism**

Subject

To be a subject is to be in the world in a way that couples consciousness with that which is not conscious, for example, nature, objectivity. This mode of being is quite distinct from mythopoetic forms of life that do not take the world as an object distinct from themselves, as with, for example, Australian Aborigines or Ancient Egyptians. The break towards the distinction in consciousness between subject and object is embedded in the construction of nature, an event that can certainly be traced back to Thales and no doubt predates that. Even so, subjectivity and self were conceived as being dispersed as a form of inter-subjectivity in the tribe, household and polis. *Phrenes*, mind, is, in Ancient Greece, frequently referred to as located partly outside the individual; as dispersed. That dispersion was collected and focused in the Christian idea of the person, as made in the image of God, as standing before God as an individual, as seeking justification before God as an individual, not as a member of a tribe or household or polis or other form of group. The inward reflexivity present in Greek thought became radically reflexive, turning in on itself to produce an inward life centred on the 'I'. This 'I', expressed as an individual entity in St Augustine, became the single centre of consciousness in Descartes. Descartes, in seeking certainty, in seeking a rational ground for incorrigible knowledge, found it in the contents of his consciousness, a consciousness that he regarded as transparent, clear all the way down. Incorrigibility had been placed firmly in individuality by St Augustine. Looking for certainty about his own existence he proclaimed that this was not something about which he could be mistaken, 'If I am mistaken I exist', he proclaimed in *The City of God*. If I think, argued Descartes, then I must be a thinking substance: '*cogito ergo sum*', 'I think, therefore I am'. The argument of course is deeply flawed; the fact that there is thinking does not warrant the existence of the singular 'I', and that 'I' cannot guarantee its transparency, indeed its transparency seems completely undercut by the possibility of self-deception. That one can deceive oneself is possible only against the idea of the self as having layers of being, hidden behind various guises and disguises. Even the presence and placing of the veils is concealed from the inquiring self. In Kant the subject became the centre of the knowing universe and the centre of the moral universe. Against Hume, who thought the self a bundle of impressions, he argued that a merely passive self was impossible, knowledge was active, and experience ordered by the categories of the mind. Morality resided in autonomous action, unconditioned by any desire. The subject was complete, individual, and sovereign over their actions. As Kant put it in *The Groundwork of the Metaphysics of Morals*, this was only an ideal, but it was a persuasive ideal, though also a problematic one. The Augustinian–Cartesian–Kantian subject was detached from the world as subject to object, as self against nature. For Descartes, that there was a world outside subjectivity was a certainty dependent upon the existence of God. If the certainty in God collapsed, the certainty of the world collapsed. For Kant, the world appeared as phenomena ordered by the categories of the mind. The difficulty here is how does one know that there is a world that matches the categories? Perhaps there are merely disordered, chaotic apperceptions, whose order is merely an apparency of the mind. Perhaps the apperceptions are merely properties of the mind and there is no world at all. Kant had no clear answer to this kind of difficulty. The relation between the mind and the world was, he said, 'a happy coincidence', an answer that, while not notable for its fulsomeness, is perhaps the best available given the basic perspective. That perspective is given by the *problématique* of the Western philosophical and theological tradition. Once subjectivity was assumed as something individual and as something distinct from the world as object, the sceptical line of reasoning began. Scepticism can never answer its own questions. A premise of solipsism will never yield a successful argument to the conclusion that there are other people in the world. The Platonic–Augustinian–Cartesian–Kantian tradition shifted from scepticism to solipsism; to the problem of other minds. How, it was asked, could one know the existence of other minds? Neither Descartes nor Kant had entirely satisfactory answers to that. For Descartes the solution rested on an argument to the existence of God, but if that failed, as it must, for there are no rational arguments to that conclusion, so other minds, other people, faded from reality and became mere appearances. For **Hobbes**, for Leibniz, for Hume and for Kant nothing outside the individual mind could be certain, there was no certainty of an external world and no assurance of people in it. What began as an exercise in expelling the poets from the Republic ended with no one left in the Republic other than a lone, sceptical and peculiarly autonomous philosopher king. (*See also* **Foucault, Michel**.)

Paul Barry Clarke and Andrew Linzey, *Dictionary of Ethics, Theology and Society* (Routledge, 1996). Religion Online. Taylor & Francis. www.routledgereligiononline.com:80/Book.aspx?id=w004 (accessed 15 October 2009).

Subject, the death of the

The death of the subject is linked to the death of God. The phrase is Nietzsche's. It is, perhaps, prophetic; it is clearly born of acute observation; it is remarkably insightful. Nietzsche observed the tendencies towards secularization, the way in which God had disappeared from the practical lives of much of humanity, the way that in many cases, including many religious movements, God had been placed in the service of man. Feuerbach made a similar point, saying that God was a projection of human need, the product of alienation. The insight was seized upon by Marx in his attempt to turn Hegel around. Hegel had started with the assumption of God as a self-positing being seeking absolute knowing. Human culture was following that discovery, and Western history was the medium through which Geist would express itself. In a very real sense Western history was God-given. Marx took Feuerbach's claim that God was a projection of human need, understood as material and economic emmiseration, as alienation from world, from the objects of one's labour, from self and from others as one's species being. Religion, he argued, was 'the sigh of the oppressed, the heart of the heartless, the opium of the people'. To be emancipated required not freedom to worship as one pleased, but to be emancipated from religion. The end of alienation would itself end the psychological and material conditions in which the projection that was God had developed; it would lead, in Nietzsche's phrase, to the death of God. As intellectual arguments made against the **rationalism** and logocentrism of Western philosophy these arguments are powerful, if not decisive. However, set against the backdrop of a general decline in religious belief, the decline of religious authority—indeed the decline of authority in general—and mere secularization, they gave intellectual weight to those willing to be persuaded or those wavering on the edge of religious disconviction.

Coupled with the death of the **subject** and the death of God is one of the most powerful and decisive shifts in intellectual understanding in the Western tradition of thought: the linguistic turn. Its effects can hardly be overstated. The linguistic turn has no single source, it is a part of a general shift in Western thought. It rests on a single and devastatingly simple observation: words, statements and terms in language refer to words, statements, terms and concepts in language. This seems so basic, so obvious, indeed almost so trite, that its revolutionary impact takes some grasping. It is this: if linguistic terms refer to other linguistic terms then they do not refer to things in the world. If that is so, then language is not about the world at all, it is always and only about language. Truth is not based on correspondence to events, or states of affairs in the world. The assumptions underlying this remarkable claim, remarkable only for its late appearance in Western intellectual life, comes from distinct sources, one philosophical, the other linguistic.

Paul Barry Clarke and Andrew Linzey (eds), *Dictionary of Ethics, Theology and Society* (Routledge, 1996). Religion Online. Taylor & Francis. www.routledgereligiononline.com:80/Book.aspx?id=w004 (accessed 15 October 2009).

Sustainability/sustainable development

Sustainability has become a key slogan of the environmental movement in recent years. The concept of sustainable development (SD) is best associated with the 1987 World Commission on Environment and Development's report titled Our Common Future, which defined it as, to paraphrase: development that meets the needs of the present generation without compromising the ability of future generations to do the same. SD became the subject of heated debates between environmentalists in north and south, and is now considered both highly contested and a plastic buzzword that can be used for all kinds of ostensibly 'green' reasons. Sustainability is a less contentious concept, that is often used as a kind of 'vanishing point' in the future toward which societies need to move if they are to avoid complete environmental, economic and social breakdown. For an overview of feminist perspectives on sustainability see MacGregor 2003.

Sherilyn MacGregor

T

Torture

Torture is the deliberate infliction of physical or mental suffering, and is normally thought of as being perpetrated by some agency of the state. The neutrality of this definition is necessitated because there are two rather separate aspects to torture historically, and correspondingly two different theoretical objections to it. In popular fiction torture is most often associated with the idea of hurting someone to make them confess to a crime, or to extract information that the state needs or wants. However, torture can also cover the use of pain and suffering as a punishment in itself. The distinction is important because, historically, torture in the first sense has more often been disapproved of than in its second sense. Torture, often referred to as judicial torture, to make a suspect confess to a crime or to gain information about, for example, a suspect's accomplices, though recently used, has, historically, equally frequently been objected to and banned. Under the Roman Republic torture could only be used on slaves, and even when, under the Empire, it was allowed on citizens, its use was restricted to cases of suspected treason. In fact this limitation to treason, the most heinous of crimes in many penal codes, was usual.

Among the most dramatic of historic uses of torture, by the Inquisition in their attempts to discover and eliminate heresy, gained its legitimacy during the 13th century when the Roman use of torture in treason cases was incorporated into canon law, heresy being seen as a directly equivalent crime. Torture as a legitimate part of canon law was not abolished until a Papal Bull of 1816. Only later, and following this precedent, did most continental European systems adopt torture and, in adopting it, often expand the range of suspected crimes for which it could be applied. By the early modern era torture was legitimated widely in Europe, though particularly so in the Italian and Germanic states. Torture was not outlawed in these states until various dates during the 18th century: the French abolished it at the Revolution, for example. Torture lingered on as a legitimate weapon of state in some places, being abolished in Naples as late as 1860.

English common law was never comfortable with judicial torture, and although it was practised from time to time, this was almost always under special prerogative writs from the monarch and, again, only in cases of treason. The well-known use of torture under Elizabeth I is an example of both these points. As early as 1628 the English judiciary declared torture illegal when it was proposed to use it on the assassin of the Duke of Buckingham to find the identity of his accomplices. The theoretical argument against torture is a combination of a due process argument (the state should not be allowed to manufacture evidence) and a straightforward humanitarian objection to the infliction of pain on possibly innocent people. This is why the concept becomes complex, because humanitarian objections to the infliction of pain on guilty people are much more recent in origin. Many forms of punishment depended primarily on the infliction of pain, as with flogging, were indifferent to the incidental infliction of pain, as with branding, or used pain to add further emphasis to the horror of the punishment for deterrent effects, as with the classically horrifying forms of execution used for some crimes. It is really only in the 20th century that sensitivities have developed to the point where even the punishment of the guilty is regarded as not justifying any avoidable physical or mental suffering, although the beginning of this trend is found in the 18th century with doctrines like the ban on cruel and unusual punishment in the US Constitution.

Torture in both senses is banned by a series of international civil rights covenants, such as the UN **Convention against Torture and Other Cruel, Inhuman or Degrading Treatment or Punishment** (1984) and the European Convention for the Prevention of Torture and Inhuman or Degrading Treatment or Punishment (1969).

David Robertson, *A Dictionary of Human Rights*, second edn (Europa Publications, 2004). Politics Online. Taylor & Francis. www.routledgepoliticsonline.com:80/Book.aspx?id=w006 (accessed 15 October 2009).

Travel writing

There was a distinct revival of interest in women's travel writing at the end of the 20th century, especially after the English publisher Virago reissued a series of British women's travel texts written in the 19th and early 20th centuries. Many of these travel texts had not been reprinted since their original publication, and they surprised many late 20th-century readers by the range of possible forms of behaviour that were available to Victorian women travellers. Rather than the shy, retiring stereotype of the middle-class British feminine women, there emerged a vision of individuals who defied conventions and travelled in conditions that would challenge most women from the turn of the 21st century.

Women travellers have conventionally been viewed as curious eccentrics, oddities who transgressed the boundaries of behaviour judged acceptable for women to maintain their femininity. Victorian notions of femininity determined that women should centre their lives largely on the private sphere of the home. When middle-class women went outside the home or when they were in the company of men, it was common for them to be chaperoned—to have with them an older female companion who would protect their 'honour'. Women who travelled, particularly those who travelled alone, were judged to be 'adventuresses', a term that has strong sexual connotations.

Women have written about their travels since at least the 14th century, beginning with the English mystic Margery Kempe (c.1373–c.1440). She travelled to the major Christian shrines—Jerusalem, Santiago de Compostela and Rome—as well as travelling extensively to holy sites in Great Britain. There were a fair number of women travellers in the 18th century, for example, Celia Fiennes (1662–1741), Mary Wollstonecraft (1759–97), and Lady Mary Wortley Montagu (1686–1762); these women were drawn mainly from the aristocracy. It was only in the 19th century that a greater number of women travelled, and the majority were from the middle classes. They travelled on missionary work, for exploration, to write reports for newspapers, to accompany their husbands, to collect botanical specimens, or simply for the sake of travelling.

Many of them write strikingly original texts. For example, Mary Kingsley (1862–1900) wrote an account of her travels, *Travels in West Africa* (1897), that is witty and idiosyncratic, stressing the problems she encountered and the humorous ways she managed to solve them. For example, unlike many male travellers, she focused on the mistakes she made and the accidents she had, pointing out how ridiculous she looked to the inhabitants of the regions: 'Going through a clump of *shenja*, I slipped, slid and finally fell plump through the roof of an unprotected hut. What the unfortunate inhabitants were doing I don't know, but I am pretty sure they were not expecting me to drop in, and a scene of great confusion occurred'. She represented herself as an indomitable spirit while not transgressing the image of herself as feminine. Characterizing her account as a 'well-intentional word-swamp', she struggled to find a form of expression that was appropriate to recounting her adventures while not offending her reading public.

Other women travel writers, such as Alexandra David-Neel (1868–1969), chose other methods to deal with the constraints of writing about other countries. In her book *My Journey to Lhasa* (1927), David Neel adopted a straightforwardly heroic masculine adventurer role model in describing her attempt to be the first Western woman to reach the 'forbidden city' of Lhasa, Tibet, in a narrative that is full of suspense and intrigue. In contrast, writers such as Nina Mazuchelli (1832–1914) stressed the efforts that they made to retain their femininity and preserve British decorum while in foreign countries. Thus, what is most notable about women's travel writing is the great variety of strategies that individual authors adopted in order to navigate the complex networks of socially acceptable behaviour.

Most of the critics who have written on this subject have been interested primarily in women who travelled to countries that were subject to British imperial rule, because of interest in colonial and postcolonial discourse and theory (Blunt 1994). However, some of them have been concerned with women who travelled to Europe and to the USA. Some of the critics have taken a rather straightforwardly biographical approach to these writings, but many of the critics in the 1990s subjected them to close textual analysis to see whether these women wrote within the same textual confines as did their male counterparts, or whether in fact women travel writers developed a different voice to describe their travels.

Some critics see women's travel writing as necessarily different textually from men's writing; so, for example, it is asserted that women travellers write about the domestic environment and about details of dress and customs more than men do. Some critics, such as Catherine Stevenson (1982), also see them as necessarily critical of the colonial powers; she views this critique as a displacement of their potential **feminism**. While in many cases that is true, it cannot be assumed that women are any less racist or less involved in colonialism simply because they are female.

Women travel writers, particularly within the colonial setting, experienced a range of pressures on their writings; they were pulled in different directions by their allegiance or resistance to colonial or imperial rule and by their need to present or challenge an acceptable feminine persona for their reading public. The writings of women travellers challenge readers to rethink stereotyped views about women travellers as eccentric and abnormal. In some sense, they force us to reconsider the way that women managed to carve out spaces for themselves within a social system that repressed women's freedoms.

Sarah Mills in Cheris Kramarae and Dale Spender (eds), *Routledge International Encyclopedia of Women: Global Women's Issues and Knowledge* (New York and London: Routledge, 2000), 1,969–70.

U

United Nations Environment Programme (UNEP)

The UNEP was established in 1972 by the UN General Assembly, following recommendations of the 1972 UN Conference on the Human Environment, in Stockholm, Sweden, to encourage international co-operation in matters relating to the human environment.

UNEP represents a voice for the environment within the UN system. It is an advocate, educator, catalyst and facilitator, promoting the wise use of the planet's natural assets for sustainable development. It aims to maintain a constant watch on the changing state of the environment; to analyse the trends; to assess the problems using a wide range of data and techniques; and to undertake or support projects leading to environmentally sound development. It plays a catalytic and co-ordinating role within and beyond the UN system. Many UNEP projects are implemented in co-operation with other UN agencies, particularly the UN Development Programme (UNDP), the World Bank group, FAO, UNESCO and WHO. About 45 intergovernmental organizations outside the UN system and 60 international non-governmental organizations (NGOs) have official observer status on UNEP's Governing Council and, through the Environment Liaison Centre in Nairobi, UNEP is linked to more than 6,000 non-governmental bodies concerned with the environment. UNEP also sponsors international conferences, programmes, plans and agreements regarding all aspects of the environment.

In February 1997 the Governing Council, at its 19th session, adopted a ministerial declaration (the Nairobi Declaration) on UNEP's future role and mandate, which recognized the organization as the principal UN body working in the field of the environment and as the leading global environmental authority, setting and overseeing the international environmental agenda. In June a special session of the UN General Assembly, referred to as 'Rio + 5', was convened to review the state of the environment and progress achieved in implementing the objectives of the UN Conference on Environment and Development (UNCED—known as the Earth Summit), which had been held in Rio de Janeiro, Brazil, in June 1992. UNCED had adopted Agenda 21 (a programme of activities to promote sustainable development in the 21st century) and the 'Rio + 5' meeting adopted a Programme for Further Implementation of Agenda 21 in order to intensify efforts in areas such as energy, freshwater resources and technology transfer. The meeting confirmed UNEP's essential role in advancing the Programme and as a global authority promoting a coherent legal and political approach to the environmental challenges of sustainable development. An extensive process of restructuring and realignment of functions was subsequently initiated by UNEP, and a new organizational structure reflecting the decisions of the Nairobi Declaration was implemented during 1999. UNEP played a leading role in preparing for the World Summit on Sustainable Development (WSSD), held in August–September 2002 in Johannesburg, South Africa, to assess strategies for strengthening the implementation of Agenda 21. Governments participating in the conference adopted the Johannesburg Declaration and WSSD Plan of Implementation, in which they strongly reaffirmed commitment to the principles underlying Agenda 21 and also pledged support to all internationally agreed development goals, including the UN Millennium Development Goals adopted by governments attending a summit meeting of the UN General Assembly in September 2000. Participating governments made concrete commitments to attaining several specific objectives in the areas of water, energy, health, agriculture and fisheries, and biodiversity. These included a reduction by one-half in the proportion of people world-wide lacking access to clean water or good sanitation by 2015, the restocking of depleted fisheries by 2015, a reduction in the ongoing loss in biodiversity by 2010, and the production and utilization of chemicals without causing harm to human beings and the environment by 2020. Participants determined to increase usage of renewable energy sources and to develop integrated water resources management and water efficiency plans. A large number of partnerships between governments, private sector interests and civil society groups were announced at the conference.

In May 2000 UNEP's first annual Global Ministerial Environment Forum (GMEF), was held in Malmo, Sweden, attended by environment ministers and other government delegates from more than 130 countries. Participants reviewed policy issues in the field of the environment and addressed issues such as the impact on the environment of population growth, the depletion of earth's natural resources, **climate change** and the need for fresh water supplies. The Forum issued the Malmo Declaration, which identified the effective implementation of international agreements on environmental matters at national level as the most pressing challenge for policy-makers. The Declaration emphasized the importance of mobilizing domestic and international resources and urged increased co-operation from civil society and the private sector in achieving sustainable development. The GMEF was subsequently convened annually.

United Nations Environment Programme (UNEP), in Europa World online (London: Routledge), www.europaworld.com/entry/io-un.1441 (accessed 1 June 2009).

United Nations peace-keeping

UN peace-keeping operations have been conceived as instruments of conflict control. The UN has used these operations in various conflicts, with the consent of the parties involved, to maintain international peace and security, without prejudice to the positions or claims of parties, in order to facilitate the search for political settlements through peaceful means such as mediation and the good offices of the UN Secretary-General. Each operation is established with a specific mandate, which requires periodic review by the **UN Security Council**. In 1988 the UN Peace-keeping Forces were awarded the Nobel Peace Prize.

UN peace-keeping operations fall into two categories: peace-keeping forces and observer missions. Peace-keeping forces are composed of contingents of military and civilian personnel, made available by member states. These forces assist in preventing the recurrence of fighting, restoring and maintaining peace, and promoting a return to normal conditions. To this end, peace-keeping forces are authorized as necessary to undertake negotiations, persuasion, observation and fact-finding. They conduct patrols and interpose physically between the opposing parties. Peace-keeping forces are permitted to use their weapons only in self-defence.

Military observer missions are composed of officers (usually unarmed), who are made available, on the Secretary-General's request, by member states. A mission's function is to observe and report to the Secretary-General (who, in turn, informs the Security Council) on the maintenance of a cease-fire, to investigate violations and to do what it can to improve the situation. Peace-keeping forces and observer missions must at all times maintain complete impartiality and avoid any action that might affect the claims or positions of the parties.

A UN Stand-by Arrangements System (UNSAS) became operational in 1994; participating countries make available specialized civilian and military personnel as well as other services and equipment. In January 1995 the UN Secretary-General presented a report to the Security Council, reassessing the UN's role in peace-keeping. The document stipulated that UN forces in conflict areas should not be responsible for peace-enforcement duties, and included a proposal for the establishment of a 'rapid reaction' force which would be ready for deployment within a month of being authorized by the Security Council. During 2000–09 the multinational UN Stand-by Forces High Readiness Brigade (SHIRBRIG), based in Denmark, was available to the UN, deploying troops to the UN Mission in Ethiopia and Eritrea (in 2000) and the UN Mission in Liberia (2003), and assisting (during 2005) with preparations for the deployment of the UN Mission in Sudan. In March 2003 a SHIRBRIG team supported ECOWAS in planning the deployment of a peace-keeping mission to Côte d'Ivoire; in 2007 it was asked to assist the **African Union** in planning its mission in Sudan (AMISOM) and, during 2008, it was assisting the African Union with the development of an African Standby Force. The Danish Government announced in late 2008 that SHIRBRIG was to be disbanded at the end of June 2009.

In August 2000 a report on UN peace-keeping activities prepared by a team of experts appointed by the Secretary-General assessed the aims and requirements of peace-keeping operations and recommended several measures to improve the performance of the Department of Peace-keeping Operations (DPKO), focusing on its planning and management capacity from the inception of an operation through to post-conflict peace-building activities, and on its rapid response capability. Proposed reforms included the establishment of a body to improve co-ordination of information and strategic analysis requirements; the promotion of partnership arrangements between member states (within the context of UNSAS) enabling the formation of several coherent multinational brigades, and improved monitoring of the preparedness of potential troop contributor nations, with a view to facilitating the effective deployment of most operations within 30 days of their authorization in a Security Council resolution; the adoption of 'on-call' reserve lists to ensure the prompt deployment of civilian police and specialists; the preparation of a global logistics support strategy; and a restructuring of the DPKO to improve administrative efficiency. The study also urged an increase in resources for funding peace-keeping operations and the adoption of a more flexible financing mechanism, and emphasized the importance of the UN's conflict prevention activities. In November the Security Council, having welcomed the report, adopted guidelines aimed at improving its management of peace-keeping operations, including providing missions with clear and achievable mandates. In June 2001 the Council adopted a resolution incorporating a Statement of principles on co-operation with troop-contributing countries, which aimed to strengthen the relationship between those countries and the UN and to enhance the effectiveness of peace-keeping operations. A new Rapid Deployment Level within UNSAS was inaugurated in July 2002. In 2004 the Department established a Special Investigation Team, at the request of the UN Secretary-General, which, in November, visited the Democratic Republic of the Congo to examine allegations of sexual exploitation and abuse committed by peace-keeping personnel. In July 2007 a new Department of Field Operations was established within the UN Secretariat to provide expert support and resources to enhance personnel, budget, information and communication technology and other logistical aspects of UN peace-keeping operations in the field. At the same time a restructuring of the DPKO was initiated. A new Office for the Rule of Law and Security Institutions was established, and the Military Division was reconstituted as the Office of Military Affairs.

The UN's peace-keeping forces and observer missions are financed in most cases by assessed contributions from member states of the organization. In recent years a significant expansion in the UN's peace-keeping activities has been accompanied by a perpetual financial crisis within the organization, as a result of the increased financial burden and some member states' delaying payment. At 31 December 2008 outstanding assessed contributions to the peace-keeping budget amounted to some US $2,880m.

By the end of January 2009 the UN had deployed a total of 63 peace-keeping operations, of which 13 were authorized in the period 1948–88 and

50 since 1988. At 31 January 2009 120 countries were contributing some 91,049 uniformed personnel to the ongoing operations, of whom 77,804 were peace-keeping troops, 10,766 civilian police and 2,479 military observers.

In 2009 the DPKO was directly supporting two political and peace-building missions (in addition to those maintained by the Department of Political Affairs): the UN Assistance Mission in Afghanistan (established in March 2002) and the UN Integrated Office in Burundi (established in January 2007, as a successor to the UN peace-keeping operation).

United Nations Peace-keeping, in Europa World online (London: Routledge), www.europaworld.com/entry/int-org_un-peace (accessed 1 June 2009).

United Nations Population Fund (UNFPA)

Created in 1967 as the Trust Fund for Population Activities, the UN Fund for Population Activities (UNFPA) was established as a Fund of the UN General Assembly in 1972 and was made a subsidiary organ of the UN General Assembly in 1979, with the UNDP Governing Council (now the Executive Board) designated as its governing body. In 1987 UNFPA's name was changed to the United Nations Population Fund (retaining the same acronym).

UNFPA aims to promote health, in particular **reproductive health**, and **gender** equality as essential elements of long-term sustainable development. It aims to assist countries, at their request, to formulate policies and strategies to reduce poverty and support development and to collect and analyse population data to support better understanding of their needs. UNFPA's activities are broadly defined by the Programme of Action adopted by the International Conference on Population and Development (ICPD), which was held in Cairo, Egypt, in September 1994. The Programme's objectives envisaged universal access to reproductive health and family planning services, a reduction in infant, child and maternal mortality, a reduction in the rate of HIV infection, improving life expectancy at birth, and universal access to primary education for all children by 2015. The Programme also emphasized the necessity of empowering and educating women, in order to achieve successful sustainable human development. A special session of the UN General Assembly (entitled ICPD + 5, and attended by delegates from 177 countries) was held in June–July 1999 to assess progress in achieving the objectives of the Cairo Conference and to identify priorities for future action. ICPD + 5 adopted several key actions for further implementation of the Programme of Action. These included advancing understanding of the connections between poverty, gender inequalities, health, education, the environment, financial and human resources, and development; focusing on the economic and social implications of demographic change; greater incorporation of gender issues into social and development policies and greater involvement of women in decision-making processes; greater support for HIV/AIDS prevention activities; and strengthened political commitment to the reproductive health of adolescents. Several new objectives were adopted by the special session, including the achievement of 60% availability of contraceptives and reproductive health care services by 2005, 80% by 2010, with universal availability by 2015. The ICPD objectives were incorporated into the Millennium Development Goals (MDGs), agreed in September 2000 by a summit of UN heads of state or government and, increasingly, are included in national development frameworks and poverty reduction strategies. The overall objective for UNFPA's strategic plan for the period 2008–11 was to accelerate the progress and national ownership of the ICPD Programme of Action, in order to help countries to achieve the MDGs.

Gender Equality: A fundamental aspect of UNFPA's mission is to achieve gender equality, in order to promote the basic **human rights** of women and, through the empowerment of women, to support the elimination of poverty. Incorporated into all UNFPA activities are efforts to improve the welfare of women, in particular by providing reproductive choice, to eradicate gender **discrimination**, and to protect women from sexual and domestic violence and coercion. UNFPA's Strategic Framework on **Gender Mainstreaming** and Women's Empowerment for 2008–11 focuses on the following six priority areas: setting policy for the ICPD Programme of Action and the MDGs; reproductive health; ending gender-based violence; adolescents and youth; emergency and post-emergency situations; and men and boys: partners for equality. The Fund aims to encourage the participation of women at all levels of decision- and policy-making and supports programmes that improve the access of all girls and women to education and that grant women equal access to land, credit and employment opportunities. UNFPA aims to eradicate traditional practices that harm women, and works jointly with UNICEF to advance the eradication of female genital mutilation. UNFPA actively participates in efforts to raise awareness of and implement Resolution 1325 of the **UN Security Council**, adopted in October 2000, which addresses the impact of armed conflict on women and girls, the role of women in peace-building, and gender dimensions in peace processes and conflict resolution. Other activities are directed at particular issues concerning girls and adolescents and projects to involve men in reproductive health care initiatives. (*See also* **Reproductive health and rights**.)

Gender Equality (United Nations Population Fund—UNFPA), in Europa World online (London: Routledge), www.europaworld.com/entry/io-un.1637 (accessed 1 June 2009).

United Nations Security Council (UNSC)

The Security Council was established as a principal organ under the United Nations Charter; its first meeting was held on 17 January 1946. Its task is to promote international peace and security in all parts of the world.

Permanent members are the People's Republic of China, France, the Russian Federation, the United Kingdom and the USA. The remaining 10 members are normally elected (five each year) by the General Assembly for two-year periods (five countries from Africa and Asia, two from Latin America, one from Eastern Europe, and two from Western Europe and others).

The Security Council has the right to investigate any dispute or situation that might lead to friction between two or more countries, and such disputes or situations may be brought to the Council's attention either by one of its members, by any member state, by the General Assembly, by the Secretary-General or even, under certain conditions, by a state that is not a member of the UN.

The Council has the right to recommend ways and means of peaceful settlement and, in certain circumstances, the actual terms of settlement. In the event of a threat to or breach of international peace or an act of aggression, the Council has powers to take 'enforcement' measures in order to restore international peace and security. These include severance of communications and of economic and diplomatic relations and, if required, action by air, land and sea forces.

All members of the UN are pledged by the Charter to make available to the Security Council, on its call and in accordance with special agreements, the armed forces, assistance and facilities necessary to maintain international peace and security.

As the UN organ primarily responsible for maintaining peace and security, the Security Council is empowered to deploy UN forces in the event that a dispute leads to fighting. It may also authorize the use of military force by a coalition of member states or a regional organization. A summit meeting of the Council convened during the Millennium Summit in September 2000 issued a declaration on ensuring an effective role for the Council in maintaining international peace and security, with particular reference to Africa. In June 2006 an intergovernmental advisory UN Peace-building Commission was inaugurated as a subsidiary advisory body of both the Security Council and the General Assembly, its establishment having been authorized by the Security Council and General Assembly, acting concurrently, in December 2005. The annual reports of the Commission were to be submitted to the Security Council for debate. During 2009 the Security Council continued to monitor closely all existing **UN peace-keeping** and political missions and the situations in countries where missions were being undertaken, and to authorize extensions of their mandates accordingly. The Council authorized the establishment of the UN Integrated Peace-building Office in Sierra Leone in October.

Europa World online (London: Routledge), www.europaworld.com/entry/io-un.448 (accessed 1 June 2009).

Virginity testing

This refers to the practice of examining the genitalia of unmarried girls or women in order to determine whether they are virgins. Virginity tests are practised in different cultural contexts all over the world. They are usually embedded in patriarchal traditions that frame the virginity of brides as central to the honour and prestige both of her and the groom's family. Actual procedures may vary; in some societies, for example, young couples are expected to produce a blood-stained sheet from their marital bed, while in others girls and women are examined by doctors or elderly, respected women before the wedding. Most virginity tests are based on the belief that a torn hymen can only result from sexual intercourse, notwithstanding the fact that this fragile part of the vagina can very easily be injured in other physical activities. In many cultures girls/women who are believed to have failed a virginity test risk negative consequences such as being disowned of or even shunned from their family or community.

Lena J. Kruckenberg

War rape – *see* **Rape, war**

Women and postcolonialism

Many women have been attracted to postcolonial theory and criticism because of the parallels between the recently decolonized nation and the situation of women within patriarchal culture. In a number of ways, both take the perspective of a socially marginalized subgroup in their relationship to the dominant culture, and this shared perspective manifests itself in narratives.

Fictions of counter-colonial resistance often draw upon the many different indigenous local and hybrid processes of self-determination to defy, erode and sometimes supplant the prodigious power of imperial cultural knowledge. Consequently, groups of people who are marginalized, whether sexually or racially, often have access to a number of different subject positions, each imbued with its own historicity and each a potentially adversarial and revolutionary agency relative to the dominant mode that holds it in captivity. The fiction of Ntozake Shang, the journalism of Mirta Vidal, and the poetry of Viola Correa, for example, highlight the impossibility of constructing a narrative or discourse that is heterogeneous, multiple and differential enough to offer fair representation to multiple subject positions. Perhaps the best known of postcolonial theorists, Edward Said, argues that ultimately the tragedy of any postcolonial question lies in the constitutive limitation imposed on any attempt to deal with relationships that are polarized, radically uneven and remembered differently. The spheres, the site of intensity, the agenda and the constituencies in the metropolitan and ex-colonized world, argues Said, overlap only partially.

Ailbhe Smythe in Cheris Kramarae and Dale Spender (eds), *Routledge International Encyclopedia of Women: Global Women's Issues and Knowledge* (New York & London: Routledge, 2000), 1,645.

Women in development

Since its creation in 1945, the UN has sought to alleviate poverty and to improve the standard of living of the world's poorest states. The overall strategy has been to fund a wide range of aid and development programmes. However, until the 1970s none of these programmes specifically took into account the role of women in the development process. In recognition of this problem, the UN embarked on a vigorous campaign to advance the position of women within the development community. This included measures to improve their access to funding, to make **gender** equity a priority, and to ensure that UN development programmes would lead to more gender-sensitive outcomes for women. To facilitate this, special units were set up within institutions such as the World Bank. Moreover, foreign aid began to target women's issues, and women began to have more input at the strategic planning level.

The most important initiative was the *International Decade for the Advancement of Women*. Lasting from 1976 to 1985, the Decade helped to open up a space for dialogue and debate about issues of concern to women. It did this in at least three ways.

First, a number of conferences were held during the period, which provided women with an opportunity to discuss their individual experiences, to take part in workshops and develop information networks. Second, two specialized agencies within the UN were established: the United Nations Development Fund for Women (UNIFEM) and the United Nations International Research and Training Institute for the Advancement of Women (INSTRAW). Third, the Decade provided an important impetus for an emerging feminist literature on Women in Development (WID).

Much of this literature remains highly critical of the UN for the gender-biased character of its aid and development programmes that allegedly fail to take account of issues central to women's lives, such as reproduction, health and child-rearing. Moreover, the programmes have done little to overcome the large inequalities between men and women in the Third World. The WID literature argues that women are integral to development but that they rarely benefit from it, largely because of a lack of access to markets, funding, decision-making and education. The goal of the WID literature, therefore, is to highlight the importance of women's roles and to help establish strategies to reduce gender inequality. The WID critique has helped to establish a presence for women within the development debate, as well as in the planning and decision-making process. In this sense, the WID literature has made a lasting contribution to Third World development and towards correcting the institutional bias against women in the UN and elsewhere. In addition, the WID literature was an important starting point for feminist incursions into development studies and international political economy. It was the first body of literature to draw attention to the need of women for better access to aid and development, gender equity and gender-sensitive development planning. (*See also* **Gender and development**.)

Craig Warkentin in Martin Griffiths and Terry O'Callaghan (eds), *International Relations: The Key Concepts*, second edn (Routledge, 2002). Politics Online. Taylor & Francis. www.routledgepoliticsonline.com:80/Book.aspx?id=w043 (accessed 15 October 2009).

Bibliography

Bibliography

Abrams, K., 'The Legal Subject in Exile', *Duke Law Journal* 51, 2001, www.law.duke.edu/journals/dlj/articles/dlj51p27.htm.

Acker, J., 'Rewriting Class, Race and Gender: Problems in Feminist Rethinking', in Myra Marx Ferree et al. (eds), *Revisioning Gender* (London: Sage Publications, 1998).

Action Alert, 'Women's Bodies as a Battleground: Sexual Violence Against Women and Girls During the War in the Democratic Republic of Congo, South Kivu (1996–2003)' (London, 2004).

Adams, C.J., *The Sexual Politics of Meat: A Feminist-Vegetarian Critical Theory*, second edn (New York: Continuum, 1999).

Adler, E., 'Seizing the middle ground: constructivism in world politics', *European Journal of International Relations* 3:3, 1997, 319–63.

——'Constructivism and International Relations', in W. Carlsnaes, T. Risse and B. Simmons, *Handbook of International Relations* (London: Sage, 2002), 95–118.

Adorno, T. and M. Horkheimer, 'The culture industry: Enlightenment as mass deception', 1944 (transcribed by Andy Blunden, 1998, www.marxists.org).

——*Mass Culture* (London: Politico, 2005).

Africa Centre for the Constructive Resolution of Disputes, 'Women, Peace & Security', Special Edition, *Conflict Trends* 3, 2003, www.accord.org.za/ct/2003-3.htm (accessed 20 December 2008).

African Gender Institute, 'Gender and Women's Studies for Africa's Transformation', www.gwsafrica.org.

African Union, *Constitutive Act of the African Union* (Lome, Togo: African Union, 11 July 2000).

——*Protocol to the African Charter on Human and Peoples' Rights on the Rights of Women in Africa* (Assembly/AU, Maputo: African Union, 11 July 2003).

——*Solemn Declaration on Gender Equality* (Assembly/AU/Decl.12(III), Addis Ababa: African Union, 6–8 July 2004).

——*Chairperson's Second Progress Report on the Implementation of the Solemn Declaration on Gender Equality in Africa* (Addis Ababa: African Union, 2006).

Agamben, G., *State of Exception* (Chicago, IL and London: University of Chicago Press, 2005).

Agarwal, A., *Environmentality: Technologies of Government and the Making of Subjects* (Durham, NC: Duke University Press, 2005).

Agarwal, B., 'The gender and environment debate', in R. Keil, D.V.J. Bell, P. Penz and L. Fawcett (eds), *Political Ecology: Global and Local* (New York: Routledge, 1998), 192–219.

Ahmed, L., *Women and Gender in Islam: Historical Roots of a Modern Debate* (New Haven, CT: Yale University Press, 1992).

——'Migrants in the Mistress's House: Other Voices in the "Trafficking" Debate', *Social Politics* 2005, 12:1, 96–117.

Ahmed-Ghosh, H., 'A History of Women in Afghanistan: Lessons Learned for the Future', *Journal of International Women's Studies* 2003, 4:3 (online).

Akhavan, N., G. Bashi, M. Kia and S. Shakhsari, 'A Genre in the Service of Empire: An Iranian Feminist Critique of Diasporic Memoirs', 2007, fanonite.org/?s = Reading+Lolita+in+Tehran (accessed 13 October 2008).

Alaimo, S., *Undomesticated Ground: Recasting Nature as Feminist Space* (Ithaca, NY: Cornell University Press, 2000).

——'Ecofeminism without nature? Questioning the relation between feminism and environmentalism', *International Feminist Journal of Politics* 10:3, 2008, 299–304.

Albertson Fineman, M. and N. Thomadsen (eds), *At the Boundaries of Law: Feminism and Legal Theory* (New York: Routledge, 1991).

Alcoff, L., 'Culture Feminism Versus Post-Structuralism: The Identity Crisis in Feminist Theory', *Signs: Journal of Women in Culture and Society* 12:3, 1988.

Allan, S., *News Culture*, second edn (Maidenhead: Open University Press, 2005).

Allen, B. *Rape Warfare: The Hidden Genocide in Bosnia-Herzegovina and Croatia* (Minneapolis, MN: University of Minnesota Press, 1996).

Alston, P., 'The UN's Human Rights Record: From San Francisco to Vienna and Beyond', *Human Rights Quarterly* 16, 1994, 365–90.

Althusser, L. and E. Balibar, *Reading Capital* (London: NLB, 1970).

Amnesty International, *Human Rights are Women's Right* (London: Amnesty International, 1995).

——*Reservations to the Convention on the Elimination of all Forms of Discrimination against Women – Weakening the Protection of Women from Violence in the Middle East and North Africa Region* (2004), web.amnesty.org/library/index/engior510092004.

Andermahr, S., T. Lovell and C. Wolkowitz, *A Glossary of Feminist Theory* (London: Arnold, 1997).

Anderson, B. and J. O'Connell Davidson, *Trafficking – a Demand Led Problem?* (Stockholm: Save the Children, 2002).

Anderson, P., *Lineages of the Absolutist State* (London: New Left Books, 1974).

——*The Origins of Post-modernity* (London: Verso, 1998).

Anthias, F. and N. Yuval-Davis, *Racialized Boundaries: Race, Nation, Gender, Colour and Class in the Anti-Racist Struggle* (London: Routledge, 1992).

Asad, T., *Formations of the Secular. Christianity, Islam and Modernity* (Stanford, CA: Stanford University Press, 2003).

Ashley, R.K. and R.B.J. Walker, 'Reading Dissidence/Writing the Discipline: Crisis and the Question of Sovereignty in International Studies', *International Studies Quarterly* 34:3, September 1990a, 367–416.

——'Speaking the Language of Exile: Dissident Thought in International Studies', *International Studies Quarterly* 34:3, September 1990b, 259–68.

Askin, K.D., *War Crimes Against Women: Prosecution in International War Crimes Tribunals* (The Hague: Kluwer Law International Law, 1997).

Askin, K.D. and D.M. Koenig, *Women and International Human Rights*, Vol. 1 (New York: Transnational Publishers, Inc., 1998).

Askola, H., *Legal Responses to Trafficking in Women for Sexual Exploitation in the European Union* (Oxford: Hart, 2007).

Asylum Aid, *Romani Women from Central and Eastern Europe: A 'Fourth World', or Experience of Multiple Discrimination* (London: Refugee's Women Research Project, 2002), www.asylumaid.org.uk/publications.php (accessed 5 February 2009).

Aubrey, J., in A. Clark (ed.) *Brief Lives, Chiefly of Contemporaries* (Oxford: 1898).

Azoy, W., *Buzkashi, Game and Power in Afghanistan* (Prospect Heights, IL: Waveland Press, 2002 [1982]).

Baden, S. and A.M. Goetz, 'Who Needs (Sex) When You Can Have (Gender)? Conflicting Discourse on Gender at Beijing', *Feminist Review* 56, Summer 1997, 3–25.

Baksh, R., L. Etchart, E. Onubogu and T. Johnson (eds), *Gender Mainstreaming in Conflict Transformation: Building Sustainable Peace* (London: Commonwealth Secretariat, 2005).

Baldick, C., *The Social Mission of English Criticism, 1848–1932* (Oxford: Clarendon Press, 1983).

Bancroft, A., 'Closed Spaces, Restricted Places: Marginalisation of Roma in Europe', *Space & Polity* 5:1, 2001, 145–57.

Banerjee, D. and M. Mayerfeld Bell, 'Ecogender: locating gender in the environmental social sciences', *Society and Natural Resources* 2007, 20, 3–19.

Barrett, M., *Women's Oppression Today: The Marxist/Feminist Encounter* (London and New York: Verso, 1988).

Barry, K., *Female Sexual Slavery* (New York: Avon, 1979).

——*The Prostitution of Sexuality: The Global Exploitation of Woman* (New York: New York University Press, 1995).

Bartky, S., 'Foucault, Femininity and Patriarchal Power', in I. Diamond and L. Quinby (eds), *Feminism and Foucault* (Boston, MA: Northeastern University Press, 1988).

Basu, A. (ed.), *The Challenge of Local Feminisms* (Boulder, CO: Westview, 1995).

Bauman, Z., *Modernity and the Holocaust* (Cornell, NY: Cornell University Press, 1989).

de Beauvoir, S., *The Second Sex*, trans. by H.M. Parshley (New York: Knopf, 1952).

——*The Second Sex* (New York: Vintage Press, 1973).

Baxter, J., *De Niro: A Biography* (London: HarperCollins, 2002).

Beasley, C., *What is Feminism Anyway? Understanding Contemporary Feminist Thought* (London: Allen and Unwin, 1999).

Beechey, V., 'Rethinking the Definition of Work', in Jensen, Hagen and Reddy (eds), *Feminization of the Labor Force* (New York: Oxford University Press, 1988), 45–62.

Behar, R. and D. Gordon, *Women Writing Culture* (Berkeley, CA: University of California Press, 1992).

Bell, V., 'Beyond the "Thorny Question": Feminism, Foucault and the Desexualisation of Rape', *International Journal of the Sociology of Law* 19, 1991, 83–100.

Bem, S. Lipositz, *The Lenses of Gender: Transforming the debate in sexual inequality* (New Haven, CT: Yale University Press, 1993).

Beneria, L. and R. Blank, 'Women and the Economics of Military Spending', in A. Harris and Y. King (eds), *Rocking the Ship of State* (Boulder, CO: Westview, 1989).

Benhabib, S., 'On Contemporary Feminist Theory', *Dissent* 36, Summer 1989, 369–78.

——'Sexual Difference on Collective Identities: The New Global Constellation', in S. James and S. Palmer (eds), *Visible Women: Essays on Feminist Legal Theory and Political Philosophy* (Oxford and Portland, OR: Hart Publishing, 2002), 137–58.

——*The Rights of Others* (Cambridge: Cambridge University Press, 2004).

Benhabib, S. and D. Cornell (eds), *Feminism as Critique: Essays on the Politics of Gender in Late-Capitalist Societies* (Cambridge: Polity Press, 1986).

Benjamin, W., *Illuminations* (London: Routledge, 1992).

Berg, A. and M. Lie, 'Feminism and Constructivism: Do Artefacts Have a Gender?', Special Issue: Feminist and Constructivist Perspectives on New Technology, *Science, Technology and Human Values* 20:3, Summer 1995, 332–51.

Beveridge, F. and S. Mullally, 'International Human Rights and Body Politics', in J. Bridgeman and S. Millns (eds), *Law and Body Politics: Regulating the Female Body* (Aldershot: Dartmouth Publishing, 1995), 240–72.

Biersteker, T.J. and C. Weber (eds), *State Sovereignty as Social Construct* (Cambridge: Cambridge Press, 1996).

Birkett, D. and S. Wheeler, *Amazonian: Penguin Book of New Women's Travel Writing* (London: Penguin, 1998).

Bitu, N., *The Situation of Roma/Gypsy Women in Europe* (Bucharest: Roma Centre for Social Intervention and Studies, 1999), www.romawomen.ro/pages/reports.htm (accessed 5 February 2009).

——'The Challenges of and for Romani Women', *Roma Rights* No. 4, 2003, www.errc.org/Romarights_index.php (accessed 5 February 2009).

——'Romani Women: No Longer Willing to Wait on Gender', *Open Society News* Summer/Fall 2005, 10–11.

Blau, F. and M. Ferber, *The Economics of Women, Men, and Work*, second edn (Englewood Cliffs, NJ: Prentice Hall, 1992).

Blunt, A., *Travel, Gender and Imperialism: Mary Kingsley and West Africa* (London: Routledge, 1994).

Bly, R., *Iron John: A Book About Men* (New York: Addison-Wesley, 1990).

Boehmer, E., *Empire Writing. An Anthology of Colonial Literature 1870–1918* (Oxford: Oxford University Press, 1998).

Bonacich, E., 'Class Approaches to Ethnicity and Race', *Insurgent Sociologist* 10:2, 1980, 9–23.

Bond, J.E., 'Intersecting Identities and Human Rights: The Example of Romani Women's Reproductive Rights', in *Georgetown Journal of Gender and the Law* Vol. 5, 2004, 897–916.

Bordo, S., *Unbearable Weight: Feminism, Western Culture and the Body* (Berkeley, CA: University of California Press, 1993).

Bottomley, A., *Feminist Perspectives on the Foundational Subjects of Law* (London: Cavendish Publishing, 1996).

Bottomley, A. and J. Conaghan (eds), *Feminist Theory and Legal Strategy* (Oxford: Blackwell, 1993).

Boyle, J., 'Ideals and Things: International Legal Scholarship and the Prison-House of Language', *Harvard International Law Journal* 26, 1985, 327–59.

——'Is Subjectivity Possible? The Postmodern Subject in Legal Theory', *University of Colorado Law Review* 62, 1991, 28pp, www.law.duke.edu/boylesite/Subject.htm.

Bretherton, C., 'Global environmental politics: putting gender on the agenda?' *Review of International Studies* 1998, 24, 1–85.

——'Movements, networks, hierarchies: a gender perspective on global environmental governance', *Global Environmental Politics* 2003, 3:2, 103–19.

Bridgeman, J. and S. Millns (eds), *Law and Body Politics: Regulating the Female Body* (Aldershot: Dartmouth Publishing, 1995).

——*Feminist Perspectives on Law: Law's Engagement with the Female Body* (London: Sweet & Maxwell 1998).

Brophy, J. and C. Smart (eds), *Women-in-Law Explorations in Law, Family and Sexuality* (London: Routledge & Kegan Paul, 1985).

Brown, S., 'Feminism, International Theory, and International Relations of Gender Inequality', *Millennium: Journal of International Studies* 17:3, 1998, 461–75.

Brown, W., *States of Injury Power and Freedom in Late Modernity* (Princeton, NJ: Princeton University Press, 1995a).

——'Rights and Identity in Late Modernity: Revisiting the "Jewish Question"', in A. Sarat and T.R. Kearns (eds), *Identities, Politics and Rights* (Ann Arbor, MI: Michigan University Press, 1995b), 85–130.

Brownmiller, S., *Against Our Will: Men, Women and Rape* (London: Penguin, 1975).

Buckingham, S., 'Ecofeminism in the 21st century', *The Geographical Journal* 170:2, 2004, 146–54.

Buckingham-Hatfield, S., *Gender and Environment* (London: Routledge, 2000).

Buss, D. and A. Manji (eds), *International Law: Modern Feminist Approaches* (Oxford and Portland, OR: Hart Publishing, 2005).

Butegwa, C., *The International Criminal Court: A Ray of Hope for the Women of Darfur?* (Kampala: Darfur Consortium, 2006), www.pambazuka.org (accessed 20 December 2008).

Butler, J., 'Variations on Sex and Gender: Beauvoir, Witig and Foucault', in S. Benhabib and D. Cornell (eds), *Feminism as Critique* (Cambridge: Polity, 1986), 128–42.

——*Gender Trouble: Feminism and the Subversion of Identity* (London: Routledge, 1990).

——'Contingent Foundations: Feminism and the Question of "Postmodernism"', in J. Butler and J.W. Scott (eds), *Feminists Theorize the Political* (London: Routledge, 1992), 3–21.

——*Bodies That Matter: On the Discursive Limits of 'Sex'* (London: Routledge, 1993).

——*Excitable Speech: A Politics of the Performative* (London: Routledge, 1997).

——*Gender Trouble: Feminism and the Subversion of Identity* (New York: Routledge, 1999 [1990]).

——'Giving an Account of Oneself', *Diacritics* 31:3, Winter 2001, 22–40.

——*Precarious Life: The Power of Mourning and Violence* (London & New York: Verso, 2004).

Butler, J. and J.W. Scott (eds), *Feminists Theorize the Political* (London: Routledge, 1992).

Cahn, C., 'The Unseen Powers: Perception, Stigma and Roma Rights', *Roma Rights* No. 3, 2007, www.errc.org/Romarights_index.php (accessed 5 February 2009).

Callinicos, A., *Against Post-modernism: A Marxist Critique* (Cambridge: Polity Press, 1991).

Carlassare, E., 'Essentialism in ecofeminist discourse', in C. Merchant (ed.), *Ecology: Key Concepts in Critical Theory* (Atlantic Highlands, NJ: Humanities Press, 1994), 220–34.

Carpenter, C. Rharli, 'Gender Theory in World Politics: Contributions of a Non-Feminist Standpoint?', *International Studies Review* 4, 3, 2003a, 153–66.

——'Stirring Gender into the Mainstream: Constructivism, Feminism and the Uses of IR Theory', *International Studies Review* 5:2, June 2003b, 297–302.

Carver, T., '"Public Man" and the Critique of Masculinities', *Political Theory* 24, 1996, 673–86.

——'Gender/Feminism/IR', *International Studies Review* 5:2, June 2003, 288–90.

——'War of the Worlds/Invasion of the Body Snatchers', *International Affairs* 80:1, 2004, 92–94.

Carver, T., M. Cochran and J. Squires, 'Gendering Jones', *Review of International Studies* 24:2, 1998, 283–98.

Cassese, A., 'The Statute of the International Criminal Court: Some Preliminary Reflections', *European Journal of International Law* 10, 1999, 144–71.

Centre for Reproductive Rights and Poradna, *Body and Soul: Forced Sterilization and Other Assaults on Roma Reproductive Freedom in Slovakia* (New York: CRR, 2003), www.reproductiverights.org/pub_bo_slovakia.html (accessed 5 February 2009).

Centre for Roma Initiatives, *Istraživanje. On the Position of Roma Women in Niksic* (2005), www.osim.org.me/fosi_rom_en/frame_publications.htm (accessed 5 February 2009).

——*Virginity does not determine whether a Roma girl is worthy or not* (2006), www.osim.org.me/fosi_rom_en/frame_publications.htm (accessed 5 February 2009).

Chamallas, M., *Introduction to Feminist Legal Theory* (Gaithersburg and New York: Aspen Law Business, 1999).

Chapman, R.A., '*Leviathan* Writ Small: Thomas Hobbes on the Family', *American Political Science Review* 1975, 69, 76–90.

Charles, N., *Feminism, the State and Social Policy* (London: MacMillan Press, 2000).

Charlesworth, H., 'Feminist Methods in International Law', *American Journal of International Law* 93, 1999, 379–94.

——'Not Waving but Drowing: Gender Mainstreaming and Human Rights in the United Nations', *Harvard Human Rights Journal* 18, 2005, 1–18.

Charlesworth, H. and C. Chinkin, 'Violence Against Women: A Global Issue', in M. A. Chen, 'Engendering World Conference: The International Women's Movement and the United Nations', *Third World Quarterly* 16:3, 1995, 477–93.

——*The Boundaries of International Law: A Feminist Analysis* (Manchester: Manchester University Press, 2000).

Charlesworth, H., C. Chinkin and S. Wright, 'Feminist Approaches to International Law', *American Journal of International Law* 85, 1991, 613–45.

Charny, I.W., 'The Whitaker Report', in I.W. Charny (ed.), *Encyclopedia of Genocide* (Santa Barbara, CA: ABC-CLIO, 1999), 581–87.

Chesterman, S., 'Human Rights as Subjectivity: The Age of Rights and the Politics of Culture', *Millennium: Journal of International Studies* 27:1, 1998, 97–118.

Chiang, L., 'Trafficking in Women: An International Issue', in K. Askin and D. Koenig (eds), *Women's International Human Rights: A Reference Guide*, Vol. 1 (Ardsley, NY: Transnational Press, 1999), 321–64.

Chinkin, C., 'Rape and Sexual Abuse of Women in International Law', *European Journal of International Law* 5:3, 1994, 326–41.

——'A Critique of the Public/Private Distinction', *European Journal of International Law* 10:2, 1999, 387–96.

di Chiro, G., 'Living environmentalisms: coalition politics, social reproduction and environmental justice', *Environmental Politics* 17:2; 2008, 276–98.

Chowdhry, G. and S. Nair, *Power, Postcolonialism and International Relations: Reading Race, Gender and Class* (London: Routledge, 2002).

Christiansen, T., K.E. Jorgensen and A. Wiener (eds), *The Social Construction of Europe* (London: Sage, 2001).

Chunn, D.E. and D. Lacombe (eds), *Law as a Gendering Practice* (Oxford: Oxford University Press, 2000).

Cixous, H., 'Sorties' in H. Cixous (with C. Clement), *La jeune née* [The newly born woman] (Paris: Union Générale d'Editions, 1975).

Cockburn, C., *The Space Between Us: Negotiating Gender and National Identities in Conflict* (London: Zed, 1998).

Code, L., *What can she know?: feminist theory and the construction of knowledge*, second edn (Ithaca, NY: Cornell University Press, 1991).

——(ed.), *Encyclopaedia of Feminist Theories* (London and New York: Routledge, 2000).

Cole, E.R., 'Coalitions as a Model for Intersectionality: From Practice to Theory', *Sex Roles* 59, 2008, 443–53.

Collier, J.F. and S. Yanagisako (eds), *Gender and Kinship: Essays Toward a Unified Analysis* (Stanford, CA: Stanford University Press, 1987).

Collins, P. Hill, *Black Feminist Thought: Knowledge, Consciousness and the Politics of Empowerment* (London: Routledge, 2000).

Commission on Human Rights, *Study of the Legal Validity of the Undertakings Concerning Minorities*, 70 UN Doc. E/CN.4/367 (1950), 70–71.

Commission on Security and Co-operation in Europe, *Accountability and Impunity Investigations into Sterilization Without Informed Consent in the Czech Republic and Slovakia* (Washington: CSCE, 2006), www.csce.gov/index.cfm?Fuseaction=Files.Download&FileStore_id = 647 (accessed 5 February 2009).

Connell, R., *Gender and Power* (Cambridge: Polity, 1995).

Connell, R.W., 'Gender and the state', in K. Nash and A. Scott (eds), *The Blackwell Companion to Political Sociology* (Oxford: Blackwell, 2001), 117–26.

Connolly, W.E., 'Identity and Difference in Global Politics', in J. Der Derian and M.J. Shapiro (eds), *International/Intertextual Relations: Postmodern Readings of World Politics* (New York: Lexington Books, 1989), 323–42.

Connor, S., 'US climate policy bigger threat to world than terrorism', *The Independent* 9 January 2004, www.independent.co.uk/news/world/americas/us-climate-policy-bigger-threat-to-world-than-terrorism-572493.html (accessed 24 May 2007).

Cook, R., 'Reservations to the Convention on the Elimination of All Forms of Discrimination against Women', *Virginia Journal of International Law* 30, 1990, 643–716.

——(ed.), *Human Rights of Women: National and International Perspectives* (Philadelphia, PA: University of Pennsylvania Press, 1994).

Cooke, M. and A. Woollacott (eds), *Gendering War Talk* (Princeton, NJ: Princeton University Press, 1993).

Copelon, R., 'Surfacing gender: Reconceptualizing crimes against women in time of war', in L.A. Lorentzen and J. Turpin (eds), *The Women & War Reader* (New York: New York University Press, 1998), 63–79.

Copjec, J., 'The Orthopsychic Subject: Film Theory and the Reception of Lacan', in *Read My Desire: Lacan against the Historicists* (Cambridge, MA: MIT Press, 1994), 15–38.

Cornell, D., 'The Philosophy of the Limit: Systems Theory and Feminist Legal Reform', in D. Cornell, M. Rosenfeld and D. Gray Carlson (eds), *Deconstruction and the Possibility of Justice* (New York/London: Routledge, 1992a), 68–94.

——*The Philosophy of Limit* (London: Routledge, 1992b).

——*Transformations* (New York and London: Routledge, 1993).

Cornwall, A. and N. Lindisfarne, *Dislocating Masculinity. Comparative Ethnography* (London and New York: Routledge, 1994).

Council of Europe and EUMC (European Monitoring Centre on Racism and Xenophobia), *Breaking the Barriers—Romani Women and Access to Public Health Care* (Luxembourg: Office for Official Publications of the European Communities, 2003), fra.europa.eu/fraWebsite/material/pub/ROMA/rapport-en.pdf (accessed 5 February 2009).

Crenshaw, K. Williams, 'Mapping the Margins: Intersectionality, Identity Politics, and Violence against Women of Color', *Stanford Law Review* 1991, 43:6, 1,241–99.

——'Whose story is it anyway? Feminst and antiracist appropriations of Anita Hill', in T. Morrison (ed.), *Race-ing justice, en-gendering power: Essays on Anita Hill, Clarence Thomas and the construction of social reality* (New York: Pantheon, 1992), 402–40.

——'Demarginalizing the Intersection of Race and Sex: A Black Feminist Critique of Antidiscrimination Doctrine, Feminist Theory and Antiracist Politics', in D.K. Weisberg (ed.), *Feminist Legal Theory: Foundations* (Philadelphia, PA: Temple University Press, 1993), 383–98.

Crowe, D., 'Conclusion', in D. Crowe and J. Kolsti (eds), *The Gypsies of Eastern Europe* (London: M.E. Sharpe, 1991), 151–58.

——*A History of the Gypsies of Eastern Europe and Russia* (Houndmills: Palgrave Macmillan, 2007).

Dabashi, H., 'Native Informers and the Making of the American Empire', *Al-Ahram* No. 797, 1–7 June 2006.

Dallmeyer, D.G. (ed.), *Reconceiving Reality: Women and International Law* (Massachusetts: American Society of International Law, 1993).

Danielsen, D. and K. Engle (eds), *After Identity: A Reader in Law and Culture* (New York/London: Routledge, 1995).

Dankelman, I., 'Climate change: learning from gender analysis and women's experience of organizing for sustainable development', *Gender and Development* 10:2, July 2002, 21–29.

Dankelman, I., K. Alam, W. Bashar Ahmed, Y. Diagne Gueye, N. Fatema and R. Mensah-Kutin, *Gender, Climate Change and Human Security: Lessons from Bangladesh, Ghana and Senegal*, report prepared for ELIAMEP (Women's Environment and Development Organization—WEDO with ABANTU for Development in Ghana, ActionAid Bangladesh and ENDA in Senegal, 2008), www.wedo.org/files/HSN%20Study%20Final%20May%2020,%202008.pdf (accessed 24 November 2008).

Davies, M., 'Talking the Inside Out: Sex and Gender in the Legal Subject', in N. Naffine and R.J. Owens (eds), *Sexing the Subject of Law* (Sydney: Law Book Co., 1997), 25–46.

Davion, V., 'Is Ecofeminism Feminist?', in K.J. Warren (ed.), *Ecological Feminism* (New York: Routledge, 1994), 8–28.

Deforges, L., R. Jones and M. Woods, 'New Geographies of Citizenship', *Citizenship Studies* 9:5, 2005, 439–51.

Delzotto, A.C., 'Weeping Women, Wringing Hands: How the Mainstream Media Stereotyped Women's Experiences in Kosovo', *Journal of Gender Studies* 2002, 11, 91–108.

Denton, F., 'Climate change vulnerability, impacts and adaptation: why does gender matter?', *Gender and Development* 10:2, 2002, 10–20.

Department of Trade and Industry, *GM Nation? Findings of the public debate* (UK: Department of Trade and Industry, 2003).

Derrida, J., 'Force of Law: the "Mystical Foundation of Authority"', in D. Cornell, M. Rosenfeld and D. Gray Carlson (eds), *Deconstruction and the Possibility of Justice* (New York and London: Routledge, 1992), 3–67.

Deutscher, P., *Yielding Gender: Feminism, Deconstruction and the History of Philosophy* (London: Routledge, 1997).

Diez, T., 'Social Constructivism', in J. Steans and L. Pettiford (with T. Diez), *International Relations: Perspectives and Themes* (London: Pearson, 2005).

Dietz, T., A. Dan and R. Shwom, 'Support for climate change policy: social psychological and social structural influences', *Rural Sociology* 2007, 72:2, 185–214.

Dietz, T., L. Kalof and P. Stern, 'Gender, values and environmentalism', *Social Science Quarterly* 83:1, 2002, 353–64.

Dirks, T., 'Raging Bull' (2009), www.filmsite.org/ragi.html.

Doane, M.A., 'Film and the Masquerade: Theorising the Female Spectator', *Scree* 23, 3–4, 1987, 77–87.

Dobson, A., *Justice and the Environment: Conceptions of Environmental Sustainability and Theories of Distributive Justice* (Oxford: Oxford University Press, 1998).

——*Citizenship and the Environment* (Oxford: Oxford University Press, 2003).

Doezema, J., 'Forced to Choose. Beyond the Voluntary v. Forced Prostitution Dichotomy', in K. Kempadoo and J. Doezema (eds), *Global Sex Workers* (London: Routledge, 1998).

——'Ouch! Western Feminists' "Wounded Attachment" to the Third World Prostitute', *Feminist Review* 2001, 67, 16–38.

——'Now You See Her, Now You Don't: Sex Workers at the UN Trafficking Protocol Negotiations', *Social and Legal Studies* 2005, 14:1, 61–90.

Donald, J. and S. Hall, *Politics and Ideology* (Milton Keynes: Open University Press, 1986).

Donnelly, J., *The Concept of Human Rights* (New York: St Martin's Press, 1985).

Douzinas, C., *Human Rights and Empire: The Political Philosophy of Cosmopolitanism* (New York: Routledge-Cavendish, 2007).

Dryzek, J., *The Politics of the Earth: Environmental Discourses*, second edn (Oxford: Oxford University Press, 2005).

Duberly, F.I., *Comprising Experiences in Rajpootnana and Central India During the Suppression of the Mutiny 1857–58*, 1859.

Dupree, L., *Afghanistan* (Princeton, NJ: Princeton University Press, 2002 [1973]).

Eagleton, T., *Literary Theory: An Introduction* (Oxford: Blackwell, 1996).

economalliance.org/about

Edkins, J., *Poststructuralism & International Relations: Bringing the Political Back In* (Boulder and London: Lynne Rienner, 1999).

Edkins, J., N. Persram and V. Pin-Fat (eds), *Sovereignty and Subjectivity* (Boulder, CO and London: Lynne Rienner, 1999).

Edkins, J. and M. Zehfuss, 'Generalising the International', *Review of International Studies* 31:3, July 2005, 451–72.

Ehrenreich, B. and A. Russell Hochschild (eds), *Global Woman: Nannies, Maids, and Sex Workers in the New Economy* (New York: Metropolitan Books, 2003).

Ehrlich, Paul, *The Population Bomb* (New York: Sierra Club, 1968).

Eisenstein, Z., *Against Empire: Feminisms, Racism and the West* (Melbourne: Spinifex Press, 2004).

Ekberg, G., 'The International Debate about Prostitution and Trafficking in Women: Refuting the Arguments', paper presented at seminar on the effects of legalization of prostitution activities (Swedish Government, 2002).

Ekiyor, T., *Women's Empowerment in Peacebuilding: A Platform for Women's Participation in Decision Making*, African Women's Development Fund (AWDF), Annual Briefing Paper (2004), www.awdf.org/awdf/?pid=32&cid = 36 (accessed 21 December 2008).

el-Bushra, J., 'Transforming Conflict: Some Thoughts on a Gendered Understanding of the Conflict Process', in S. Jacob, R. Jacobson and J. Marhband (eds), *States of Conflict: Gender, Violence, and Resistance* (London: Zed, 2000), 66–86.

Elliott, L., *The Global Politics of the Environment*, second edn (London: Macmillan, 2004).
Elshtain, J.B., *Women and War* (New York: Basic Books, 1987).
——*Public Man, Private Women: Women in Social and Political Thought*, second edn (Princeton, NJ: Princeton University Press, 1993).
Eminova, E., 'Raising New Questions about an Old Tradition', *Open Society News* Summer/Fall 2005, 13.
——'Negotiations: Feminism, racism and difference', *Development* 49:1, 2006, 35–37.
Engels, F., *The Origin of the Family, Private Property, and the States* (London: Lawrence and Wishart, 1972).
Engle, K., 'After the Collapse of the Public/Private Distinction: Strategizing Women's Rights', in D.G. Dallmeyer (ed.), *Reconceiving Reality: Women and International Law* (Massachusetts: American Society of International Law, 1993), 143–56.
——'Feminism and Its (Dis)Contents: Criminalizing Wartime Rape in Bosnia and Herzegovina', *American Journal of International Law* 99:4, October 2005, 778–815.
——'Calling in the Troops: The Uneasy Relationship Among Women's Rights, Human Rights, and Humanitarian Intervention', *Harvard Human Rights Journal* 20:189, 2007.
Enloe, C., *Bananas, Beaches and Bases: Making Feminist Sense of International Politics* (Berkeley, CA: University of California Press, 1989).
——*The Morning After: Sexual Politics at the End of the Cold War* (Berkeley, CA: University of California Press, 1993).
——*Maneuvers: The International Politics of Militarizing Women's Lives* (Berkeley, CA: University of California Press, 2000).
——'Interview with Professor Cynthia Enloe', carried out at the February 2001 Annual Convention of the ISA in Chicago, IL, *Review of International Studies* 27, 2001, 649–66.
——'"Gender" is Not Enough: The Need for a Feminist Consciousness', *International Affairs* 80:1, 2004, 95–97.
Epstein, C., 'In Praise of Women Warriors', *Dissent* 38, 1991, 421–22.
——*The Power of Words in International Relations: Birth of an Anti-Whaling Discourse* (Cambridge, MA: MIT Press, 2008).
ERRC (European Roma Rights Centre), *A Special Remedy: Roma and Schools for the Mentally Handicapped in the Czech Republic* (Budapest: ERRC, 1999), www.errc.org/cikk.php?cikk=115 (accessed 5 February 2009).
——*'Romani women in Romani and majority societies*', *Roma Rights* 1, 2000, www.errc.org/Romarights_index.php (accessed 5 February 2009).
——*Statement at the OSCE Human Dimension Implementation Meeting: Romani Women's Rights* (Budapest: ERRC, 2006), www.errc.org/cikk.php?cikk=2637 (accessed 5 February 2009).
——*The Glass Box. Exclusion of Roma from Employment* (Budapest: Westimprint, 2007), www.errc.org/db/02/14/m00000214.pdf (accessed 5 February 2009).
——*Statement of the European Roma Rights Centre on the Occasion of the OSCE Human Dimension Implementation Meeting: Tolerance and Non-Discrimination* (Budapest: ERRC, 2008), www.osce.org/item/33776.html (accessed 5 February 2009).
Etchart, L. and R. Baksh, 'Applying a Gender Lens to Armed Conflict, Violence and Conflict Transformation', in *Gender Mainstreaming in Conflict Transformation: Building Sustainable Peace* (London: Commonweath Secretariat, 2005).

European Commission, *The Situation of Roma in an Enlarged European Union* (Luxembourg: European Commission Directorate-General for Employment and Social Affairs, 2004), ec.europa.eu/employment_social/publications/index_en.htm (accessed 5 February 2009).

Fearon, J. and A. Wendt, 'Rationalism v. Constructivism: A Sceptical View', in W. Carlsnaes, T. Risse and B. Simmonds (eds), *Handbook of International Relations* (London: Sage, 2002).

Fekete, L., 'Enlightened fundamentalism? Immigration, feminism and the Right', *Race and Class* 2006, 48:1, 1–22.

Feldblum, M., *Reconstructing Citizenship: The Politics of Nationality Reform and Immigration in Contemporary France* (New York: SUNY Press, 1999).

Femmes Africa Solidarité, *Gender is my Agenda: Civil Society's Guidelines and Mechanism for the Implementation, Monitoring and Evaluation of the Solemn Declaration on Gender Equality in Africa* (Dakar and Geneva: FAS, 2005).

Ferber, M. and J. Nelson (eds), *Beyond Economic Man* (Chicago, IL: University of Chicago Press, 1993).

Filmer, Sir R., '*Patriarcha*', in J. Sommerville (ed.), *Sir Robert Filmer: 'Patriarcha' and Other Writings* (Cambridge: Cambridge University Press, 1991).

Fineman, M.A. and N. Sweet Thomadsen (eds), *At the Boundaries of Law: Feminism and Legal Theory* (London and New York: Routledge, 1991).

Firestone, S., *The Dialectic of Sex* (London: Women's Press, 1979).

FitzGerald, S., 'Putting Trafficking on the Map: The Geography of Feminist Complicity', in V. Munroe and M. Della Giusto (eds), *Critical Perspectives on Prostitution* (Aldershot: Ashgate, 2008).

Fitzgerald, T., *Religion and the Secular. Historical and Colonial Formations* (London: Equinox, 2007a).

——*Discourse on Civility and Barbarity* (Oxford: Oxford University Press, 2007b).

Flax, J., 'Postmodernism and Gender Relations in Feminist Theory', *Signs* 12, 1987, 621.

—'The End of Innocence', in J. Butler and J.W. Scott (eds), *Feminists Theorize the Political* (London: Routledge: 1992), 445–63.

Fleming, M., 'Genocide and the Body Politic in the Time of Modernity', in R. Gellately and B. Kiernan (eds), *The Specter of Genocide: Mass Murder in Historical Perspective* (Cambridge: Cambridge University Press, 2003), 97–113.

Forsdick, C., C. Fowler and L. Kostova (eds), *Travel writing and ethics: theory and practice* (London: Routledge, 2010).

Foster, S. and S. Mills (eds), *An anthology of women's travel writing* (Manchester: Manchester University Press, 2002).

Foucault, M., *The Order of Things: An Archaeology of the Human Sciences* (London: Tavistock, 1970 [1966]).

——*The Archaeology of Knowledge* (London: Tavistock, 1972 [1969]).

——*The History of Sexuality: Volume 1: An Introduction*, trans. by R. Hurley (London: Penguin, 1976).

——*Power/Knowledge – Selected Interviews and Other Writings 1972–1977* (Brighton: Harvester Press, 1980).

——'Governmentality', *Ideology and Consciousness* 6, Summer 1986, 5–21.

Foundation for Community Development, *Implementation of the Solemn Declaration on Gender Equality in Africa: Achievements and Challenges – Drawing Lessons from Mozambique* (Maputo: FDC, 2008).

Fouskas, V.K., 'An interview with Donald Sassoon', *Journal of Southern Europe and the Balkans* 8:3, 2006.

Fowler, C., 'Journalists in Feminist Clothing: Men and Women Reporting Afghan Women during Operation Enduring Freedom, 2001', *Journal of International Women's Studies* 2006a, 8:2, 4–19.

——'Recuperating Narratives With Troublesome Titles. A critical meta-commentary on the problem of reading Beatrice Grimshaw's *From Fiji to the Cannibal Islands*', *Ecloga* 2006b, 5, 40–41.

——*Chasing Tales: travel writing, journalism and the history of British ideas about Afghanistan* (Amsterdam and New York: Rodopi, 2007a).

——'Replete with danger: The Legacy of British Travel Narratives to News Media Coverage of Afghanistan', *Studies in Travel Writing* 2007b, 11, 155–75.

Fraser, A., *The Gypsies* (Oxford: Blackwell, 1992).

——'Becoming Human: The Origins and Development of Women's Human Rights', *Human Rights Quarterly* 21, 1999, 853–906.

Fraser, N., *Scales of Justice: Reimagining Political Space in a Globalizing World* (Oxford: Polity Press, 2007).

Fraser, S. (ed.), *The Bell Curve Wars: Race, Intelligence, and the Future of America* (London: HarperCollins, 1995).

Freeman, M.A., 'The Human Rights of Women in the Family: Issues and Recommendations for Implementation of the Women's Convention', *Journal of International Affairs* 5:1, 1997, 149–66.

——'International Institutions and Gender Justice', *Journal of International Affairs* 52:2, Spring 1999, 513–43.

Freeman, S. (ed.), *The Cambridge Companion to Rawls* (Cambridge: Cambridge University Press, 2003).

Freudenberg, W. and D. Davidson, 'Nuclear families and nuclear risks: The effects of gender, geography, and progeny on attitudes toward a nuclear waste facility', *Rural Sociology* 2007, 72:2, 215–43.

Frug, M.J., *Postmodern Legal Feminism* (New York: Routledge, 1992).

Fukuyama, F., 'Women and the Evolution of World Politics', *Foreign Affairs* 77:5, 1998, 24–40.

Fuller, G., 'Which way is up?', *Sight and Sound* 16:9, September 2006.

Fuss, D., *Essentially Speaking: Feminism, Nature and Difference* (New York: Routledge, 1989).

GAATW, *Human Rights Standards for the Treatment of Trafficked Persons* (1999), www.inet.co.th/org/gaatw/gendercc.

——'A Rights Based Approach to Trafficking', *Alliance News* 22, 2004.

——*Collateral Damage: The Impact of Anti-Trafficking Measures on Human Rights around the World* (Bangkok: Amarin Printing and Publishing, 2007).

Gervais, M., 'Human Security and Reconstruction Efforts in Rwanda: Impact on the Lives of Women', in H. Afshar and D. Eade (eds), *Development, Women, and War: Feminist Perspectives* (Oxford: Oxfam, 2004).

Geuss, R., *Political Philosophy and Real Politics* (Princeton, NJ: Princeton University Press, 2008).

Gheorghe, N. and T. Acton, 'Citizens of the world and nowhere: Minority, ethnic and human rights for Roma during the last hurrah of the nation state', in W. Guy (ed.), *Between past and future—the Roma of Central and Eastern Europe* (Hatfield: Hertfordshire Press, 2001), 54–70.

Ghose, I., *The Power of the Female Gaze. Women Travellers in Colonial India* (Oxford: Oxford University Press, 1998a).

——*Memsahibs Abroad: Writings by Women Travellers in Nineteenth Century India* (Oxford: Oxford University Press, 1998b).

Gibbs, S.T., 'Theme Panel III: Multiple Tiers of Sovereignty: The Future of International Governance', *American Society of International Law, Proceedings of the 88th Annual Meeting* (Washington, DC, April 1994).

Gibson, S., 'On Sex, Horror and Human Rights', *Women: and Cultural Review* 4:3, 1993a, 251–60.

——'The Discourse of Sex/War: Thoughts on Catherine Mackinnon's 1993 Oxford Anmesty Lecture', *Feminist Legal Studies* 2:2, 1993b, 179–88.

Gibson-Graham, J.-K., *The End of Capitalism (as We Knew It): A Feminist Critique of Political Economy* (Cambridge, MA and Oxford: Blackwell, 1996).

Gilligan, C., *In a Difference Voice: Psychological Theory and Women's Development* (Massachusetts: Harvard University Press, 1982).

——*In a Different Voice: Psychological Theory and Women's Development* (Cambridge, MA: Harvard University Press, 1993).

Gil-Robles, A., *Final Report on the Human Rights Situation of the Roma, Sinti, and Travellers in Europe* (Strasbourg: Council of Europe, 2006), ec.europa.eu/employment_social/fundamental_rights/roma/rpub_en.htm (accessed 5 February 2009).

Ginty, M., 'Women at Center of Consumer Eco-Push', *Women's e-news* (online), 2006 womensenews.org/article.cfm/dyn/aid/2784 (accessed 25 November 2008).

Gökay, B., 'Tectonic Shifts and Systemic Faultlines', Inaugural Lecture, Keele University, October 2009.

Goldberg, D.T., *Racist Culture: Philosophy and the Politics of Meaning* (Oxford: Blackwell, 1993).

Goldhagen, D.J., *Hitler's Willing Executioners: Ordinary Germans and the Holocaust* (New York: Vintage, 1997).

Goldstein, J., 'Feminism', in *International Relations* (London: Pearson, 2005).

Goldstein, J.S., *War and Gender: How Gender Shapes the War System and Vice-Versa* (Cambridge: Cambridge University Press, 2001).

Goodey, J., 'Sex Trafficking in Women From Central and East European Countries; Promoting a "Victim-Centred" and "Woman-Centred" Approach to Criminal Justice Intervention', *Feminist Review* 2004, 76, 26–46.

Gordon, C. (ed.), *The Power/Knowledge: Selected Interviews and Other Writings 1972–1977* (Brighton: Harvester Press, 1980).

Gowan, P., 'Crisis in the heartland', *New Left Review* January 2009.

Gramsci, A., *Antonio Gramsci Pre-Prison Writings* (Cambridge: Cambridge University Press, 1994).

——*Further selections from The Prison notes*, ed. and trans. Derek Boothman (Minneapolis, MN: University of Minnesota Press, 1995).

Grant, R. and K. Newland (eds), *Gender and International Relations* (Milton Keynes: Open University Press, 1991).

Green, L.L., 'Gender Hate Propaganda and Sexual Violence in the Rwandan Genocide: An Argument for Intersectionality in International Law', *Columbia Human Rights Law Review* 33, summer 2002, 733–76.

Grewal, I., *Home and Harem: Nation, Gender, Empire, and the Cultures of Travel* (London: Leicester University Press, 1996).

Grimshaw, B., *From Fiji to the Cannibal Islands* (London: Eveleigh Nash, 1907).

Grosz, E., 'A Thousand Tiny Sexes: Feminism and Rhizomatics', in C.V. Boundas and D. Olkowski (eds), *Gilles Deleuze and the Theater of Philosophy* (New York/ London: Routledge, 1994a), 187–210.

Grosz, L., *Volatile Bodies, Towards a Corporeal Feminism* (London: Routledge, 1994b).

Guillebaud, J., *Youthquake: Population, fertility and environment in the 21st century* (Optimum Population Trust, 2007), www.optimumpopulation.org/opt.sub.briefing.climate.population.May07.pdf (accessed 14 November 2008).

Gunew, S., *Haunted Nations: The Colonial Dimensions of Multiculturalisms* (New York: Routledge, 2004).

Guy, W., 'Ways of Looking at Roma: the Case of Czechoslovakia', in F. Rehfisch (ed.), *Gypsies, Tinkers and Other Travellers* (London: Academic Press, 1975), 201–16.

Guy, W. (ed.), *Between past and future – the Roma of Central and Eastern Europe* (Hatfield: Hertfordshire Press, 2001).

de Haas, H., *The myth of invasion: Irregular migration from West Africa to the Maghreb and the European Union*, IMI research report, October 2007.

Habermas, J., 'Modernity versus post-modernity', *New German Critique* winter 1981, 22, 3–14.

Hajer, M., *The politics of environmental discourse: ecological modernization and the policy process* (Oxford: Oxford University Press, 1995).

Hall, S., 'New Ethnicities', in K. Mercer (ed.), *Black Film/British Cinema* (London: Institute of Contemporary Art, 1988), 27–31.

Halley, J., *Split Decisions: How and Why To Take A Break from Feminism* (Princeton, NJ: Princeton University Press, 2006).

Halliday, F., 'Hidden From International Relations; Women and the International Arena', *Millennium: Journal of International Studies* 17:3, 1988, 419–28.

Hammond, A., '"The Unending Revolt": Travel in an Era of Modernism', *Studies in Travel Writing* 2003, 7, 169–89.

Hancock, I.F., *The Pariah Syndrome: An Account of Gypsy Slavery* (Ann Arbor: Karoma Publishers, 1987), www.geocities.com/~Patrin/pariah-contents.htm (accessed 5 February 2009).

——*We are the Romani people* (Hatfield: Hertfordshire Press, 2002).

Handrahan, L., 'Conflict, Gender, Ethnicity and Post-Conflict Reconstruction', *Security Dialogue* 35, 2004, 429–45.

Hannertz, U., *Foreign News. Exploring the World of Foreign Correspondents* (Chicago, IL: Chicago University Press, 2004).

Haraway, D., 'A Manifesto for Cyborgs: Science, Technology, and Socialist Feminism in the 1980s', in L.J. Nicholson (ed.), *Feminism/Postmodernism* (New York: Routledge, 1990).

——*Simians, Cyborgs, and Women: The Reinvention of Nature* (New York: Routledge, Chapman and Hall, 1991).

——*Modest-Witness, Second-Millennium: Femaleman Meets Oncomouse: Feminism and Technoscience* (New York: Routledge, 1996).

Hardin, G., 'The Tragedy of the Commons', *Science* 162:3858, December 1968.

Harding, S., 'Is there a Feminist Method?', in S. Harding (ed.), *Feminism and Methodology* (Bloomington: Indiana University Press, 1987), 1–14.

——*Whose Science? Whose Knowledge? Thinking From Women's Lives* (Ithaca, NY: Cornell University Press, 1991).

——'Can Feminist Thought Make Economics More Objective?', *Feminist Economics* 1:1, 1995, 7–32.

Hargreaves, I., J. Lewis and T. Speers, *Towards a Better Map: Science, the Public and the Media* (Swindon, UK: Economic and Social Research Council, 2003).

Hartmann, B. and A. Hendrixson, 'Pernicious peasants and angry young men: the strategic demography of threats', in B. Hartmann, B. Subramaniam and C. Zerner (eds), *Making Threats: Biofears and Environmental Anxieties* (Lanham, MD: Rowman and Littlefield, 2005).

Hartsock, N.C., *Money, Sex and Power: Toward a Feminist Historical Materialism* (Boston, MA: Northeastern University Press, 1983).

——'Foucault on Power: A Theory for Women?', in L.J. Nicholson (ed.), *Feminism/ Postmodernism* (New York/London, 1990), 157–75.

——'Gender and Sexuality: Masculinity, Violence, and Domination', in M. Steger and N.S. Lind (eds), *Violence and Its Alternatives: An Interdisciplinary Reader* (New York: St Martin's Press, 1999), 95–112.

Haskell, M., *From Reverence to Rape: The Treatment of Women in the Movies* (Chicago, IL: University of Chicago Press, 1974).

Hawkes, G., *A Sociology of sex and sexuality* (Buckingham: Open University Press, 1996).

Healy, G. and M. O'Connor, *The Links between Prostitution and Sex Trafficking: A Brief Handbook* (Stockholm: CATW/ EWL, 2006).

Hennessy, R., *NAFTA From Below: Maquiladora Workers, Campesinos, and Indigenous Communities Speak Back* (New York: Routledge, 2006).

Hernandez-Truyol, B.E., 'Human Rights Through a Gendered Lens: Emergence, Evolution, Revolution', in K.D. Askin and D.M. Koenig (eds), *Women and International Human Rights Volume 1* (New York: Transnational Publishers, 1998), 3–39.

Hevener, N.K., 'An Analysis of Gender Based Treaty Law: Contemporary Developments in Historical Perspective', *Human Rights Quarterly* 8, February 1986, 70–88.

Hinton, R.W.K., 'Husbands, Fathers, and Conquerors I', *Political Studies* 1967, 15, 291–300.

——'Husbands, Fathers, and Conquerors II', *Political Studies* 1968, 16, 55–67.

Hirschmann, N.J., *Rethinking Obligation: A feminist method for political theory* (Ithaca, NY: Cornell University Press, 1992).

——*Gender, Class and Freedom in Modern Political Theory* (Princeton, NJ: Princeton University Press, 2007).

Hirsi Ali, A., *The Caged Virgin, An Emancipation Proclamation for Women and Islam* (Free Press, 2006).

Hobbes, T., *Human Nature and De Corpore Politico (the Elements of Law)* (Oxford: Oxford University Press, 1994).

——*Leviathan*, revised edn (Cambridge: Cambridge University Press, 1996).

——*Thomas Hobbes: the Correspondence*, two vols (Oxford: Oxford University Press, 1997).

——*On the Citizen*, trans. by Michael Silver (Cambridge: Cambridge University Press, 1998).

Hoffman, J., *Gender and Sovereignty: Feminism, the State and International Relations* (Basington: Palgrave, 2001).

Homer-Dixon, T.F., *Environment, Scarcity, and Violence* (Princeton, NJ: Princeton University Press, 1999).

——*The Upside of Down: Catastrophe, Creativity and the Renewal of Civilisation* (Toronto: Random House, 2006).

Honig, B., *Democracy and the Foreigner* (Princeton, NJ: Princeton University Press, 2001).

Honig, J.W. and N. Both, *Srebrenica: Record of a War Crime* (London: Penguin, 1996).

Hoogensen, G. and K. Stuvoy, 'Gender Resistance and Human Security', *Security Dialogue* 37:2, 2006, 207–28.

hooks, b., 'Black Women Shaping Feminist Thought', in *Feminist Theory: From the Margin to the Center* (Boston, MA: South End Press, 1984), 1–16.

——*Yearning: Race, Gender, and Cultural Politics* (Toronto: Between the Lines, 1990).

Hooper, C., *Masculinities, International Relations and Gender Politics* (New York: Columbia University Press, 2001).

Hopf, T., 'The Promise of Constructivism in International Relations Theory', *International Security* 23:1, Summer 1998, 171–200.

Hudson, H., *A Feminist Perspective on Human Security in Africa* (Pretoria: Institute for Security Studies, 2006), www.issafrica.org.

Hughes, D.M., 'The "Natasha" Trade: The Transnational Shadow Market of Trafficking in Women', *Journal of International Migration* 53:2, 2000, 625–51.

Human Rights Watch, *The Nature of the Abuse, in A Modern Form of Slavery: Trafficking of Burmese Women and Girls into Brothels in Thailand* (New York: Human Rights Watch, 1993), 3–7.

Hunt, A., *Explorations in Law and Society: Toward a Constitutive Theory of Law* (New York and London: Routledge, 1993).

Hunt, A. and G. Wickham, *Foucault and Law: Towards Sociology of Law as Governance* (London and Boulder: Pluto Press, 1994).

Hunter, L., A. Hatch and A. Johnson, 'Cross-national gender variation in environmental behaviours', *Social Science Quarterly* 2004, 85:3, 677–94.

Hunter, R., 'Gender in Evidence: Masculine Norms v Feminist Reforms', *Harvard Women's Law Journal* 19, 1996, 127–282.

Huntington, S.P., 'The Clash of Civilizations?', *Foreign Affairs* 72:3, Summer 1993, 22–49.

Hutchings, K., 'Towards a feminist international ethics', *Review of International Studies*, 26:5, 2000, 111–30.

Hynes, H.P., 'Taking population out of the equation: reformulating I = PAT', in J. Silliman and Y. King (eds), *Dangerous intersections: feminist perspectives on population, environment and development* (Cambridge, MA: South End Press, 1999), 39–73.

Intergovernmental Panel on Climate Change, *Fourth Assessment Report Climate Change 2007* (2007).

Irigaray, L., *Speculum de l'autre femme* [Speculum of the other woman] (Paris: Minuit, 1974).

——*The Ethics of Sexual Difference*, trans. by G.C. Gill (Ithaca, NY: Cornell University Press, 1988).

——*This Sex Which Is Not One* (Ithaca, NY: Cornell University Press, 1985).

Ishay, M., *The History of Human Rights: From Ancient Times to the Globalization Era* (Berkeley and Los Angeles, CA: University of California Press, 2004).

Islam, S., *The Ethics of Travel. From Marco Polo to Kafka* (Manchester: Manchester University Press, 1996).

Izsak, R., *The European Romani Women's Movement – International Roma Women's Network* (AWID's Building Feminist Movements and Organisations (BFEMO), 2008), www.awid.org/eng/About-AWID/AWID-News/Changing-Their-World (accessed 5 February 2009).

Jacobsen, J., *The Economics of Gender* (Oxford: Blackwell, 1994).

Jagger, A.M. and P.S. Rothenberg (eds), *Feminist Framework* (New York: MacGraw Hill, 1984).

James, N., *Heat* (London: BFI Publishing, 2002).

James, S. and S. Palmer (eds), *Visible Women: Essays on Feminist Legal Theory and Political Philosophy* (Oxford and Portland, OR: Hart Publishing, 2002).

James, S.M., 'Viewpoint – Shades of Othering: Reflections on Female Circumcision/ Genital Multilation', *Sign: Journal of Women in Culture and Society* 23:4, 1998, 1,031–48.

Jenkins, R., *Social Identity* (London: Routledge, 1996).

——*Rethinking Ethnicity: Arguments and Explorations* (London: Sage, 1997).

Johnsson-Latham, G., 'A study on gender equality as a prerequisite for sustainable development. What we know about the extent to which women globally live in a more sustainable way than men, leave a smaller ecological footprint and cause less climate change', report to the Environment Advisory Council (Sweden: Ministry of the Environment, 2007).

Johnston, C., 'Women's cinema as counter-cinema', in B. Nichols (ed.), *Movies and Methods: an anthology*, Vol. 1 (University of California Press, 1976 [1973]), 208–17.

Jones, A., 'Does "Gender" Make the World Go Around? Feminist Critiques of International Relations', *Review of International Studies* 22:4, 1996, 405–29.

——'Engendering Debate', *Review of International Studies* 24: 2, 1998, 299–303.

Jones, J.P. III, H.J. Nast and S.M. Roberts (eds), *Thresholds in Feminist Geography: Difference, Representation, Methodology* (Lanham, MD: Rowman and Littlefield, 1997).

Jordon, A.D., 'Human Rights or Wrongs? The Struggle for a Rights-Based Response to Trafficking in Human Rights', *Gender and Development* 2002, 10:1, 28–37.

Juma, M., 'The African Union Machinery and Gender Justice in Post-conflict Societies', *Africa Policy Brief* (African Policy Institute) 2005, No.8.

Kabbani, R., *Europe's Myth of Orient* (London: Pandora, 1986).

Kapur, R., 'The Tragedy of Victimization Rhetoric: Resurrecting the "Native" Subject in International/Post Colonial Feminist Legal Politics', *Harvard Human Rights Journal* 15, spring 2002, 1–38, www.law.harvard.edu/students/orgs/hrj/iss15/index.shtml.

——*Erotic Justice: Law and the New Politics of Postcolonialism* (London: Glasshouse Press / New Delhi: Permanent Black, 2005).

——'Revisioning the Role of Law in Women's Human Rights Struggles', in S. Meckled-Garcia and B. Cali (eds), *The Legalization of Human Rights: Multidisciplinary Perspectives on Human Rights and Human Rights Law* (London and New York: Routledge, 2006), 101–16.

Kapur, R. and B. Cossman (eds), *Subversive Sites: Feminist Engagements with Law in India* (London: SAGE Publications, 1996).

Keller, E.F., *Reflections on Gender and Science* (New Haven, CT: Yale University Press, 1985).

Kelly, L., '"You Can Find Anything You Want": A Critical Reflection on Research on Trafficking in Persons within and into Europe', *International Migration* 2005, 43:1/2, 235–65.

Kelly, M.P., *Martin Scorsese: A Journey* (New York: Avalon, 2004).

Kempadoo, K., 'Women of Color and the Global Sex Trade: Transnational Feminist Perspectives', *Meridians, Feminism, Race, Transnationalism* 1:2, 2001, 28–51.

Kennedy, D., 'The International Human Rights Movement: Part of the Problems?', *Harvard Human Rights Journal* 15, Spring 2000, 101–26, www.law.harvard.edu/students/orgs/hrj/iss15/index.shtml.

Keohane, R., 'International Relations Theory: Contributions of a Feminist Standpoint', *Millennium: Journal of International Studies* 18, 2, summer, 1989.

——'Beyond Dichotomy: Conversations Between International Relations and Feminist Theory', *International Studies Quarterly* 42:1, 1998, 193–98.

Kerrigan, W., 'Modernism and the Politics of Literary Taste', *Raritan* 2003, 22:4.

Kiljunen, P., *Energy Attitudes 2006: Public Opinion in Finland*, results of a follow-up study concerning Finnish attitudes towards energy issues 1983, 2006, www.sci.fi/~yhdys/eas_06/english/eng_luku2–4.htm (accessed 2 December 2008).

King, M., 'What Difference does it make? Gender as a Tool in Building Peace', in D. Rodriguez and Natukunda-Togboa (eds), *Gender and Peacebuilding in Africa* (San Jose: University for Peace, 2005).

Kingdom, E., 'Body Politics and Rights', in J. Bridgeman and S. Millns (eds), *Law and Body Politics: Regulating the Female Body* (Aldershot: Dartmouth Publishing, 1995), 1–21.

Kingsley Kent, S., *Gender and Power in Britain, 1640–1990* (London: Routledge, 1999).

Kinsella, H., 'For a Careful Reading: The Constructivism of Gender Constructivism', *International Studies Review* 5:2, June 2003, 294–97.

Kipling, R., *The Man Who Would Be King And Other Stories* (Oxford: Oxford Paperbacks, 1999).

——*Kim* (London: Penguin, 2007).

——*Barrack Room Ballads* (London: Standard Publications, 2008).

Kligman, G., 'On the social construction of "otherness" identifying "the Roma" in post-socialist communities', *Review of Sociology* 7:2, 2001, 61–78.

Klímová-Alexander, I., *The Romani Voice in World Politics: The United Nations and Non-State Actors* (Aldershot: Ashgate, 2005).

Koen, K., 'Claiming space: Reconfiguring women's roles in post-conflict situations', Institute for Security Studies, Occasional Paper 121, February 2006, www.issafrica.org (accessed 20 December 2008).

Kofman, E., 'Citizenship, Migration and the Reassertion of National Identity', *Citizenship Studies* 2005, 9:5, 453–67.

Kohn, E.A., 'Rape as a Weapon of War: Women's Human Rights During the Dissolution of Yugoslavia', *Golden Gate University Law Review* 24:199, 1994, 199–221.

Koji, T., 'Emerging Hierarchy in International Human Rights and Beyond: From the Perspective of Non-Derogable Rights', *European Journal of International Law* 12:5, 2001, 917–42.

Koskenniemi, M., *From Apology to Utopia* (Helsinki: Finnish Lawyers Publication Co., 1990).

Kouvo, S., 'The United Nations and Gender Mainstreaming', in D. Buss and A. Manji (eds), *International Law: Modern Feminist Approaches* (Oxford and Portland, OR: Hart Publishing, 2005), 237–52.

Kovats, M., *The politics of Roma identity: between nationalism and destitution* (2003), www.opendemocracy.net/node/1399/pdf (accessed 5 February 2009).

Kramarae, C. and D. Spender (eds), *Routledge International Encyclopedia of Women* (New York and London: Routledge, 2000).

Krane, J., J. Oxkan-Martinez and K. Ducey, 'Violence Against Women and Ethno-racial Minority Women: Examining Assumptions about Ethnicity and Race', *Canadian Ethnic Studies* 32:3, 2000, 1–18.

Krasner, S., 'Structural Causes and Regime Consequences: Regime as Intervening Variable', *International Organisation* 36:2, 1982, 185–205.

Krause, J., 'The International Dimensions of Gender Inequality and Feminist Politics: a "New Direction" for International Political Economy?', in J. MacMillan and A. Linklater (eds), *Boundaries in Question: New Directions in International Relations* (London and New York: Pinter Publishers, 1995), 128–43.

Kristeva, J., *La Révolution du langage poétique* [Revolution in general linguistics] (Paris: Seuil).

Kuhn, A., *Women's pictures: feminism and cinema* (London and Boston, MA: Routledge & K. Paul, 1982).

Kundnani, A., 'Integrationism: the politics of anti-Muslim racism', *Race and Class* 2007, 48:2, 24–44.

Kymlicka, W., *Multicultural Citizenship* (Oxford: Oxford University Press, 1995).

Lacey, N., *Unspeakable Subjects: Essays in Legal and Social Theory* (Oxford: Hart Publishing, 1998).

——'Violence, Ethics and Law: Feminist Reflections on a Familiar Dilemma', in S. James and S. Palmer (eds), *Visible Women Essays on Feminist Legal Theory and Political Philosophy* (Oxford and Portland, OR: Hart Publishing, 2002), 117–36.

Lamb, C., *The Sewing Circles of Herat* (HarperCollins, 2002).

Lamont, M. and V. Molnár, 'The Study of Boundaries in the Social Sciences', *Annual Review of Sociology* Vol. 28, 2002, 167–95.

LaMotta, J., *My Story* (Englewood Cliffs, NJ: Da Capo Press, 1997 [1970]).

di Lampedusa, G.T., *Il gattopardo* (Milan: Feltrinelli, 1986 [1956]).

Lapsley, R. and M. Westlake, *Film Theory: An Introduction* (Manchester: Manchester University Press, 1989).

Lattanzi, F. and W.A. Schabas (eds), *Essays on the Rome Statute of the International Criminal Court*, Vol. I (Ripa Fagnano Alto: Editrice il Sirente, 2000).

de Lauretis, T., *Technologies of gender: Essays on theory, film, and fiction* (Bloomington, IN: Indiana University Press, 1987).

Lefort, C., *The Political Forms of Modern Society*, trans. by J.B. Thompson (Cambridge: Polity Press, 1986).

di Leonardo, M., 'Gender, Culture and Political Economy: Feminist Anthropology in Historical Perspective', in *Gender at the Crossroads of Knowledge: Feminist Anthropology in the Postmodern Era* (Berkeley, CA: University of California Press, 1991).

Levene, M., *Genocide in the Age of the Nation-State, Volume I: The Meaning of Genocide* (London: I.B. Taurus, 2005).

Lewchuk, W., 'Men and Monotony', *Journal of Economic History* 53:4, 1993, 824–56.

Lewis, R., *Gendering Orientalism: Race, Femininity and Representation* (London: Routledge, 1995).

——*Rethinking Orientalism: Women, Travel and the Ottoman Harem* (New Brunswick: Rutgers University Press, 2004).

Liégeois, J.-P., *Roma in Europe* (Strasbourg: Council of Europe Publishing, 2007).

Littig, B., *Feminist Perspectives on Environment and Society* (Harlow: Prentice Hall, 2001).

Locke, J., *Two Treatises of Government* (Cambridge: Cambridge University Press, 1988).

de Londras, F., 'Telling Stories and Hearing Truths: Providing an Effective Remedy to Genocidal Sexual Violence Against Women', in R. Henham and P. Behrens (eds), *The Criminal Law of Genocide: International, Comparative and Contextual Aspects* (Hampshire: Ashgate, 2007).

Longino, H., *Science as Social Knowledge* (Princeton, NJ: Princeton University Press, 1990).

Lovell, T., *Consuming Fiction: Questions for Feminists* (London: Verso, 1987).

Lowe, L., *Critical Terrains: French and British Orientalism* (Cornell, NY: Cornell University Press, 1991).

Lukacs, G., *History and Class Consciousness* (London: Merlin Press, 1971).

Luke, T.W., 'On environmentality: geo-power and eco-knowledge in the discourses of contemporary environmentalism', *Cultural Critique* (Fall) 1995, 31, 57–81.

Lyotard, J.-F., *The Post-modern Condition; A Report on Knowledge* (Minneapolis, MI: University of Minnesota Press, 1979).

——*The Differend: Phrases in Dispute* (Minneapolis, MN: University of Minneapolis Press, 1988 [1983]).

MacCormack, C. and M. Strathern (eds), *Nature, Culture and Gender* (Cambridge: Cambridge University Press, 1980).

MacGregor, S., 'Feminist Perspectives on Sustainability', in D. Bell and A. Cheung (eds), *Introduction to Sustainable Development, UNESCO Encyclopedia of Life Support Systems* (Oxford: EOLSS Publishers, 2003).

——'From care to citizenship: calling ecofeminism back to politics', *Ethics and the Environment* 9:1, Spring 2004, 56–84.

——'The Public, the Private, the Planet and the Province: Women's Quality of Life Activism in Urban Southern Ontario', in M. Hessing, R. Raglan and C. Sandilands (eds), *This Elusive Land: Women and the Canadian Environment* (Vancouver: University of British Columbia Press, 2005).

——*Beyond Mothering Earth: Ecological Citizenship and the Politics of Care* (Vancouver: University of British Columbia Press, 2006).

——'Natural allies, perennial foes? On the trajectories of feminist and green political thought', *Contemporary Political Theory* 8:3, 2009, 329–39.

Mackenzie, S., *Visible Histories: Women and Environments in a Post-War British City* (Montreal: McGill-Queen's University Press, 1989).

MacKinnon, C., 'Feminism, Marxism, Method and the State: An Agenda for Theory', *Signs* 7:3, 1982, 515–44.

——'Feminism, Marxism, method and the state: toward feminist jurisprudence', *Signs* 8:2, 1983, 638–58.

——'Pornography, Civil Rights, and Speech', *Harvard Civil Rights – Civil Liberties Law Review* 20, 1985, 70.

——*Feminism Unmodified* (Harvard, MA: Harvard University Press, 1987).

——*Toward a Feminist Theory of the State* (London: Harvard University Press, 1989).

——'Crimes of War, Crimes of Peace', in S. Shuite and S. Harley (eds), *The Oxford Amnesty Lectures* (New York: Basic Books, 1993a), 83–109, 230–43 (notes).

——'Difference and Dominance: On Sex Discrimination', in D. Kelly Weisberg (ed.), *Feminist Legal Theory: Foundations* (Philadelphia, PA: Temple University Press, 1993b), 276–87.

——'Genocide's Sexuality', in *Are Women Human? And Other International Dialogues* (Cambridge, MA: Harvard/Belknap, 2006).

Macpherson, C.B., *The Political Theory of Possessive Individualism: Hobbes to Locke* (Oxford: Clarendon Press, 1962).

Macrory, P. (ed.), *A Journal of the First Afghan War* (Oxford: Oxford University Press, 2002 [1969]).

Mahmood, S., *The Politics of Piety: The Islamic Revival and the Feminist Subject* (Princeton, NJ: Princeton University Press, 2005).

Marchand, M., 'Different Communities/Different Realities/Different Encounters: A Reply to J. Ann Tickner', *International Studies Quarterly* 42:1, March 1998, 199–204.

Marchand, M.H. and A.S. Runyan, 'Feminist Sighting of Global Restructuring: Conceptualizations and Reconceptualizations', in M.H. Marchand and A.S. Runyan (eds), *Gender and Global Restructuring: Sightings, Sites and Resistance* (New York: Routledge, 2000).

Marcuse, H., *One Dimensional Man* (London: Routledge and Kegan Paul, 1964).

Marshall, L., 'Militarism and Violence Against Women', *Z-Magazine*, 17:4, April 2004, 1.

Martinich, A.P., *Hobbes: a Biography* (Cambridge: Cambridge University Press, 1999).

——*Hobbes* (London: Routledge, 2005).

Marushiakova, E. and V. Popov, 'Historical and Ethnographic Background: Gypsies, Roma, Sinti', in W. Guy (ed.), *Between past and future – The Roma of Central and Eastern Europe* (Hatfield: Hertfordshire Press, 2001), 33–53.

Masika, R., 'Editorial', *Gender and Development* (special issue on climate change) 2002, 10:2, 2–9.

Mason, A., 'Multiculturalism and the critique of essentialism', in A.S. Laden and D. Owen (eds), *Multiculturalism and Political Theory* (Cambridge: Cambridge University Press, 2007), 221–43.

Massey, D., *Spatial Divisions of Labour: Social Structures and the Geography of Production* (London: Macmillan/ New York: Methuen, 1994).

——*Space, Place, and Gender* (Cambridge: Polity Press/ Minneapolis: University of Minnesota Press, 1994).

Matrix, *Making Space: Women and the Man-Made Environment* (London: Pluto Press, 1984).

Mattingly, D. and K. Falconer al-Hindi (eds), 'Should Women Count? The Role of Quantitative Methodology in Feminist Geographic Research', *Professional Geographer* 47:4, 1995, 427–66.

McClintock, A., *Imperial Leather: Race, Gender and Sexuality in the Colonial Context* (London: Routledge, 1995).

McClintock, C., '"No longer a future heaven": Nationalism, Gender, and Race', in G. Eley and R. Suny (eds), *Becoming National: A Reader* (New York: Oxford University Press, 1996).

McClure, K., 'The Issue of Foundations: Scientized Politics, Politicized Science, and Feminist Critical Practice', in J. Butler and J.W. Scott (eds), *Feminists Theorize the Political* (London: Routledge, 1992), 341–68.

——'Taking Liberties in Foucault's Triangle: Sovereignty, Discipline, Governmentality, and the Subject of Rights', in A. Sarat and T.R. Kearns (eds), *Identities, Politics and Rights* (Ann Arbor: Michigan University Press, 1995), 149–92.

McColgan, A., *Women Under the Law: The False Promise of Human Rights* (Singapore: Longman, 2000).

McDowell, L., 'Space, Place and Gender Relations, Part 1: Feminist Empiricism and the Geography of Social Relations', *Progress in Human Geography* 17:2 (1993a), 157–79.

——'Space, Place and Gender Relations, Part 2: Identity, Difference, Feminist Geometries and Geographies', *Progress in Human Geography* 17:3 (1993b), 305–18.

——*Gender, Identity and Place: Understanding Feminist Geographies* (Cambridge: Polity Press/ Minneapolis: University of Minnesota Press, 1999).

McLean, J., 'Political Theory, International Theory and Problems of Ideology', *Millennium: Journal of International Studies* 10:2, 1981, 102–25.

McNay, L., *Foucault and Feminism: Power, Gender and the Self* (Cambridge: Polity Press, 1992).

——*Gender and Agency: Reconfiguring the Subject in Feminist and Social Theory* (London: Sage Publications, 2001).

Meckled-Garcia, S. and B. Cali (eds), *The Legalization of Human Rights: Multi-disciplinary Perspectives on Human Rights and Human Rights Law* (London and New York: Routledge, 2006).

Mehta, *U.S., Liberalism and Empire: A Study in Nineteenth Century British Liberal Thought* (Chicago, IL: Chicago University Press, 1999).

Meijer, I. Cosstera and B. Prins, 'How Bodies Come to Matter: An Interview with Judith Butler', *Signs* 23:1, winter 1998, 275–86.

Meintjes, S. et al., 'There is no Aftermath for Women', in S. Meintjes, A. Pillay and M. Tursen (eds), *The Aftermath: Women in Post Conflict Transformation* (London: Zed Books, 2001).

Mellor, M., *Feminism and Ecology* (Washington Square, NY: New York University Press, 1997).

Melman, B., *Women's Orients: English Women and the Middle East, 1718–1918. Sexuality, Religion and Work* (Michigan: University of Michigan Press, 1992).

Memedova, A., 'Romani Men and Romani Women Roma Human Rights Movement: A Missing Element', in *Roma Rights* No.4, 2003, www.errc.org/Romarights_index.php (accessed 5 February 2009).

Merchant, C., *The Death of Nature* (New York: Continuum, 1980).

——*Earthcare: Women and the Environment* (New York: Routledge, 1996).

Metodieva, M., *The Roma Women's Movement Takes on Technology* (2003), www.techsoup.org/learningcenter/consultants/archives/page10297.cfm (accessed 5 February 2009).

Mettraux, G., *International Crimes and the ad hoc Tribunals* (Oxford: Oxford University Press, 2005).

Mies, M. and V. Bennholdt-Thomsen, *The Subsistence Perspective: Beyond the Globalized Economy* (London: Zed Books, 1999).

Mies, M. and V. Shiva, *Ecofeminism* (London: Zed Books, 1993).

Mihalache, I., 'Romani Women's Participation in Public Life', in *Roma Rights* 4, 2003, www.errc.org/Romarights_index.php (accessed 5 February 2009).

Millennium: Journal of International Studies, Special Issue: 'Women in International Relations', *Millennium: Journal of International Studies* 17:3, 1988.

Mill, J.S. *The Subjection of Women: A Critical Edition*, Edward Alexander ed., (New Brunswich, NJ: Transaction Publisher, 2001).

Mills, C.W. 'Multiculturalism as/and/or anti-racism?', in A.S. Laden and D. Owen (eds), *Multiculturalism and Political Theory* (Cambridge: Cambridge University Press, 2007), 89–114.

Mills, S., *Discourses of Difference* (London: Routledge, 1992).

Milmo, C., 'Prince Charles jets in to US to collect environment award', *The Independent* 27 January 2007, www.independent.co.uk/environment/prince-charles-jets-in-to-us-to-collect-environment-award-433823.html (accessed 4 February 2007).

Mitchell, E., 'Apes and essences; some sources of significance in the American gangster films', in P. Lehman (ed.), *Wide Angle,* 1976, 1:1.

Mittleman, J., *The Globalisation Syndrome* (Princeton, NJ: Princeton University Press, 2000).

Mohanty, C.T., 'Under Western Eyes: Feminist Scholarship and Colonial Discourses', in C.T. Mohanty, A. Russo and L. Torres (eds), *Third World Women and the Politics of Feminism* (Bloomington and Indianapolis: Indiana University Press, 1991), 51–80.

——'Feminist Encounters: Locating the Politics of Experience', in A. Philips (ed.), *Feminism & Politics* (New York: Oxford University Press, 1998), 254–72.

——*Feminism without Borders: Decolonising Theory, Practicing Solidarity* (Durham and London: Duke University Press, 2003).

Mohanty, C.T., A. Russo and L. Torres (eds), *Third World Women and the Politics of Feminism* (Bloomington and Indianapolis: Indiana University Press, 1991).

Momsen, J. Henshall, 'Gender differences in environmental concern and perception', *Journal of Geography* 2000, 99, 47–56.

Monk, J. and S. Hanson, 'On Not Excluding Half of the Human in Human Geography', *Professional Geographer* 34/1, 1982, 11–23.

Moon, W. and S. Balasubramanian, 'Public attitudes toward agrobiotechnology: the mediating role of risk perceptions on the impact of trust, awareness, and outrage', *Review of Agricultural Economics* 2004, 26:2, 186–208.

Moore, H., *Feminism and Anthropology* (Cambridge: Polity Press, 1988).

Morokvasic, M., '"Settled in Mobility": Engendering Post Wall Migration in Europe', *Feminist Review* 2004, 77, 7–25.

Morris, M. and O'Connor, L. (eds), *The Virago Book of Women Travellers* (London: Virago Press, 2004).

Moss, P. (ed.), 'Feminism as Method', *Canadian Geographer* 37:1, 1993, 48–61.

——'Discussion and Debate: Symposium on Feminist Participatory Research', *Antipode* 27, 1995, 71–101.

Moutz, A., 'Embodying the Nation-State: Canada's Response to Human Smuggling', *Political Geography* 2004, 23:3, 323–45.

Mukhopadhyay, M. and N. Singh (eds), *Gender Justice, Citizenship and Development* (New Delhi: Zuuban, 2007).

Mulvey, L., 'Visual Pleasure and Narrative Cinema', *Screen* 16, 3:6, 1975, 6–18.

——'Notes on Sirk and Melodrama', *Movie* 25, 1977, 53–56.

Munro, V., 'A Tale of Two Servitudes: A Domestic Response to Trafficking of Women for Prostitution in the UK and Australia', *Social and Legal Studies* 2005, 14: 1, 91–114.

——'Exploring Exploitation: Trafficking in Sex, Work, and Sex Work', in V. Munroe and M. Della Giusta (eds), *Critical Perspectives on Prostitution* (Aldershot: Ashgate, 2008).

Murphy, C., 'Seeing Women, Recognizing Gender, Recasting International Relations', *International Organisation* 3:5, 1996, 513–38.

Murray, A., 'Debt Bondage and Trafficking: Don't Believe the Hype', in K. Kempadoo and J. Doesema (eds), *Global Sex Workers: Rights, Resistance and Redefinition* (London: Routledge, 1998), 51–68.

Musse, F., 'War crimes against Women and Girls' in J. el-Bushra and J. Gardeber (eds), *Somalia–the Untold Story: The War through the Eyes of Somali Women* (London: Catholic Institute for International Relations (CIIR), 2004).

Naffine, N., *Law and the Sexes* (London: Unwin and Hyman, 1990).

——'Possession: Erotic Love in the Law of Rape', *Modern Law Review* 57, 1994, 10–37.

——'Can Women be Legal Persons?', in S. James and S. Palmer (eds), *Visible Women: Essays on Feminist Legal Theory and Political Philosophy* (Oxford and Portland OR, 2002a), 69–90.

——*Gender and Justice* (Aldershot: Dartmouth Publishing, 2002b).

Naffine, N. and R.J. Owens (eds), *Sexing the Subject of Law* (Sydney: Law Book Co., 1997).

Nast, H.J. (ed.), 'Women in the Field', *Professional Geographer* 46:1, 1994, 54–102.

Nelson, B.J. and N. Chowdhury, *Women and Politics Worldwide* (New Haven, CT and London: Yale University Press, 1994).

Nelson, J., 'Feminism and Economics', *Journal of Economic Perspectives* 9:2, 1995, 131–48.

Nelson, L.H., *Who Knows: From Quine to a Feminist Empiricism* (Philadelphia, PA: Temple University Press, 1990).

Neufeld, M., *The Restructuring of International Relations Theory* (Cambridge: Cambridge University Press, 1995).

New Scientist, 'Are men to blame for global warming?', *New Scientist* 10 November 2007, Issue 2629, www.newscientist.com/article/mg19626293.600-are-men-to-blame-for-global-warming.html (accessed 12 September 2008).

Newey, G., *Hobbes and 'Leviathan'* (London: Routledge, 2008).

Nicholson-Lord, D., *Why the UK Needs a Population Policy: An Optimum Population Trust Briefing* (2006), www.optimumpopulation.org/opt.sub.briefing.whypoppolicy.Jul06.pdf (accessed 4 March 2007).

Nielsen Company, 'Climate change and influential spokespeople: a global Nielsen on-line survey', Nielsen Company and the Oxford University Environmental Change Institute, June 2007, www.eci.ox.ac.uk/publications/downloads/070709nielsen-celeb-report.pdf (accessed 12 November 2008).

Nietzsche, F., *On the Genealogy of Morals* (London: Dover Publications, 2003).

Nowak, M., 'The Prohibition of Gender-Specific Discrimination under the International Covenant on Civil and Political Rights', in W. Benedek, E.M. Kisaakye and G. Oberleitner (eds), *Human Rights of Women: International Instruments and African Experiences* (London/New York: Zed Books, 2002), 105–18.

Nussbaum, M., *Sex and Social Justice* (Oxford: Oxford University Press, 1999a).

——'The Professor of Parody: The Hip Defeatism of Judith Butler', *The New Republic* 22 February 1999b, 37–45.

——*Upheavals of Thought: The Intelligence of Emotions* (New York: Routledge, 2001).

——'Rawls and Feminism', in S. Freeman (ed.), *The Cambridge Companion to Rawls* (Cambridge: Cambridge University Press, 2003).

NWP/OSI (Network Women's Program/Open Society Institute), *Bending the Bow: Targeting Women's Human Rights and Opportunities* (New York: Open Society Institute, 2002), www.soros.org/initiatives/women/articles_publications/publications (accessed 5 February 2009).

——*A Place at the Policy Table. Report of the Roma Women's Forum* (Budapest: Open Society Institute, 2003), www.soros.org/initiatives/women/articles_publications/publications (accessed 5 February 2009).

Oakley, A., *Gender on Planet Earth* (Cambridge: Polity Press, 2002).

Ochieng, R., 'The Scars on Women's Minds and Bodies: Women's Roles in Post-Conflict Reconstruction in Uganda', *Canadian Women Studies* 2003, 22:2.

ODIHR (OSCE Office for Democratic Institutions and Human Rights), EUMC and Council of Europe, *International conference on the implementation and harmonization of national policies for Roma, Sinti and Travellers. Guidelines for a common vision. Bucharest, Romania, 2–6 May 2006*, Draft Conference Report, HDIM.ODIHR/523/06 (2006), www.osce.org/documents (accessed 5 February 2009).

O'Donovan, K., *Sexual Divisions in Law* (London: Weidenfeld and Nicolson, 1985).

——'With Sense, Consent, or Just a Con? Legal Subjects in the Discourse of Autonomy', in N. Naffine and R.J. Owens (eds), *Sexing the Subject of Law* (Sydney: Law Book Co., 1997), 47–64.

Ogata, S. and A. Sen, *Human security now: final report of the Commission on Human Security* (New York: United Nations, 2003).

Ohmae, K., *The Borderless World* (New York: HarperCollins, 1991).

Okin, S., 'Justice and Gender', *Philosophy and Public Affairs* 1987, 16, 3–46.

——*Justice, Gender and the Family* (New York: Basic Books, 1989).

Okin, S., *Is Multiculturalism Bad for Women?* (Princeton, NJ: Princeton University Press, 1999).

O'Neill, O., 'Justice, Gender and International Boundaries', *British Journal of Political Science* 20, 1989, 439–59.

Ophuls, W., 'Unsustainable Liberty, Sustainable Freedom', in D. Pirages (ed.), *Building Sustainable Societies: A Blueprint for a Post-Industrial World* (New York: M.E. Sharpe, 1996), 33–44.

Oprea, A., 'The Exclusion of Romani Women in Statistical Data: Limits of the Race-versus-Gender Approach', *Online Journal of the EU Monitoring and Advocacy Programme*, 2003, www.eumap.org/journal/features/2003 (accessed 5 February 2009).

——'Re-Envisioning Social Justice from the Ground Up: Including the Experiences of Romani Women', *Essex Human Rights Review* Vol. 1, No. 1, 2004, projects.essex.ac.uk/ehrr/vol1no1.html (accessed 5 February 2009).

——'The Arranged Marriage of Ana Maria Cioaba, Intra-Community Oppression and Romani Feminist Ideals', *European Journal of Women's Studies* 12:2, 2005, 133–48.

Orford, A. (ed.), *International Law and Its Others* (Cambridge: Cambridge University Press, 2006).

Ortner, S. and H. Whitehead (eds), *Sexual Meanings: The Cultural Construction of Gender and Sexuality* (Cambridge: Cambridge University Press, 1981).

Outshoorn, J., *The Politics of Prostitution: Women's Movements, Democratic States and the Globalisation of Sex Commerce* (Cambridge: Cambridge University Press, 2004).

Ovalle, O.M., 'Romani Women's Rights', *Roma Rights* 4, 2006, www.errc.org/Romarights_index.php (accessed 5 February 2009).

Paglia, C., *Sex, Art and American Culture* (London: Penguin, 1992).

Pateman, C., *The Sexual Contract* (Oxford: Polity Press, 1988).

——'"God Hath Ordained to Man a Helper": Hobbes, Patriarchy and Conjugal Right', in M. Lyndon Shanley and C. Pateman (eds), *Feminist Interpretations and Political Theory* (University Park, PA: Pennsylvania State University Press, 1991), 53–73.

Pateman, C. and E. Crosz (eds), *Feminist Challenges: Social and Political Theory* (Boston, MA: Northeastern University Press, 1986).

Pearson, R., *Heredity and Humanity: Race, eugenics and modern science* (Washington, DC: Scott-Townsend, 1996).

Perrons, D., *Globalization and Social Change* (London: Routledge, 2004).

Petchesky, R.P. and K. Judd (eds), *Negotiating reproductive rights: women's perspectives across countries and cultures* (New York: Zed Books, 1988).

Peter, J. and A. Wolper (eds), *Women's Rights Human Rights: International Feminist Perspectives* (London: Routledge, 1995).

Peterson, V.S., 'Transgressing Boundaries: Theories of Knowledge, Gender and International Relations', *Millennium: Journal of International Studies* 21:2, 1992a, 183–206.

——(ed.), *Gendered States: Feminists (Re)Visions of International Relations Theory* (Boulder, CO: Lynne Reinner Publishers, 1992b).

——'Gendered Nationalism', *Peace Review* 6:1, 1994, 77–83.

Peterson, V.S. and L. Parisi, 'Are Women Human? It's Not an Academic Question', in T. Evans (ed.), *Human Rights Fifty Years On: A Reappraisal* (Manchester: Manchester University Press, 1998), 132–60.

Peterson, V.S. and A.S. Runyan (eds), *Global Gender Issues* (Boulder, CO: Westview Press, 1993).

——'Gender as a lens on world politics', in *Global Gender Issues*, second edn (Boulder, CO: Westview Press, 1999), 21–68.

Petrova, D., 'The Roma: Between a Myth and the Future', in *Roma Rights* 3, 2004, www.errc.org/Romarights_index.php (accessed 5 February 2009).

Pettman, J., *Wording Women: A feminist international politics* (London: Routledge, 1996).

Philips, A., *Feminism & Politics* (New York: Oxford University Press, 1998).

——'Feminism and the Politics of Difference. Or, Where Have All the Women Gone?', in S. James and S. Palmer (eds), *Visible Women: Essays on Feminist Legal Theory and Political Philosophy* (Oxford and Portland, OR: Hart Publishing, 2002), 11–28.

——'More on Culture and Representation', *Social Legal Studies* 17:4, 2008, 555–58.

Phillips, D., 'The Good, the Bad and the Ugly: the Many Faces of Constructivism', *Educational Researcher* 24:7, October 1995, 5–12.

Pionke, A., 'Representations of the Indian Mutiny in Victorian Higher Journalism', in *The Victorian Web. Literature, history and culture in the age of Victoria* (2008), usp.nus.edu.sg/Victorian/history/empire/1857/intro.html (accessed 1 September 2008).

Plumwood, V., *Feminism and the Mastery of Nature* (New York: Routledge, 1993).

——*Environmental Culture: The Ecological Crisis of Reason* (New York: Routledge, 2002).

Pogány, I., *The Roma Café* (London: Pluto Press, 2004).

Ponzio, R., 'Theme Panel IV: The End of Sovereignty?', at the American Society of International Law *Proceedings* of the 88th Annual Meeting (Washington, DC, 1994).

Pratt, M.L., *Imperial Eyes. Studies in Travel Writing and Transculturation* (London: Routledge, 1982).

——'Fieldwork in Common Places', in J. Clifford and G.E. Marcus (eds), *Writing Culture. The Poetics and Politics of Ethnography* (Berkeley, CA: University of California Press, 1986).

Prentice, S., 'Taking sides: what's wrong with eco-feminism?', *Women and Environments* 10:3, 1988, 9.

Price, L.S., 'Finding the Man in the Soldier-Rapist: Some Reflections on Comprehension and Accountability', *Women's Studies International Forum* 2:2, 2001, 211–27.

Puechguirbal, N., 'Gender and Peacebuilding in Africa: Analysis of Some Structural Obstacles', in D. Rodriguez and Natukunda-Togboa (eds), *Gender and Peacebuilding in Africa* (San Jose: University for Peace, 2005).

Purdie-Vaughns, V. and R.P. Eibach, 'Intersectional Invisibility: The Distinctive Advantages and Disadvantages of Multiple Subordinate-Group Identities', *Sex Roles* Vol. 59, 2008, 377–91.

Rabinow, P. (ed.), *The Foucault Reader: An Inroduction to Foucault's thought* (Harmondsworth: Penguin, 1987).

Ramazanoglu, C. (ed.), *Up Against Foucault: Explorations of some Tensions between Foucault and Feminism* (London and New York: Routledge, 1993).

Rawls, J., *A Theory of Justice* (Oxford: Oxford University Press, 1972).

——*Political Liberalism* (New York: Columbia University Press, 1993).

Raymond, J.G., 'Prostitution as Violence against Women: NGO Stonewalling in Beijing and Elsewhere', *Women's Studies International Forum* 1998, 21:1, 1–9.

Rees, W. and L. Westra, 'When consumption does violence: can there be sustainability in environmental justice in a resource-limited world?', in J. Agyeman, R. Bullard and B. Evans (eds), *Just Sustainabilities: Development in an Unequal World* (London: Earthscan, 2003), 99–124.

Rehn, E. and E. Johnson Sinead, 'Women and Peace Operations', in *Women, War and Peace: The Independent Expert's Assessment on the Impact of Armed Conflict on Women and Women's Role in Peace-building* (New York: UNIFEM, 2002), www.unifem.undp.org/resources/assessment/index.html (accessed 20 December 2008).

Rhodes, S., 'Heat', www.all-reviews.com.

Richa, A., 'Compulsory heterosexuality and lesbian existence', in *Bread, blood, and poetry: Selected prose 1979–1985* (New York: Norton, 1986).

Richardson, D., *Women, Motherhood and Childrearing* (London: Macmillan, 1993).

Richey, K.C., 'Several Steps Sideways: International Legal Developments Concerning War Rape and the Human Rights of Women', *Texas Journal of Women & the Law* 17:109, 2007.

Riddell-Dixon, E., 'Mainstreaming Women's Human Rights: Problems and Prospects Within the Centre for Human Rights', *Global Governance* 5:2, April–June 1999, 149–72.

Ringold, D., M.A. Orenstein and E. Wilkens, *Roma in an Expanding Europe: Breaking the Poverty Cycle* (Washington, DC: World Bank, 2005).

Robinson, F., *Globalizing Care: Feminist Ethics and International Relations* (Oxford: Westview, 1999).

Robinson, J., *Unsuitable for Ladies: An Anthology of Women Travellers* (Oxford: Oxford Paperbacks, 2001).

——*Wayward Travellers: A Guide to Women Travellers* (Oxford: Oxford Paperbacks, 2005).

Rodriguez, D. and F. Natukunda-Togboa (eds), *Gender and Peacebuilding in Africa* (San Jose: University for Peace, 2005), www.upeace.org/resources/GPBA.cfm (accessed 20 December 2008).

Rooney, P., 'Methodological Issues in the Construction of Gender as a Meaningful Variable in Scientific Studies of Cognition', *Proceedings of the Biennial Meeting of the Philosophy of Science Association* 1, 1994, 109–19.

Rosaldo, M.Z. and L. Lamphere (eds), *Women, Culture and Society* (Stanford, CA: Stanford University Press, 1974).

Rose, G., *Feminism and Geography: The Limits of Geographical Knowledge* (Cambridge: Polity Press, 1993).

Rostami-Povey, E., *Afghan Women: Identity and Invasion* (London: Zed Books, 2007).

Rubin, G., 'The Traffic in Women: Notes on a "Political Economy" of Sex', in R. Rapp Reiter (ed.), *Toward an Anthropology of Women* (New York: Monthly Review Press, 1975).

Russell, M., *The Blessings of A Good Thick Skirt: Women Travellers and Their World* (London: Flamingo, 2007).

SaCouto, S., 'Advances and Missed Opportunities in the International Prosecution of Gender-Based Crimes', *Michigan State Journal of International Law* 2007, 14, 137.

Said, E.W., *Orientalism. Western Conceptions of the Orient* (London: Penguin Classics, 1978).

——'Nationalism, human rights, and interpretation', in Barbara Johnson (ed.), *Freedom and Interpretation: The Oxford Amnesty Lectures 1992* (New York: Basic Books, 1993a).

——*Culture and Imperialism* (New York: Vintage Books, 1993b).

——*The Politics of Dispossession: The Struggle for Palestinian Self-Determination 1969–1994* (London: Chatto & Windus, 1994).

Salleh, A., *Ecofeminism as Politics: Nature, Marx and the Postmodern* (London: Zed Books, 1997).

Sandilands, C., 'On "green consumerism": Environmental privatization and "family values"', *Canadian Women's Studies/Les Cahiers de la Femme* 3:3, spring 1993, 45–47.

——*The Good-Natured Feminist: Ecofeminism and the Quest for Democracy* (Minneapolis, MN: University of Minnesota Press, 1999).

Sargisson, L., 'What's wrong with ecofeminism?', *Environmental Politics* 2001, 10:1, 52–64.

Sassen, S., 'Women's Burden: Counter-Geographies of Globalisation and the Feminisation of Survival', *Nordic Journal of International Law* 2002, 71, 255–74.

Sassoon, D., *The Culture of the Europeans* (London: HarperCollins, 2006).

de Saussure, F., *Cours de linguistique générale* [Course in General Linguistics] (Paris: Pauot, 1988 [1916]).

Schabas, W.A., *Genocide in International Law: The Crimes of Crimes* (Cambridge: Cambridge University Press, 2000).

——*An Introduction to the International Criminal Court*, second edn (Cambridge: Cambridge University Press, 2004).

Schemo, D.J., 'Files in Paraguay Detail Atrocities of U.S. Allies', *New York Times* 11 August 1999, A10 C1.

Scherrer, C.P., *Genocide and Crisis in Central Africa: Conflict Roots, Mass Violence, and Regional War* (Westport, CT: Praeger, 2002).

Schiebinger, L., 'Making Natural Knowledge: Constructivism and the History of Science', *The American Historical Review* 10:5, December 1998, 1,554–55.

Schiffer, R., *Oriental Panorama: British Travellers in Nineteenth-Century Turkey* (Amsterdam and New York: Rodopi, 1999).

Schlosberg, D. and S. Rinfret, 'Ecological modernisation, American style', *Environmental Politics* 2008, 17:2, 254–75.

Schmidt, B., 'On the History and Histography of International Relations', in W. Carlsnaes, T. Risse and B. Simmonds, *Handbook of International Relations* (London: Sage, 2002).

Schmitt, C., *The Nomos of the Earth in the International Law of Jus Publicum Europaeum (1950)*, trans. by G.L. Ulmen (New York: Telos, 2006).

Schneider, M. Deutsch, 'About Women, War and Darfur: The Continuing Quest for Gender Violence Justice', *North Dakota Law Review* 83:915, 2007.

Schochet, G., 'Thomas Hobbes on the Family and the State of Nature', *Political Science Quarterly* 1967, 82, 427–45.

——*Patriarchalism in Political Thought: The Authoritarian Family and Political Speculation and Attitudes Especially in Seventeenth-Century England* (New York: Basic Books, 1975).

——'Intending (Political) Obligation: Hobbes and the Voluntary Basis of Society', in M. Deitz (ed.), *Thomas Hobbes and Political Theory* (Lawrence, KA: University Press of Kansas, 1990), 55–73.

Schriber, M.S., *Telling Tales: Selected Writings by Nineteenth-Century American Women Abroad* (Illinois: North Illinois University Press, 1995).

Schroeder, E., 'A Window of Opportunity in the Democratic Republic of the Congo: Incorporating a Gender Perspective in the Disarmament, Demobilization and Reintegration Process', *Peace, Conflict and Development* 2004, 5.

Schultz, D.L., 'An Intersectional Feminism of Their Own: Creating European Romani Women's Activism', *Journal for Politics, Gender, and Culture* 4:8/9, 2005, 243–77.

Schultz, I., 'Women and waste', *Capitalism, Nature, Socialism* 1993, 4:2, 51–63.

Schultz, I. et al., 'Research on gender, environment and sustainable development', in *Studies on Gender Impact Assessment of the Programmes of the 5th Framework Programme for Research, Technological Development and Demonstration* (Frankfurt: Institut für sozial-ökologische Forschung, 2001).

Schwarz, D.R., *Imagining the Holocaust: Narrative and Memory in Major Holocaust Literary Works* (New York: St Martin's Press, 1999).

Scicluna, H., 'Anti-Romani Speech in Europe's Public Space—The Mechanism of Hate Speech', *Roma Rights* 3, 2007, www.errc.org/Romarights_index.php (accessed 5 February 2009).

Scott, J.W., 'Experience', in J. Butler and J.W. Scott (eds), *Feminists Theorize the Political* (London: Routledge, 1992), 22–40.

——*Gender and the Politics of History* (New York: Columbia University Press, 1999 [1988]).

Seager, J., *Earth Follies: Coming to Feminist Terms with the Global Environmental Crisis* (New York: Routledge, 1993).

——'The 6-billionth baby: designated green scapegoat', *Environment and Planning A* 2000, 32, 1,711–14.

——'Pepperoni or broccoli? On the cutting wedge of feminist environmentalism', *Gender, Place and Culture* 10:2, June 2003a, 167–74.

——'Rachel Carson died of breast cancer: the coming of age of feminist environmentalism', *Signs* 28:3, 2003b, 945–72.

Sellers, S., *Language and sexual difference: Feminist writing in France* (Basingstoke: Macmillan, 1991).

Sells, M.A., *The Bridge Betrayed: Religion and Genocide in Bosnia* (Berkeley, CA: University of California Press, 1998).

Semelin, J., *Purify and Destroy: The Political Uses of Massacre and Genocide*, trans. by M.J. Dwyer (New York: Columbia, 2007).

Shackley, S., C. McLachlan and C. Gough, *The Public Perceptions of Carbon Capture and Storage*, Tyndall Centre for Climate Change Research Working Paper 44 (2004), www.tyndall.ac.uk/publications/working_papers/wp44.pdf (accessed 14 October 2008).

Shahrani, M.N., *The Kirgiz and Wakhi of Afghanistan* (Illinois: Waveland Press, 1979).

Sharpe, J., 'The Unspeakable Limits of Rape: Counter-violence and Counter-Insurgency', in P. Williams and L. Chrisman (eds), *Colonial Discourse and Post-Colonial Theory: A Reader* (Hemel Hempstead: Harvester, 1993).

Shaw, M., *War and Genocide: Organised Killing in Modern Society* (London: Polity, 2003).

——*What is Genocide?* (London: Polity, 2007).

Shields, S.A., 'Gender: An Intersectionality Perspective', *Sex Roles* 59, 2008, 301–11.

Shipler, D.K., *Arab and Jew: Wounded Spirits in a Promised Land* (New York: Penguin, 1986).

Short, P., *Pol Pot: Anatomy of a Nightmare* (New York: Henry Holt, 2004).

Shute, S. and S. Hurley (eds), *On Human Rights: the Oxford Amnesty Lectures* (New York: Basic Books, 1993).

Sideris, T., 'Rape in War and Peace: Social Context, Gender, Power and Identity', in S. Meintjes, A. Pillay and M. Tursen (eds), *The Aftermath: Women in Post Conflict Transformation* (London: Zed Books, 2001).

Silliman, J., 'Introduction', in J. Silliman and Y. King (eds), *Dangerous Intersections* (Cambridge, MA: South End Press, 1999), viii–xxiv.

Simmel, G., *On Individuality and Social Forms* (Chicago, IL: University of Chicago Press, 1971).

Simons, M. and J. Gettleman, 'Sudan Leader is Accused of Genocide', *New York Times* 15 July 2008, www.nytimes.com/2008/07/15/world/africa/15sudan.html (accessed 15 July 2008).

Singer, J., 'The Player and the Cards: Nihilism and Legal Theory', *Yale Law Journal* 94, 1984, 1–70.

Singer, P.W., *Children at War* (Berkeley, CA: University of California Press, 2005).

Sinner, S.D., *Open Wound: The Genocide of German Ethnic Minorities in Russia and the Soviet Union: 1915–1949 and Beyond* (Fargo: North Dakota State University Press, 2000).

Sjoberg, L. and C.E. Gentry, *Mothers, Monsters, Whores: Women's Violence in Global Politics* (London: Zed Books, 2007).

Skjelsbaek, I. and D. Smith (eds), *Gender, Peace and Conflict* (London: Thousand Oaks/ Sage, 2001).

Skutsch, M., 'Protocols, treaties, and action: the "climate change process" viewed through gender spectacles', *Gender and Development* 10:2, July 2002, 30–39.

Slapper, G. and D. Kelly, *Sourcebook on the English Legal System*, second edn (London: Cavendish Publishing, 2000).

Smart, C., 'Legal Subjects and Sexual Objects: Ideology, Law and Female Sexuality', in J. Brophy and C. Smart (eds), *Women-in-Law Explorations in Law, Family and Sexuality* (London: Routledge & Kegan Paul, 1985), 50–70.

——'Feminism and Law: Some Problems of Analysis and Strategy', *International Journal of the Sociology of Law* 14:2, 1986, 109–23.

——*Feminism and the Power of Law* (London: Routledge, 1989).

——'The Woman of Legal Discourse', *Social and Legal Studies* 1, March 1992, 29–44.

——*Law, Crime and Sexuality: Essays in Feminism* (London: Sage Publications, 1995).

Smart, C. and J. Brophy, 'Locating Law: A Discussion of the Place of Law in Feminist Politics', in J. Brophy and C. Smart (eds), *Women-in-Law Explorations in Law, Family and Sexuality* (London: Routledge & Kegan Paul, 1985), 1–20.

Smith, A., *Conquest: Sexual Violence and American Indian Genocide* (Boston, MA: South End Press, 2005).

Smith, D., 'The Psychocultural Roots of Genocide', *American Psychologist* 53:7, 1998, 743–53.

——*Writing the Social: Critique, Theory and Investigations* (Toronto: University of Toronto Press, 1999).

Smith, D. Clayton, 'Environmentalism, feminism and gender', *Sociological Inquiry* 71:3, 2001, 314–34.

Smith, R., 'Women and Genocide', *Holocaust and Genocide Studies* 8:3, 1994.

Smith, S., *Moving Lives: 20th Century Women's Travel Writing* (Minnesota: University of Minnesota Press, 2001).

——'The United States and the Discipline of International Relations: Hegemonic Country, Hegemonic Discipline', *International Studies Quarterly* 4:2, summer 2002, 67–86.

Solomon, L.S., D. Tomaskovic-Devey and B.J. Risman, 'Nuclear Power Perception and Sex Differences: The Case of Shearon Harris', *Sex Roles* 21:5/6, 1989, 401–19.

Sontag, S., 'Waiting for Godot in Sarajevo', *Performing Arts Journal* 47, 1994, 87–106.

Spain, D., *Gendered Spaces* (Chapel Hill: University of North Carolina Press, 1992).

Spees, P., 'Women's Advocacy in the Creation of the International Criminal Court: Changing the Landscape of Justice and Power', *Signs* 4, 2003, 1,233–55.

Spivak, G.C., *In another world: Essays in cultural politics* (London: Routledge, 1987).

——'Can the Subaltern Speak?', in C. Nelson and L. Grossberg (eds), *Marxism and the Interpretation of Culture* (London: Macmillan Press, 1988), 271–313.

——'French Feminism Revisited: Ethics and Politics', in J. Butler and J.W. Scott (eds), *Feminists Theorize the Political* (London: Routledge, 1992a), 54–87.

——'The Politics of Translations', in M. Barrett and A. Philips (eds), *Destabilising Theory: Contemporary Feminist Debates* (Stanford, CA: Stanford University Press, 1992b).

——'Subaltern Talk: Interview with the Editors (29 October 1993)', in D. Landry and G. MacLean (eds), *The Spivak Reader* (London: Routledge, 1993).

——*A Critique of Postcolonial Reason: Toward a History of the Vanishing Present* (Cambridge, MA: Harvard University Press, 1999).

Staeheli, L. et al., *Mapping Women, Making Politics: Feminist Perspectives on Political Geography* (London: Routledge, 2004).

Stanko, E.A., 'Missing the Mark? Policing Battering', in J. Hammer, J. Radford and E. Stanko (eds), *Women, Policing and Male Violence* (London: Routledge, 1989).

Stanley, L. and S. Wise, *Breaking Out* (London: Routledge and Kegan Paul, 1983).

Stanlick, N.A., 'Lords and Mothers: Silent Subjects in Hobbes's Political Theory', *International Journal of Politics and Ethics* 1, 2001, 171–82.

Staub, E., *The Roots of Evil: The Origins of Genocide and Other Group Violence* (Cambridge: Cambridge University Press, 1992).

Staudt, K., *Free Trade? Informal economies at the U.S.—Mexico border* (Philadelphia, PA: Temple University Press, 1998).

Steans, J., 'Engaging from the Margins: Feminist Encounters with the "Mainstream" of International Relations', *British Journal of Politics and International Relations* 5:3, 2003, 428–54.

——*Gender and International Relations: Issues and Debates* (Cambridge: Polity Press, 2006).

Steidle, B., *The Devil Came on Horseback: Bearing Witness to the Genocide in Darfur* (New York: Perseus, 2007).

Stevenson, C., *Victorian women travel writers in Africa* (Boston, MA: Twayne, 1982).

Stewart, M., *The Time of the Gypsies* (Oxford: Westview Press, 1997).

Stigmaye, A. (ed.), *Mass Rape: The War Against Women in Bosnia-Herzegovina* (Lincoln, NE: Nebraska University Press, 1994).

Stoparic, B., 'Women push for seats at climate policy table', *Women's e-news* 2006, www.womensenews.org/article.cfm/dyn/aid/2804 (accessed 9 November 2008).

Stover, E. and H.M. Winstein (eds), *My Neighbor, My Enemy: Justice and Community in the Aftermath of Mass Atrocity* (Cambridge: Cambridge University Press, 2004).

Strathern, M. (ed.), *Dealing with Inequality: Analysing Gender Relations in Melanesia and Beyond* (Cambridge: Cambridge University Press, 1987).

Straus, S., 'Organic Purity and the Role of Anthropology in Cambodia and Rwanda', *Patterns of Prejudice* 35:2, 2001, 47–62.

——*The Order of Genocide: Race, Power, and War in Rwanda* (Ithaca, NY: Cornell University Press, 2006).

Straus, S. and R. Lyons, *Intimate Enemy: Images and Voices of the Rwandan Genocide* (New York: Zone Books, 2006).

Strauss, L., *The Political Philosophy of Hobbes: Its Basis and its Genesis*, trans. by E. Sinclair (Chicago, IL: Chicago University Press, 1952).

Stoffell, B., 'Hobbes on Self-Preservation and Suicide', *Hobbes Studies* 4:i, 1991, 26–33.

Sturgeon, N., *Ecofeminist Natures: Race, Gender, Feminist Theory and Political Action* (New York: Routledge, 1997).

Sullivan, B., 'Trafficking in Woman', *International Feminist Journal of Politics* 2003, 5:1, 67–91.

Surdu, L. and M. Surdu, *Broadening the Agenda: The Status of Romani Women in Romania*, research report prepared for the Roma Participation Programme (New York: Open Society Institute, 2006), www.soros.org/initiatives/roma/articles_publications/publications (accessed 5 February 2009).

Svaldi, D., *Sand Creek and the Rhetoric of Extermination: A Case Study in Indian-White Relations* (Lanham, MD: University Press of America, 1989).

Swirski, P., *From Lowbrow to Nobrow* (McGill-Queen's University Press, 2005).

Sylvester, C., *Feminist Theory and International Relations in a Postmodern Era* (Cambridge: Cambridge University Press, 1994a).

——'Empathetic Cooperation: A Feminist Method for IR', *Millennium: Journal of International Studies* 23:3, 1994b, 315–34.

——'Homeless in International Relations? Women's Place in Canonical Texts and Feminist Reimaginings', in A. Philips (ed.), *Feminism & Politics* (New York: Oxford University Press, 1998), 44–66.

Szakács, J., 'Breaking the Double Chain', in *Transitions Online* 10 July 2003, www.tol.cz (accessed 5 February 2009).

Tatz, C.M., *With Intent to Destroy: Reflections on Genocide* (New York: W.W. Norton, 2003).

Temple-Raston, D., *Justice on the Grass: Three Rwandan Journalists, Their Trial for War Crimes, and a Nation's Quest for Redemption* (New York: Free Press, 2005).

Terkheimer, D., 'Recognizing and Remedying the Harm of Battering: A Call to Criminalize Domestic Violence', *The Journal of Criminal Law & Criminology* 94, 2004, 959–1,031.

Thompson, A. and K. Annan, *The Media and the Rwanda Genocide* (New York: Pluto Press, 2007).

Thornton, M. (ed.), *Public and Private: Feminist Legal Debates* (Melbourne: Oxford University Press, 1995).

Thucydides, *History of the Peloponnesian War*, trans. By Rex Warner (London: Penguin Books, 1954).

Tickner, J.A., *Gender in International Relations: Feminist Perspectives on Achieving Global Security* (New York: Columbia University Press, 1992).

——'You Just Don't Understand: Troubled Engagements Between Feminists and IR Theorists', *International Studies Quarterly* 41:4, 1997, 611–32.

——'Continuing the Conversation', *International Studies Quarterly* 42:1, 1998, 205–10.

——'Why Women Can't Run the World: International Politics According to Francis Fukuyama', *International Studies Review* 3,1999a, 3–12.

——'Searching for the Princess? Feminist Perspectives in International Relations', *Harvard International Review* 21:4, Fall 1999b, 44–48.

——*Gendering World Politics: Issues and Approaches in the Post-Cold War Era* (New York: Columbia University Press, 2001).

Tilly, C., *Durable Inequality* (Berkeley, CA: University of California Press, 1999).

——'Relational Studies of Inequality', *Contemporary Sociology* 29:6, 2000, 782–85.

——*Identities, Boundaries, and Social Ties* (London: Paradigm Publishers, 2005).

Timmerman, J., 'When Her Feet Touch the Ground: Conflict between the Roma Familistic Custom of Arranged Juvenile Marriage and Enforcement of International Human Rights Treaties', *Journal of Transnational Law and Policy* 13:2, 2004, 475–97.

Tindall, D.B., S. Davies and C. Mauboulès, 'Activism and conservation behaviour in an environmental movement: The contradictory effects of gender', *Society and Natural Resources* 2003, 16, 909–32.

Tinker, G.E., *Missionary Conquest: The Gospel and Native American Cultural Genocide* (Minneapolis, MN: Fortress Press, 1993).

Tomson, J.E., 'State Sovereignty in International Relations: Bridging the Gap Between Theory and Empirical Research', *International Studies Quarterly* 39, 1995, 213–33.

Tong, R., 'Feminist thought in transition: never a dull moment', *The Social Science Journal* 2007, 44, 23–39.

Tooze, R. and C. Murphy, 'Getting Beyond the "Common Sense" in the IPE Orthodoxy', in R. Tooze and C. Murphy (eds), *The New International Political Economy* (Boulder, CO: Lynne Rienner, 1993).

Torgovnik, J., *Intended Consequences*, mediastorm.org/0024.htm (accessed December 2008).

Totten, S. (ed.), *Genocide at the Millennium: A Critical Bibliographic Review*, 5 (New Brunswick, NJ: Transaction Publishers, 2004).

——*The Intervention and Prevention of Genocide: A Critical Bibliography* (London: Routledge, 2006).

Totten, S. and P.R. Bartrop, *Investigating Genocide: An Analysis of the Darfur Atrocities Documentation Project* (New York: Taylor and Francis, 2006).

——*Dictionary of Genocide*, two vols (Westport, CT: Greenwood, 2007).

Totten, S. and E. Markusen (eds), *Genocide in Darfur: Investigating the Atrocities in Darfur* (New York: Routledge, 2006).

Totten, S., W.S. Parsons and I. Charny (eds), *A Century of Genocide: Critical Essays and Eyewitness Accounts* (London: Routledge, 2004).

Trail, A., *Scarface* (London: Bloomsbury, 1997).

Traverso, E., *The Origins of Nazi Violence*, trans. by J. Lloyd (New York: New Press, 2003).

Treaty of Peace Between the Allied Powers and Turkey (Treaty of Sèvres), 10 August 1920, art. 230, reprinted in *American Journal of International Law* 15, 179 (Supp. 1921).

Treaty with Turkey and Other Instruments Signed at Lausanne, 24 July 1923, Decl. VIII (Declaration of Amnesty), reprinted in *American Journal of International Law* 18, 1 (Supp. 1924).

Tuana, N., *Feminism and Science* (Bloomington, IN: Indiana University Press, 1989).

Turner, H.A., Jr, *General Motors and the Nazis: The Struggle for Control of Opel, Europe's Biggest Carmaker* (New Haven: Yale University Press, 2005).

Turshen, M. and C. Twagiramariya (eds), *What Women Do in Wartime: Gender and Conflict in Africa* (London: Zed Books, 1998).

UNDP (United Nations Development Programme), *The Roma in Central and Eastern Europe: Avoiding the Dependency Trap* (Bratislava: UNDP, 2002), hdr.undp.org/en/reports/regionalreports/europethecis/name,3203,en.html (accessed 5 February 2009).

UNESCO, *Statement on Women's Contribution to a Culture of Peace*, Fourth Conference on Women: Action for Equality, Development and Peace, 4–15 September, Beijing, People's Republic of China (1995).

United Nations, UN GAOR, 6th Comm., 3d Sess., 63d-135th mtgs, UN Doc A/C.6/SR.63-A/C.6/SR.135 (1948), 78th mtg, 19 October 1948.

——*Convention on the Political Rights of Women* (New York: United Nations General Assembly, 1952).

——*Convention on the Elimination of All Forms of Discrimination Against Women (CEDAW)* (New York: United Nations General Assembly, 1979).

——*Fourth Conference on Women: Action for Equality, Development and Peace, Beijing Declaration and Platform of Action* (The Beijing Platform of Action), UN Doc. A/Conf, 177/20 (1995).

——*Security Council Resolution 1265 on the Protection of Civilians in Armed Conflict*, S/RES/1265 (New York: United Nations Security Council, 1999).

——'United Nations Conventions Against Transnational Crime' (2000a), www.odccp.org/crime_cicp_convention.html#final.

——'The Protocol to Suppress and Punish Trafficking in Persons, Especially Women and Children, Supplementing the United Nations Convention Against Transnational Organised Crime' (2000b), www.odccp.org/crime_cicp_convention.html#final.

——Optional Protocol to the Convention on the Elimination of All Forms of Discrimination against Women, 2000c, www.un.org/womenwatch/daw/cedaw/protocol (accessed 4 November 2009).

——*Security Council Resolution 1325 on the Inclusion of Women in Peace Process*, S/RES/1325 (New York: United Nations Security Council, 2000d).

——*General Recommendation 19, Violence against Women*, 11th session, 1992, General comments of the Committee on the Elimination against Women, UN Doc. HRI/GEN/1/Rev. 6 (2003).

——*Violence Against Women: Report of the Secretary-General*, A/59/281 (2004).

——*Report of the International Commission of Inquiry on Darfur to the United Nations Secretary-General Pursuant to Security Council Resolution 1564 of 18 September 2004* (Geneva: UN, 2005).

United Nations Centre for Social Development and Humanitarian Affairs, *Strategies for Confronting Domestic Violence: A Resource Manual*, UN Doc.ST/CSDHA/20 (1993).

United Nations Division for the Advancement of Women, 'History of an Optional Protocol', at www.un.org/womenwatch/daw/cedaw/protocol/history.htm.

——'Signatures to and Ratifications of the Optional Protocol', www.un.org/womenwatch/daw/cedaw/protocol/sigop.htm.

——'What is an Optional Protocol? Why an Optional Protocol?', at www.un.org/womenwatch/daw/cedaw/protocol/text.htm.

United Nations Economic and Social Council, *Report of the Special Rapporteur on Violence Against Women, Its Causes and Consequences, Ms. Radhika Coomaraswamy, submitted in accordance with Commission on Human Rights Resolution 1995/85*, E/CN.4/1996/53/Add.2 (1996a).

——*Monitoring the Implementation of the Nairobi Forward-Looking Strategies for the Advancement of Women: Other Issues, Elaboration of a Draft Optional Protocol to the Convention on the Elimination of All Forms of Discrimination Against Women*, report of the Secretary-General, E/CN.6/1996/10 (1996b).

——*Convention on the Elimination of All Forms of Discrimination Against Women, Including the Elaboration of a Draft Optional Protocol to the Convention: Comparative Summary of Existing Communications and Inquiry Procedures and Practices Under International Human Rights Instruments and Under the Charter of the United Nations*, report of the Secretary-General, E/CN.6/1997/4 (1997).

——*Integration of the Human Rights of Women and the Gender Perspective: Integrating the Human Rights of Women Throughout the United Nations System, Report of the Secretary-General*, E/CN.4/soo4/64 (2004).

United Nations Office on Drugs and Crime, *Trafficking in Persons Global Pattern* (Vienna: UNDOC, 2006).

Urban, J., 'Interrogating Privilege/Challenging the "Greening of Hate"', *International Feminist Journal of Politics* 2007, 9:2, 251–65.

Uvin, P., *Aiding Violence: The Development Enterprise in Rwanda* (West Hartford, CT: Kumarian Press, 1998).

Vajpeyi, A., 'The face of truth', (2004), www.india-seminar.com/2004/542/542%20essay.htm.

Valentine, G., 'Theorizing and Researching Intersectionality: A Challenge for Feminist Geography', *Professional Geographer* 59, 2007, 10–21.

Valentino, B.A., *Final Solutions: Mass Killing and Genocide in the 20th Century* (Ithaca, NY: Cornell University Press, 2005).

Valenzuela, M.E., J. Jaquette and S.L. Wolchik, *Women and Democracy: Latin America and Central and Eastern Europe* (Baltimore, MD: Johns Hopkins University Press, 1998).

Van Schaack, B., 'The Crime of Political Genocide: Repairing the Genocide Convention's Blind Spot', *Yale Law Journal* 106, May 1997, 2,259.

——'Darfur and the Rhetoric of Genocide', *Whittier Law Review* 2005, 26/1101.

Vasquez, J., *The Power of Power Politics* (London: Pinter, 1983).

Vermeersch, P., *The Romani Movement: Minority Politics and Ethnic Mobilisation in Contemporary Central Europe* (New York: Berghahn, 2006).

Visweswaran, K., 'Histories of Feminist Ethnography', *Annual Reviews in Anthropology* 27, 1997, 591–621.

Von Verdross, A., 'Forbidden Treaties in International Law', *American Journal of International Law* 31, 1937, 571.

Vulliamy, E., *Seasons in Hell: Understanding Bosnia's War* (New York: Simon and Schuster, 1994).

Wagner, J., 'The Systematic Use of Rape as a Tool of War in Darfur: A Blueprint for International War Crimes Prosecutions', *Georgetown Journal of International Law* 37:193, 2005.

Walgenbach, K., 'Gender als interdependente Kategorie', in K. Walgenbach, G. Dietze, A. Hornscheidt and K. Palm (eds), *Gender als interdependente Kategorie: Neue Perspektiven auf Intersektionalität, Diversität und Heterogenität* (Opladen: Verlag Barbara Budrich, 2007), 23–64.

Walker, A.J., 'Methods, Theory and the Practice of Feminist Research: A Response to Janet Chafetz', *Journal of Family Issues* 25, 2004, 990–94.

Walker, R.B.J., 'Gender and Critique in the Theory of International Relations', in S.V. Peterson (ed.), *Gendered States Feminist (Re)Vision of International Relations Theory* (London: Lynne Rienne Publishers, 1992), 179–202.

——*Inside/outside: international relations as political theory* (Cambridge: Cambridge University Press, 1993).

Wallach-Scott, J., *Gender and the Politics of History* (New York: Columbia University Press, 1999).

Waller, J., *Becoming Evil: How Ordinary People Commit Genocide and Mass Killing* (Oxford: Oxford University Press, 2002).

Wallis, A., *Silent Accomplice: The Untold Story of France's Role in the Rwandan Genocide* (London and New York: I.B. Tauris, 2007).

Walton-Roberts, M., 'Rescaling Citizenship: Gendering Canadian Immigration Policy', *Political Geography* 23:3, 2004, 267–81.

Waltz, K., *Theory of International Politics* (Reading, MA: Addison-Wesley, 1979).

Walzer, M., *Just and Unjust Wars: A Moral Argument with Historical Illustrations*, fourth edn (New York: Basic Books, 2006).

Warbrick, C. and V. Lowe (eds), *The United Nations and the Principles of International Law: Essays in Memory of Michael Akehurst* (London: Routledge, 1994).

Warner, L.R., 'A Best Practices Guide to Intersectional Approaches in Psychological Research', *Sex Roles* Vol. 59, 2008, 454–63.

Warren, K.J., *Ecofeminist Philosophy: A Western Perspective on What it is and Why it Matters* (Lanham, MD: Rowman and Littlefield, 2000).

Warren, M.A., *Gendercide: The Implications of Sex Selection* (Totowa, NJ: Rowman and Littlefield, 1985).

Weber, C., 'Good Girls, Bad Girls and Little Girls: Male Paranoia in Robert Keohane's Critique of Feminist International Relations', *Millennium: Journal of International Studies* 23:2, 1994, 337–49.

——*Simulating Sovereignty, Intervention, the State and Symbolic Exchange* (Cambridge: Cambridge University Press, 1995).

——'Gender', in *International Relations Theory: A Critical Introduction* (London: Routledge, 2001).

WEDO, *Beijing Betrayed* (New York: Women's Environment and Development, 2005), www.wedo.org/library.aspx?ResourceID1/431

Weinrib, E.J., 'Legal Formalism: On the Immanent Rationality of Law', *Yale Law Journal* 97, 1988, 949.

Weir, A., *Sacrificial Logics: Feminist Theory and the Critique of Identity* (London: Routledge, 1996).

Weisberg, K.K. (ed.), *Feminist Legal Theory: Foundations* (Philadelphia, PA: Temple University Press, 1993).

Weisburd, A.M., 'The Emptiness of the Concept of Jus Cogens, as Illustrated by the War in Bosnia-Herzegovina', *Michigan Journal of International Law* 17, 1995, 1, 19.

Weitz, E.D., *A Century of Genocide: Utopias of Race and Nation* (Princeton, NJ: Princeton University Press, 2003).

Wendell, S., *The Rejected Body: Feminist Philosophical Reflections on the Disabled Body* (London: Routledge, 1996).

Wendoh, S. and T. Wallace, 'Re-thinking Gender Mainstreaming in African NGOs and Communities', in F. Porter and C. Sweetman (eds), *Mainstreaming Gender in Development: A Critical Review* (United Kingdom: Oxfam-GB, 2005).

Wendt, A., 'Anarchy is What States Make of It: The Social Construction of Power Politics', *International Organization* 46:2, 1992, 391–425.

——'Collective Identity Formation and the International State', *American Political Science Review* 88, 1994, 384–96.

——*Social Theory of International Politics* (Cambridge: Cambridge University Press, 1999).

——'The State as Person in International Relations', *Review of International Studies* 30: 2, April 2004, 289–316.

——'How Not to Argue Against State Personhood. A Response', *Review of International Studies* 30:2, April 2005, 357–60.

West, C. and S. Fenstermaker, 'Doing difference', *Gender & Society* 1995, 9:1, 8–37.

West, C.L., 'I Ain't the Right Kind of Feminist', in C.T. Mohanty, A. Russo and L. Torres (eds), *Third World Women and the Politics of Feminism* (Bloomington, IN: Indiana University Press, 1991), ix–xii.

West, C.M. (ed.), *Violence in the Lives of Black Women: Battered, Black, and Blue* (Binghamton, NY: Haworth Press, 2004).

West, L., *Feminist Nationalisms* (London: Routledge, 1997).

West, R., 'Jurisprudence and Gender', in D.K. Weisberg (ed.), *Feminist Legal Theory: Foundations* (Philadelphia, PA: Temple University Press, 1993), 75–98.

Weston, B., 'Human Rights' *Human Rights Quarterly* 6, 1986, 257–82.

Wheatley, S., 'Democracy in International Law: A European Perspective', *International & Comparative Law Quarterly* 51, April 2002, 225–47.

Whitaker, B., 'Revised and Updated Report on the Question of the Prevention and Punishment of the Crime of Genocide', UN Doc. E/CN. 4/Sub.2/1985/6 (1985), 20.

Whitbead, J., 'Mainstreaming Gender in Conflict Reduction: From Challenge to Opportunity', in C. Sweetman (ed.), *Gender, Peacebuilding, and Reconstruction* (United Kingdom: Oxfam-GB, 2004).

White, J.V. and C. Blakeslley, 'Women or Rights: How Should Women's Rights be Conceived and Implemented?', in K.D. Askin and D.M. Keonig (eds), *Women and International Human Rights Law Volume 2: International Courts, Instruments, and Organizations and Select Regional Issues Affecting Women* (Ardsley, NY: Transnational Publishers, 2000), 51–78.

Whitworth, S., *Feminism and International Relations* (Basingstoke: Macmillan, 1994).

Wichterich, C., *The Globalised Woman* (London: Zed Books, 2000).

Wight, C., 'State Agency: Social Action Without Human Activity?', *Review of International Studies* 30:2, April 2004, 269–80.

Wijers, M., 'Women, Labour and Migration: The Position of Trafficked Women and Strategies for Support', in K. Kempadoo and J. Doesema (eds), *Global Sex Workers: Rights, Resistance and Redefinition* (London: Routledge, 1998), 69–78.

——'European Union Policies on Trafficking in Women', in M. Possilli (ed.), *Gender Policies in the European Union* (New York: Peter Lang, 2001).

Williams, P.J., 'On Being the Object of Property', in M. Albertson Fineman and N. Sweet Thomadsen (eds), *At the Boundaries of Law: Feminism and Legal Theory* (London/New York: Routledge, 1991), 22–39.

——'Alchemical Notes: Reconstructing Ideals from Deconstructed Rights', in K.K. Weisberg (ed.), *Feminist Legal Theory: Foundations* (Philadelphia, PA: Temple University Press, 1993a), 496–506.

——'Law and Everyday Life', in A. Sarat and T.R. Kearns (eds), *Law and Everyday Life* (Ann Arbor: Michigan University Press, 1993b), 171–90.

Williams, Z., 'Eco-mom: here to save the world!' *The Guardian* 17 April 2008.

Wilmer, F., 'Identity, Culture, and Historicity', *World Affairs* 160:1, 1997, 3–16.

Wilson, A., 'The forced marriage debate and the British State', *Race and Class* 2007, 29:1, 25–38.

Wing, A.K. and S. Merchan, 'Rape, Ethnicity, and Culture: Spirit Injury from Bosnia to Black America', *Columbia Human Rights Law Review* 25, 1993, 1–48.

Wishik, H.R., 'To Question Everything: The Inquiries of Feminist Jurisprudence', in K.K. Weisberg (ed.), *Feminist Legal Theory: Foundations* (Philadelphia, PA: Temple University Press, 1993), 22–31.

Wolfrum, R. (ed.), *United Nations, 2 Vols, Law, Policies and Practices* (Muenchen: C.H. Beck, 1995), 1,081–90.

Women and Geography Study Group of the Institute of British Geographers (WGSG/IBG), *Gender and Geography: An Introduction to Feminist Geography* (London: Hutchinson, 1984).

Women and Geography Study Group of the Royal Geographical Society with the Institute of British Geographers (WGSG/IBG), *Feminist Geographies: Explorations in Diversity and Difference* (Harlow, Essex: Longman, 1997).

Women's Aid, 'Comments and recommendations on the Domestic Violence, Crime and Victims Bill 2004', 2004a, www.wpmensaid.org.uk/policy&consultations/DVBill/DVAct_info_section_index.htm.

——'Women's National Commission response to the Domestic Violence, Crime and Victims Bill 2004', 2004b, www.womensaid.org.uk/policy&consultations/DVBill/WNC_response_final.htm.

Women's Environmental Network and the National Federation of Women's Institutes, *Women's Manifesto on Climate Change*, 2007, www.wen.org.uk/general_pages/reports/manifesto.pdf (accessed 14 November 2008).

Woodiwiss, A., 'The Law Cannot Be Enough: Human Rights and the Limits of Legalism', in S. Meckled-Garcis and B. Cali (eds), *The Legalization of Human Rights: Multidisciplinary Perspectives on Human Rights and Human Rights Law* (London and New York: Routledge, 2006), 32–48.

Woodward, S.L., *Balkan Tragedy: Chaos and Dissolution After the Cold War* (Washington, DC: Brookings Institution, 1995).

Wootton, D. (ed.), *Modern political thought: readings from Machiavelli to Nietzsche* (Hackett Publishing, 1996).

Wright, J.H., 'Going Against the Grain: Hobbes's Case for Original Maternal Dominion', *Journal of Women's History* 2002, 14, 123–55.

Wright, L., 'Gay Genocide as Literary Trope', in E.S. Nelson (ed.), *AIDS: The Literary Response* (New York: Twayne Publishers, 1992), 50–68.

Wyler, E., 'From "State Crime" to Responsibility for "Serious Breaches of Obligations Under Peremptory Norms of General International Law"', *European Journal of International Law* 13:5, 2002, 1,147–60.

Yannis, A., 'The Concept of Suspended Sovereignty in International Law and its Implications in International Politics', *European Journal of International Law* 13:5, 2002, 1,037–52.

Yasuaki, O., 'International Law in and with International Politics: The Functions of International Law in International Society', *European Journal of International Law* 14:1, 2003, 105–40.

Yeatman, A., 'Feminist Theory of Social Differentiation', in L.J. Nicholson (ed.), *Feminism/Postmodernism* (New York/London: Routledge, 1990), 281–99.

——*Postmodern Envisioning of the Political* (London: Routledge, 1994).

Young, I.M., 'Impartiality and the Civil Public Some Implications of Feminist Critiques of Moral and Political Theory', in S. Benhabib and D. Cornell (eds), *Feminism as Critique* (Cambridge: Polity, 1986), 56–76.

——*Justice and the Politics of Difference* (Princeton, NJ: Princeton University Press, 1990a).

——*Throwing Like a Girl* (Bloomington, IN: Indiana University Press, 1990b).

——'Gender as Seriality: Thinking About Women as a Social Collective', *Sign* 19:3, 1994, 713–38.

——*Inclusion and Democracy* (Oxford: Oxford University Press, 2000).

——'Structural injustice and the politics of difference', in A.S. Laden and D. Owen (eds), *Multiculturalism and Political Theory* (Cambridge: Cambridge University Press, 2007), 60–88.

Young, J.E., *Writing and Rewriting the Holocaust: Narrative and the Consequences of Interpretation* (Bloomington, IN: Indiana University Press, 1988).

——*The Texture of Memory* (New Haven: Yale University Press, 1993).

——*At Memory's Edge: After-images of the Holocaust in Contemporary Art and Architecture* (New Haven, CT: Yale University Press, 2000).

Young, K., C. Wolkowitz and R. McCullagh (eds), *Of Marriage and the Market: Women's Subordination in International Perspective* (London: Routledge and Kegan Paul, 1985 [1981])

Youngs, G., *Political Economy, Power and the Body: Global Perspectives* (London: Palgrave, 1999).

——'Feminist International Relations: A Contradiction in Terms? Or: Why Women and Gender are Essential to Understanding the World "We" Live in', *International Affairs* 80:1, 2004, 75–87.

Youngs, T., *Travellers in Africa: British travelogues 1850–1900* (Manchester: Manchester University Press, 1994).

——'Auden's travel writings', in S. Smith (ed.), *The Cambridge Companion to W. H. Auden* (Cambridge: Cambridge University Press, 2004).

Yule, A., *Al Pacino: A Life on the Wire* (London: Warner Books, 1992).

Yuval-Davis, N., *Gender & Nation* (London and New York: Sage, 1997).

Yuval-Davis, N. and P. Werbner (eds), *Women, Citizenship and Difference* (London: Zed Books, 1999).

Zalewski, M., 'The "Women/Women" Question in International Relations', *Millennium: Journal of International Studies* 23:2, 1994, 407–23.

——'Well, What is the Feminist Perspective on Bosnia?', *International Affairs* 71, 1995, 339–56.

——'Where is Woman in International Relations? To Return as a Woman and be Heard', *Millennium* 27, 1999, 847–67.

——'Feminism and International Relations: An Exhausted Conversation?', in F. Harvey and M. Brecher (eds), *Critical Paradigms in International Studies* (Michigan: University of Michigan Press, 2002).

——'"Women's Troubles" Again in IR', *International Studies Review* 5:2, June 2003, 291–94.

Zalewski, M. and J. Parpart (eds), *The 'Man' Question in International Relations* (Oxford: Westview, 1998).

Zarembo, A., 'Judgment Day: In Rwanda, 92,392 genocide suspects await trial', *Harper's Magazine* April 1997, 68–80.

Zawati, H.M. and I.M. Mahmoud, *A Selected Socio-Legal Bibliography on Ethnic Cleansing, Wartime Rape and Genocide in the Former Yugoslavia and Rwanda* (Lewiston, NY: Edwin Mellen, 2004).

Zelezny, L., P. Poh-Pheng Chua and C. Aldrich, 'Elaborating on gender differences in environmentalism', *Journal of Social Issues* 2000, 56, 443–57.

Zelinsky, W., 'The Strange Case of the Missing Female Geographer', *Professional Geographer* 25/2, 1973, 151–67.

Žižek, S., *The Puppet and the Dwarf: The Perverse Core of Christianity* (Cambridge, MA: MIT Press, 2003).

——*The Ticklish Subject: the absent centre of political ontology*, second edn (New York: Verso, 2009).

Zimbardo, P., *The Lucifer Effect: Understanding How Good People Turn Evil* (New York: Random House, 2008).

Zinn, H., 'Whose Atrocity is Bigger?', *ZNet* 25 May 1999.

Zinn, M.B. and B. Dill, 'Theorizing difference from multiracial feminism', *Feminist Studies* 22:2, 1996, 321–31.

Zlatar, Z., *Njegos's Montenegro: Epic Poetry, Blood Feud and Warfare in a Tribal Zone 1830–1851* (Boulder, CO: East European Monographs, 2005).

Zlotnik, H., 'The Global Dimension of Female Migration', *Migration Information Source* 2003, www.migrationinformation.org/Feature/display.cfm?ID=109.

Zukier, H., 'The Twisted Road to Genocide: On the Psychological Development of Evil During the Holocaust', *Social Research* 61:2, 1994, 423–55.